New Advances in Low-Energy Processes for Geo-Energy Development: 2nd Edition

New Advances in Low-Energy Processes for Geo-Energy Development: 2nd Edition

Guest Editors

Daoyi Zhu
Yibo Li
Qingyuan Chen

MDPI

Basel • Beijing • Wuhan • Barcelona • Belgrade • Novi Sad • Cluj • Manchester

Guest Editors
Daoyi Zhu
Faculty of Petroleum
China University of
Petroleum-Beijing at Karamay
Karamay
China

Yibo Li
Faculty of Petroleum and
Natural Gas Engineering
Southwest Petroleum
University
Chengdu
China

Qingyuan Chen
Faculty of Chemistry and
Chemical Engineering
Southwest Petroleum
University
Chengdu
China

Editorial Office
MDPI AG
Grosspeteranlage 5
4052 Basel, Switzerland

This is a reprint of the Special Issue, published open access by the journal *Processes* (ISSN 2227-9717), freely accessible at: https://www.mdpi.com/journal/processes/special_issues/T10D3JDIPB.

For citation purposes, cite each article independently as indicated on the article page online and as indicated below:

Lastname, A.A.; Lastname, B.B. Article Title. *Journal Name* **Year**, *Volume Number*, Page Range.

ISBN 978-3-7258-8151-2 (Hbk)
ISBN 978-3-7258-8152-9 (PDF)
https://doi.org/10.3390/books978-3-7258-8152-9

Contents

About the Editors

Daoyi Zhu

Daoyi Zhu holds a Bachelor of Science in Applied Chemistry from Yangtze University (2011), as well as a Master's degree in Chemical Engineering (2014) and a Ph.D. in Oil and Gas Field Development Engineering (2018) from the China University of Petroleum (Beijing). His academic career spans a visiting scholar appointment at Missouri University of Science and Technology's Department of Petroleum Engineering (2016–2018), as well as his position as an Associate Professor of Petroleum Engineering at China University of Petroleum-Beijing at Karamay, which he has held since 2021. As an active researcher in energy engineering, Dr. Zhu specializes in oilfield chemistry, reservoir conformance control, CCUS technologies, and geo-energy development. He has made significant scholarly contributions, including publishing 40+ peer-reviewed SCI journal papers, authoring four textbooks (*Oilfield Chemistry, Literature Search and Technical Writing,* *CO_2 and Geo-Energy Engineering,* and *Petroleum Reservoir Engineering*), and translating multiple technical works. His editorial leadership extends to roles as Guest Editor and Executive Board Member for prominent international journals in the field.

Yibo Li

Yibo Li is a Professor at the State Key Laboratory of Oil and Gas Reservoir Geology and Exploitation, School of Petroleum and Natural Gas Engineering, Southwest Petroleum University, China. His research interests mainly focus on enhanced oil recovery (EOR) technologies and theories, chemical flooding, heavy oil thermal recovery, gas injection, reservoir conformance control, and unconventional oil and gas development. He has long been engaged in fundamental and engineering research related to complex oil and gas reservoirs and enhanced oil recovery processes. Prof. Li has undertaken and participated in a series of major national and industrial research projects, including projects funded by the National Natural Science Foundation of China, the National Key Research and Development Program of China, the National Basic Research Program (973 Program), and key collaborative projects with petroleum enterprises. He has published more than 68 academic papers, including over 30 papers as the first or corresponding author in internationally recognized journals such as *SPE Journal, Fuel, Journal of Petroleum Science and Engineering,* and *Journal of Thermal Analysis and Calorimetry.* His publications have been cited more than 1500 times in Web of Science with an H-index of 24. He has also authored and co-authored academic monographs related to CCUS and petroleum engineering, obtained 18 invention patents including international patents, and delivered presentations at international academic conferences. In addition, he serves as Vice Director of the Natural Gas Engineering Center of the Ministry of Education, Visiting Professor at Kazan Federal University, Academic Committee Member of the World Thermal EOR Conference, and Associate Editor of SCI and EI indexed journals.

Qingyuan Chen

Qingyuan Chen is currently an Assistant Research Fellow at the College of Chemistry and Chemical Engineering, Southwest Petroleum University. Holding a doctoral degree in Chemical Engineering and Technology from Southwest Petroleum University, his research focus lies in enhanced oil recovery, heavy oil viscosity reduction, amphiphilic carbon nanotube emulsification and drag reduction, active polymer flooding, and the development and mechanism research of chemical flooding systems for low-permeability and strongly heterogeneous reservoirs. His research undertakings include presiding over the Youth Science Fund Project of the National Natural Science Foundation of China, and multiple open funds from national and provincial key laboratories and university academic initiation programs, while also participating in the NSFC General Project. Additionally, he has multiple high-level papers published in authoritative journals such as *Langmuir*, *Physics of Fluids*, *Colloids and Surfaces A*, and *Energy & Fuels*, with several Chinese invention patents authorized in the fields of nano viscosity reducers, profile control and flooding systems, and functional nanomaterial preparation. Relevant research provides important theoretical and technical support for the efficient development of heavy oil and unconventional oil and gas reservoirs.

processes

MDPI

Editorial

New Advances in Low-Energy Processes for Geo-Energy Development: 2nd Edition

Daoyi Zhu

Petroleum School, China University of Petroleum-Beijing at Karamay, Karamay 834000, China; zhudaoyi@cupk.edu.cn

Geo-energy systems, including oil, gas, and geothermal ones, are accelerating toward a synergistic development stage characterized by high efficiency, low energy consumption, and low carbon emissions [1]. Guided by the concept of low-energy processes, the industry is re-engineering the entire workflow chain, including formation stimulation [2,3], injection–production control [4], wellbore engineering, surface safety, and digital intelligence [5–9]. The goal is to achieve higher recovery factors, more robust operational safety, and improved environmental performance with less energy input and material consumption [10–12]. Achieving this goal relies on multi-scale coupled understanding (pore–fracture–wellbore–pipeline–surface) [13–15], process intensification and material innovation [16,17], and data-driven adaptive optimization control.

The development of unconventional reservoirs and hard-to-recover resources—such as those with high water cut, deep burial, and strong heterogeneity—remains the primary focus for enhancing production capacity and reducing unit energy consumption [18–21]. New technologies and materials are addressing key challenges, including proppant transport and placement during fracturing [2,22,23], flowback optimization in acidizing operations, structural reinforcement for well completion and stability, and high-performance materials for deep conformance improvement and water plugging [24–26]. These advancements significantly reduce treatment pressures and the frequency of workover operations [27], as well as enable more precise flow-path control to minimize ineffective circulation and repeated energy expenditure [28,29]. Simultaneously, artificial intelligence and big data are transforming static design into a closed-loop system of dynamic perception–prediction–control. This reduces trial-and-error costs and energy waste. For surface operations, safety assessments for CO_2 transport and storage, along with accurate characterization of the thermal conductivity of fluid-bearing rocks [30], provide low-energy design bases and safety boundaries for geothermal utilization, CCUS, and multi-energy complementarity [31].

This Special Issue, "New Advances in Low-Energy Processes for Geo-Energy Development: 2nd Edition", focuses on fundamental and methodological innovations. It compiles ten papers that systematically present recent progress in low-energy processes across three key areas: unconventional reservoir stimulation and wellbore engineering, data- and intelligence-driven production optimization, and subsurface storage/transport and thermal property characterization. The following paragraphs provide an overview and commentary on the core findings of these papers, organized according to the three themes above.

The first category focuses on low-energy stimulation and wellbore process intensification. Lu et al. [32] addressed the fundamental unit of complex fracture networks—asymmetric branch fractures—and constructed an experiment using a rough-walled dual-branch slot. It is shown that structural asymmetry significantly induces non-uniform proppant transport and irregular proppant bed morphology, and that reducing branch height and width increases local flow velocity, inhibits settling, and delivers more proppant to the far end. However, injection pressure also increases; a height ratio of 0.25 easily causes screen-out and ineffective placement; and the upper branch also struggles to form a proppant bed and is prone to closure. This research provides geometric optimization and operational parameter boundaries for achieving more effective propping with lower pumping energy in unconventional fracturing. Wang et al. [33] targeted the high energy consumption in acidizing deep, high-pressure, and sensitive offshore Dongying Formation reservoirs. An integrated acidizing–flowback tubing string is proposed, which enables reverse acid injection through the casing–tubing annulus and rapid pump recovery of spent acid/byproducts, reducing single-trip time and secondary contamination. Multi-physics analysis shows maximum deformations of 1.4, 1.9, 0.18, 2.7, 1.8, and 2.5 m under buoyancy, the piston effect, viscosity, helical buckling, temperature difference, and expansion, respectively, and total deformation is less than 3 m. Under different well depth conditions, axial force is 400–600 kN, stress is 260–350 MPa, and the safety factor is greater than 3.0. This process ensures wellbore safety while reducing operational energy and cost through process integration.

Zhao et al. [34] proposed a one-trip shape-memory thermal-sensitive screen pipe, which uses a specific temperature to spontaneously expand and reinforce the wellbore wall. Finite element calculations and theoretical evaluation show that the radial compaction force after pipe–wall contact optimizes near-wellbore stress distribution and reduces sand production risk. It also effectively enhances bottomhole flowing pressure and depressurization capacity in shale oil and gas completions. This passive deployment–active reinforcement single-trip completion method is expected to reduce multiple workovers and energy-intensive sand control operations. Lv et al. [35] addressed the issue of uneven stage and cluster placement caused by the chaotic–strongly heterogeneous internal structure of clastic glutenite reservoirs. A gravel distribution characterization model is established based on imaging logs, identification analysis software is developed, and conventional-imaging log correlation prediction and Kriging point–line–surface interpolation methods are constructed. Model accuracy exceeds 80% under validation from drilled cores, borehole cameras, and imaging logs. It helps avoid high-gravel zones, and wellhead pressure is lower than in offset wells, with reduced fluid volume per well and better production. The precise formation perception–cluster design reduces ineffective fracturing energy.

Xu et al. [36] used amphiphilic carbon dots as a multifunctional modifier. Carbon dot-enhanced preformed particle gels (CD-PPGs) were prepared via in situ polymerization with dual crosslinking from hydrogen bonding and hydrophobic association. Temperature resistance (up to 110 °C), salt tolerance (up to 15×10^4 mg/L), and mechanical properties are significantly improved, and in-depth self-aggregation and targeted high-strength plugging characteristics are demonstrated. Under permeability conditions of 539.0–2988.6 mD, a 0.5 PV (pore volume) injection achieves a plugging rate greater than 95%, and in heterogeneous simulations, total oil recovery reaches 52.6%, which is 20.5% higher than that from waterflooding alone. This material aids low-energy conformance improvement with lower dosage and longer-lasting deep regulation. He and Wang [37] used pyrolyzed biochar from agricultural and forestry waste as a functional enhancer, and constructed an inorganic composite gel system for high-temperature, high-salinity fractured-vuggy carbonate reservoirs. The 0.5 wt% biochar significantly optimizes the matrix pore structure. Through micro-aggregate filling and interfacial chemical bonding, the cured body achieves a compressive strength greater than 2 MPa, and no significant strength decay is observed after aging at 130 °C for 30 days. An adsorption slow-release mechanism stabilizes hydration kinetics at high temperature, extending the setting time to 15 h, which solves the flash setting challenge in deep well pumping. The system exhibits excellent shear-thinning and fracture

https://doi.org/10.3390/pr14111706

plugging capabilities, and also shows a tough yield-reconstruction plugging characteristic, balancing durability and low-energy injection.

The second category emphasizes data- and intelligence-driven low-energy production optimization. Xin et al. [38] proposed a method combining data augmentation (noise perturbation and sliding window) with Bayesian hyperparameter optimization for downhole pressure prediction. A CNN-BiGRU-multi-head attention deep network is constructed to address challenges of limited samples and strong operational disturbances. The proposed model outperforms mainstream methods on metrics such as MAE and R^2, achieving an R^2 of 0.9831, and demonstrates strong generalization and engineering practicality. This provides a predictive core for intelligent production management and energy optimization under conditions of reduced trial-and-error and shutdowns. Xin et al. [39] further proposed a TCN-LSTM-AVOA intelligent prediction workflow, which integrates temporal convolutional networks and long short-term memory networks. The nonlinear swarm intelligence algorithm African Vulture Optimization is introduced for automatic key hyperparameter tuning, which effectively captures the time-lag and nonlinear characteristics of high-dimensional time series. On 2D three-phase heterogeneous reservoir data, performance metrics of RMSE 7.0806, MAE 3.4780, R^2 0.9975, and MAPE 1.81% are achieved, and prediction accuracy and robustness are significantly improved. This supports adaptive optimization of injection–production strategies for energy consumption minimization.

The third category focuses on subsurface storage/transport safety and geothermal/thermal property characterization. Li et al. [40] developed crack arrest assessment software for supercritical CO_2 pipelines containing impurities based on the Battelle two-curve model (BTCM). A three-layer architecture using Python (v.3.12.4) and PyQt6 (v.6.10.0) integrates resistance curve and decompression wave models, and a property database covering pure and multicomponent systems (including PR, HEOS, and GERG-2008 equations) is embedded. The software generates pressure drop, decompression, and resistance curves, and quickly identifies pressure plateaus and determines whether crack arrest has occurred. Compared with existing experiments, full-scale burst test data, and the HLP model, the error is within 10%. This tool provides efficient, low-cost safety assessment capability for CCUS transportation, indirectly reducing comprehensive energy consumption from redundant safety margins. Chen et al. [41] used a segmented differential effective medium (SDEM) framework coupled with high-precision experiments. An integrated measurement-modeling study is conducted on the effects of total porosity, grain size, and saturation on the thermal conductivity (λ) of porous rocks, using a self-developed one-dimensional divided-bar apparatus and ultrasonic pulse transmission. Experiments are performed on Fontainebleau sandstone and quartz sand packs of different grain sizes under dry and saturated conditions up to 2000 psi. The SDEM provides significantly better predictions of λ for saturated sand packs: fine-grained materials exhibit the highest λ and compressional wave velocity and are more stress-sensitive. Shear wave velocity is greater in coarse-grained materials. The model fits both dry and saturated data with a single parameter, and also enables a thermal fluid substitution analogous to Gassmann's equation. This enhances in situ thermal conductivity prediction capability based on acoustic logs and provides a basis for the low-energy design of geothermal and reservoir thermal management systems.

This Special Issue, "New Advances in Low-Energy Processes for Geo-Energy Development: 2nd Edition", focuses on key advances in the following areas: low-energy stimulation and wellbore process intensification for unconventional and deep complex reservoirs (including proppant transport geometry optimization, integrated acidizing–flowback, one-trip thermal-sensitive screen completion, intelligent characterization and cluster optimization of gravel distribution, deep high-performance conformance control, and inorganic composite water-plugging materials); AI-driven prediction and control for the entire lifecycle of oil

 https://doi.org/10.3390/pr14111706

and gas production; and safety and thermophysical property modeling for CCUS and geothermal utilization. The related work provides reusable methods and tools for reducing pumping energy and circulating fluid consumption, lowering tripping and shutdown frequency, improving fluid and energy use efficiency, and strengthening safety boundaries. Looking ahead, more innovations are expected in cross-scale mechanism and data integration, closed-loop field trials and digital twins, as well as low-energy processes for CCUS/hydrogen storage and transportation and comprehensive geothermal utilization, to continuously advance geo-energy development toward higher efficiency, safety, and sustainability.

Conflicts of Interest: The author declares no conflict of interest.

References

1. Yin, Y.; Zhang, L.-W.; Mei, K.-Y.; Cheng, X.-W.; Gan, M.-G.; Wang, Y.; Zhang, C.-M. CT investigation on oilwell cement deterioration caused by H_2S along a leaking channel under high temperature: Insights for geothermal applications. *Pet. Sci.* **2026**, *23*, 2235–2247. [CrossRef]
2. Tahir, M.; Farzaneh, A.; Steineder, D.; Mahi, M.; Enzendorfer, C.; Dardalic, M.; König, M. A Comparative Study of Acid Stimulation Techniques and Acid Recipes for a Low-Permeability and High-Temperature Dolomite Reservoir. *SPE J.* **2026**, *in press*. [CrossRef]
3. Lei, Z.; Dou, X.; Wang, Q.; Wang, R.; Ji, D.; Chen, Z.; Xing, G. A semi-analytical model of a hydraulically fractured horizontal well with pre-Darcy flow and stimulated reservoir volume in a radial composite shale reservoir. *SPE J.* **2025**, *30*, 743–761. [CrossRef]
4. Kachuma, D.; Hasanzade, R.; Tomin, P.; Thomadakis, M.E.; Franc, J.; Magri, V.A.; Byer, T.J.; Cusini, M.; Settgast, R.R.; Gross, H. High-Resolution Simulations of Geological CO_2 Injection: Application to the SPE11 Benchmark. *SPE J.* **2026**, *31*, 680–696. [CrossRef]
5. Maciel, F.S.; Galvin-Carney, J.E.; Waltrich, P.J. Low-Cost Detection of Gas Influx in Oil Well Drilling from Acoustic Logging Data. *SPE J.* **2026**, *in press*. [CrossRef]
6. Nguyen, Q.M.; Onur, M. A deep-learning-based reservoir surrogate for performance forecast and nonlinearly constrained life-cycle production optimization under geological uncertainty. *SPE J.* **2025**, *30*, 3931–3949. [CrossRef]
7. Duman, B. A real-time green and lightweight model for detection of liquefied petroleum gas cylinder surface defects based on YOLOv5. *Appl. Sci.* **2025**, *15*, 458. [CrossRef]
8. Ma, Z.; Hu, H.; Zhou, X.; Zhang, H.; Zhang, Y.; Li, G.; Tian, S.; Wang, T. Interpretable automated machine learning workflow for intelligent drilling in the petroleum industry: Case study on rate of penetration prediction. *SPE J.* **2025**, *30*, 3240–3259. [CrossRef]
9. Hui, G.; Ren, Y.; Bi, J.; Wang, M.; Liu, C. Artificial intelligence applications and challenges in oil and gas exploration and development. *Adv. Geo-Energy Res.* **2025**, *17*, 179–183. [CrossRef]
10. Lee, J.S.; Tsang, Y.F.; Kwon, E.E.; Sim, S.J. Upgrading microalgae for petroleum alternatives: CO_2 upcycling for photosynthesis-based society. *Renew. Sustain. Energy Rev.* **2025**, *222*, 116025. [CrossRef]
11. Jia, Z.; Liu, Q.; Kuang, Q.; Zou, Q.; Su, R.; Zeng, Z.; Li, L.; Ma, X.; Wu, Y. Nitrogen-doped ultramicroporous carbon spheres for efficient, low-energy CO_2 capture: Mechanical compaction assisted activation enhances ultramicroporosity and solar-driven desorption. *Chem. Eng. Sci.* **2025**, *320*, 122463. [CrossRef]
12. Busaeri, N.; Hiron, N.; Meylani, V.; Wulandana, R.; Fronitasari, D. Low-Energy Carbon Capture with Zeolite Molecular Sieves: A Study of Cascade Filter Efficiency in ccs Prototypes. *Energy Built Environ.* **2026**, *in press*. [CrossRef]
13. Long, L.; Jinrui, Z.; Qinghua, F.; Dan, Z. Algebraic multigrid method for tight oil reservoir simulation with nonlinear seepage. *Geosystem Eng.* **2025**, *28*, 112–118. [CrossRef]
14. Magri, V.; Castelletto, N.; Osei-Kuffuor, D.; Settgast, R.; Cusini, M.; Tobin, W.; Aronson, R.; Tomin, P. A Multigrid Reduction Framework for Efficient and Scalable Multiphysics Simulations. In *Proceedings of the SPE Reservoir Simulation Conference, Galveston, TX, USA, 25–27 March 2025*; SPE: Galveston, TX, USA, 2025.
15. Fu, S. Multiscale Space Enhanced Preconditioners for Compressible Multiphase Flow in Highly Heterogeneous Media. *SPE J.* **2026**, *31*, 1932–1949. [CrossRef]
16. Shah, M.; Prajapati, M.; Pardiwala, J.M. Nanotechnology origination: A development path for petroleum upstream industry. *Chem. Pap.* **2025**, *79*, 2037–2051. [CrossRef]
17. El-Masry, J.F.; Maalouf, E.; Abbas, A.H.; Bou-Hamdan, K.F. Advancements in green materials for chemical enhanced oil recovery: A review. *Pet. Res.* **2025**, *11*, 259–276. [CrossRef]

https://doi.org/10.3390/pr14111706

18. Liao, Y.-Q.; Wang, T.-T.; He, T.; Xie, D.-Z.; Xie, K.; Yang, C.-H. Petroleum recovery from salt cavern through natural gas displacement: Insights from a gas–oil two-phase flow model with gas dissolution and exsolution. *Pet. Sci.* **2025**, *22*, 4226–4239. [CrossRef]
19. Pang, X.-J.; Wang, G.-W.; Zhang, Y.-J.; Yue, D.-L.; Li, H.-B.; Kuang, L.-C.; Li, C.-L. Insights into the pore structure and hydrocarbon accumulation of lacustrine organic-rich shales. *Pet. Sci.* **2025**, *22*, 957–976. [CrossRef]
20. Xia, Y.-X.; Elsworth, D.; Xu, S.; Xia, X.-Z.; Cai, J.-C.; Lu, C. Pore-scale gas–water two-phase flow and relative permeability characteristics of disassociated hydrate reservoir. *Pet. Sci.* **2025**, *22*, 3344–3356. [CrossRef]
21. Han, S.-C.; Li, L.-L.; Hu, S.; Jing, C.-H.; Wu, X.-D.; Yang, J.-C.; Liu, T.; Xiao, C.-Y. Experimental study of acid fracturing behavior in carbonate reservoirs with different fracture-cavity development. *Pet. Sci.* **2025**, *22*, 2937–2949. [CrossRef]
22. Kozhevnikov, E.; Belkov, F.; Turbakov, M.; Ivanov, Z.; Riabokon, E.; Fedurov, D.; Guzev, M.; Gladkikh, E.; Wu, J. An Experimental Study of Proppant Transport Mechanisms to the Pump Intake in Low-Flow Oil Wells. *Pet. Res.* **2026**, *11*, 163–175. [CrossRef]
23. Yin, B.; Lou, Y.-S.; Liu, S.-Y.; Zhang, Y. Mechanism of proppant transport and deposition in rough intersecting fractures after offshore fracturing. *Pet. Sci.* **2025**, *22*, 1270–1288. [CrossRef]
24. Zhu, D.; Bai, B.; Hou, J. Polymer gel systems for water management in high-temperature petroleum reservoirs: A chemical review. *Energy Fuels* **2017**, *31*, 13063–13087. [CrossRef]
25. Zhu, D.; Peng, S.; Zhao, S.; Wei, M.; Bai, B. Comprehensive review of sealant materials for leakage remediation technology in geological CO_2 capture and storage process. *Energy Fuels* **2021**, *35*, 4711–4742. [CrossRef]
26. Zhu, D.; Hou, J.; Wei, Q.; Wu, X.; Bai, B. Terpolymer gel system formed by resorcinol–hexamethylenetetramine for water management in extremely high-temperature reservoirs. *Energy Fuels* **2017**, *31*, 1519–1528. [CrossRef]
27. Wang, J.; Cai, Y.; Liu, D.; Gao, C.; Wang, Z.; Yang, C.; Lu, J.; Hu, J. Liquid Nitrogen Cryogenic Fracturing in Coalbed Methane Reservoirs: A Comprehensive Review. *SPE J.* **2026**, *in press*. [CrossRef]
28. Guo, S.; Zhu, D. Mini-Review of Black Nanosheets for Enhanced Oil Recovery Used in Low-Permeability/Ultra-Low-Permeability Reservoirs. *Energy Fuels* **2025**, *39*, 16768–16793. [CrossRef]
29. Guo, S.; Cheng, H.-B.; Tan, H.-G.; Li, H.-Y.; Zhang, J.; Gao, Y.-Q.; Zhu, D.-Y. Huff-n-puff recovery performance and mechanism analysis of black nanosheets in low-permeability reservoirs based on NMR technology. *Pet. Sci.* **2025**, *22*, 2992–3004. [CrossRef]
30. Zhu, D.; Zhao, Q.; Chen, P.; Lu, J.; Yang, Y.; Guo, S.; Zhang, T. Laboratory evaluation of antileakage performance against CO_2 of alkali-activated gel-reinforced cement for carbon capture, utilization, and storage. *SPE J.* **2025**, *30*, 3776–3791. [CrossRef]
31. Zhang, J.; Wu, Y.; Liu, P.; Wang, C.; Liu, P.; Xi, C. Experimental investigation on enhanced oil recovery and carbon storage by multimedia synergistic electrical heating-assisted CO_2 stimulation in developing medium-deep heavy oil reservoirs. *SPE J.* **2025**, *30*, 6406–6427. [CrossRef]
32. Lu, Z.; Qu, H.; Liu, Y.; Liu, Z.; Liu, S.; Zhang, P.; You, K. Experimental Study on Proppant Transport and Distribution in Asymmetric Branched Fractures. *Processes* **2025**, *13*, 3482. [CrossRef]
33. Wang, L.; Shi, M.; Chen, Y.; Wang, T.; Wang, J. An Integrated Tubing String for Synergistic Acidizing-Flowback: Simulation and Optimization Targeting Offshore Dongying Formation. *Processes* **2025**, *13*, 3582. [CrossRef]
34. Zhao, Y.; Wei, Y.; Qin, X.; Ran, Y. A New Kind of Thermosensitive Screen Used for Wellbore Stability. *Processes* **2025**, *13*, 3674. [CrossRef]
35. Lv, Z.; Xu, J.; Liang, T.; Li, P.; Chen, X.; Cheng, H.; Zhang, Y. Quantitative Evaluation Methods and Applications for Gravel Characteristics Distribution in Conglomerate Reservoirs. *Processes* **2025**, *13*, 3911. [CrossRef]
36. Xu, G.; Li, X.; Yang, J.; Tong, C.; Wang, X.; Wang, T. Pre-Crosslinked Gel Particles Enhanced by Amphiphilic Nanocarbon Dots in Harsh Reservoirs: Synthesis and Deep Stimulation Mechanism. *Processes* **2025**, *13*, 3994. [CrossRef]
37. He, S.; Wang, T. Biochar-Enhanced Inorganic Gel for Water Plugging in High-Temperature and High-Salinity Fracture-Vuggy Reservoirs. *Processes* **2026**, *14*, 1014. [CrossRef]
38. Xin, X.; Jiang, X.; Liu, S.; Yu, G.; Jiang, X. A Method for Predicting Bottomhole Pressure Based on Data Augmentation and Hyperparameter Optimisation. *Processes* **2026**, *14*, 1194. [CrossRef]
39. Xin, X.; Xie, Z.; Liu, S.; Yu, G.; Cao, J. Smart Oil Production Forecasting Process Using Deep Learning and African Vulture Optimization Algorithm. *Processes* **2026**, *14*, 1558. [CrossRef]
40. Li, X.; Wang, D.; Jiang, X.; Yu, Y.; Xing, X. Development and Application of Software for Calculating the Crack Arrest Toughness of Impurity-Containing Carbon Dioxide Pipelines Based on the BTCM. *Processes* **2025**, *13*, 3807. [CrossRef]
41. Chen, Y.; Myers, M.T.; Hathon, L.; Unomah, G.C.; Myers, D. Staged Effective Medium Modeling and Experimental Validation for Rock Thermal Conductivity. *Processes* **2026**, *14*, 1437. [CrossRef]

 https://doi.org/10.3390/pr14111706

processes

MDPI

Article

Experimental Study on Proppant Transport and Distribution in Asymmetric Branched Fractures

Zhitian Lu [1], Hai Qu [2,*], Ying Liu [1], Zhonghua Liu [1], Su Liu [1], Pengcheng Zhang [1] and Kaige You [1]

1 School of Petroleum and Natural Gas Engineering, Chongqing University of Science and Technology, Chongqing 401331, China; lzt1604375357@163.com (Z.L.); yingliu19851204@gmail.com (Y.L.); liuzhonghua@cqust.edu.cn (Z.L.); oscarliu011101@163.com (S.L.); zhangpc021102@163.com (P.Z.); youkaige702@163.com (K.Y.)
2 School of Energy Resources, China University of Geosciences, Beijing 100029, China
* Correspondence: quhai729@163.com

Abstract

Hydraulic fracturing is a key technique for creating complex fractures in unconventional reservoirs to enhance energy recovery. Asymmetric branched fractures, as fundamental units, are widely observed in complex fracture networks. Effective proppant distribution within such structures is critical but remains poorly understood. To investigate this, a rough-walled slot with two branches was developed, where asymmetry was introduced by inserting plates with different geometries on one side. The results show that the structural asymmetry between the left and right branches can significantly induce non-uniform transport and irregular sand bed morphology. Reducing the height and width of branch fractures increases fluid velocity, limiting proppant settling within the branch. As the flow area decreases, the fluid velocity increases, driving more proppant through the branch toward the distal fracture region. Injection pressure increases as the flow area of the branch fracture decreases. At a height ratio of 0.25, sand plugging and ineffective proppant placement probably occur within the natural fracture. When the branch is located at the upper section, proppants hardly settle to form a bed, leading to closure of the fracture. The study provides new insights into optimizing proppant placement in complex fractures.

Keywords: proppant transport; branched fracture; sand bed; particle settling; hydraulic fracturing

1. Introduction

Hydraulic fracturing is a key technique for creating complex fracture systems in unconventional oil and gas reservoirs, thereby enhancing energy recovery [1]. Due to in situ stress heterogeneity and natural cracks, asymmetric branched fractures are fundamental units in the system that commonly occur and play a crucial role in fluid flow [2,3]. Ensuring effective proppant distribution within such structures is essential to sustaining long-term fracture conductivity [4,5]. Hydraulic fracturing tends to generate complex fracture networks in unconventional reservoirs because of the presence of well-developed natural fractures, as shown in Figure 1. Due to the random geometry of natural fractures, asymmetric branched fractures are typically characterized by differences in height and width between the two branches, resulting in unequal cross-sectional areas [6]. Such asymmetry disrupts local flow dynamics, leading to non-uniform proppant placement, decreased fracture conductivity, and reduced contribution from affected zones [7]. However, there is currently little research on proppant distribution in such structures, and it remains

https://doi.org/10.3390/pr13113482

poorly understood. An in-depth study of particle transport and distribution in asymmetric branched fractures become inevitably significant.

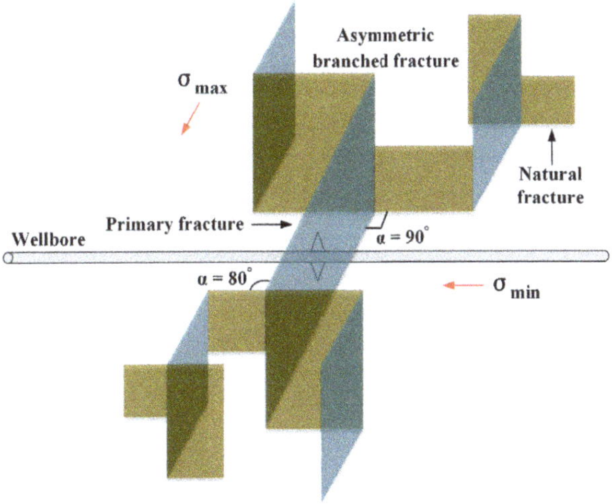

Figure 1. Complex fracture systems in the reservoir with natural fractures, including primary fractures, natural fractures and asymmetric branched fractures.

Proppant transport experiments provide an effective approach to visualizing and better understanding proppant transport across various fracture geometries, including planar [8], branched [9], and complex fracture networks [10]. Various straight smooth slots have been developed under dynamic and geometric similarity, without considering fluid leak off through the walls [11]. Model dimensions are generally under 3 m long, 0.5 m high, and 2–10 mm wide [12]. The injection parameters can be aligned with field conditions, including fluid velocity in the fracture, proppant mesh size, fluid type (slickwater or clear water), and sand concentration. Kern conducted pioneering experiments using a planar acrylic slot [13]. The sand quickly settled to the slot bottom and formed an initial bed. As more sand accumulated, the bed height would reach an equilibrium height. Then, sand transport was primarily dominated by fluidization and settling, progressively filling the fracture. The equilibrium height correlates power law with the fluid velocity, proppant properties, and proppant concentration [14]. The correlation evaluates fracturing design and treatment [15]. Attaching transparent plates with a certain level of surface roughness to the smooth walls allows for visualization and analysis of the influence of wall roughness on particle migration [16]. The roughness enhances particle–wall interactions by increasing collisions and friction, which alters particle trajectories, reduces their velocity, and may result in particles becoming trapped or deposited along the walls. The rough walls hinder particle transport, forming a higher sand bed that extends closer to the wellbore [17]. However, the planar slot has constant height and width along the flow direction, while the effects of changes in flow area on proppant transport remain largely unexplored.

Recent studies have introduced advanced models to explore proppant transport within complex fracture networks. Table 1 lists transparent complex slots. Sahai developed a large, smooth fracture model consisting of one primary slot and two secondary slots [18]. They found that dune accumulation within the primary slot strongly affects proppant entry into the secondary slots. Two distinct mechanisms were identified: (1) when the pump rate exceeds a threshold (corresponding to a critical velocity in the primary slot), proppants are carried around the slot corner into the secondary slot; and (2) gravity-driven particles fall from the primary slot into the secondary slot, occurring regardless of the pump rate. Tong and Mohanty designed a complex model featuring a primary slot intersected by a secondary slot at angles ranging from 30° to 90° [9]. Their results showed that proppant placement in the secondary slot increases with decreasing intersection angle and increasing flow rate. Wen developed a slot to simulate complex fracture networks [19]. A vortex develops at the fracture intersection, which erodes the sand bed and reduces its height at the junction. These turning points can significantly decrease the propped area. Thus, proppant transport and distribution in complex fractures differ markedly from those in straight fractures. Nevertheless, most existing models assume constant fracture height and width, which limits their ability to capture the influence of geometric variability.

Table 1. Model structure and parameters for proppant transport.

Reference	Length (m)	Height (m)	Width (mm)	Inner Wall	Intersection Angle (°)	Liquid	Proppant
Sahai, Miskimins and Olson [18]	P = 1.2 S = 0.3	0.6	5.4	Smooth	90	Water	Sand
Tong and Mohanty [9]	P = 0.38 S = 0.19	0.076	2.0	Smooth	30, 60, 90	Water	Sand
Wen, Wang and Duan [19]	P = 0.6 S = 0.3 T = 0.6	0.4	10	Smooth	90	Water Guar gum	Ceramsite
Alotaibi and Miskimins [16]	P = 1.2 S = 0.3	0.6	5.4	Rough	90	Water	Sand

P = Primary fracture; S = Secondary fracture; T = Tertiary fracture.

Existing experimental models for proppant transport generally assume constant fracture dimensions in the width and height directions. Moreover, experimental studies on asymmetric, branched fracture models are lacking, and the characteristics of proppant placement within such complex fracture geometries remain unclear. To address these gaps, this study is the first to experimentally investigate proppant transport in asymmetric, branched fractures. A rough-walled slot with two branches was developed, and plastic plates of varying heights and widths were placed on one side of the fracture to introduce controlled geometric asymmetry. The effects of fracture width, height, and branch position on proppant distribution are systematically analyzed, enabling a quantitative assessment of how fracture asymmetry influences proppant placement. The findings provide new insights into fracture-scale proppant distribution and offer guidance for optimizing hydraulic fracturing in naturally fractured reservoirs.

2. Materials and Methods

2.1. Experimental Apparatus

The T-shaped slot was constructed using acrylic sheets due to their favorable mechanical and optical properties. The sheets possess good toughness, allowing them to be bent multiple times to form the desired curvature without cracking. In addition, their high transparency enables clear observation of particle–fluid interactions during the experiments. The thickness of each acrylic sheet was 10 mm to ensure adequate strength and to prevent deformation under injection pressure conditions. Due to the difficulty of realizing fluid leak-off in transparent plates, proppant transport experiments based on flat models typically neglect the leak-off effect along the fracture walls. If leak-off were considered, the fluid velocity within the fracture would be reduced to some extent. Figure 2 shows the particle transport system. The slot is designed to simulate particle-fluid flow in an asymmetric fracture. A primary fracture (PF) intersects a natural fracture (NF) at 90°, continues to extend within it, and induces secondary fractures, also oriented at 90°. The left branch consists of NF_1 and SF_1, and the right branch consists of NF_2 and SF_2. Considering the limitations of pump capacity and the pressure tolerance of the acrylic plates, the slot height typically ranges from 0.1 to 0.6 m [10]. Both branches are of equal dimensions, each with an inner height of 0.27 m. In addition, due to the length constraint of the acrylic plates, the total internal length of the slot was kept within 3 m to facilitate assembly and ensure structural stability [18]. The PF has a length of 0.50 m, and the lengths of NF_1 and NF_2 are 0.3 m, and the secondary fractures SF_1 and SF_2 are each 0.48 m long.

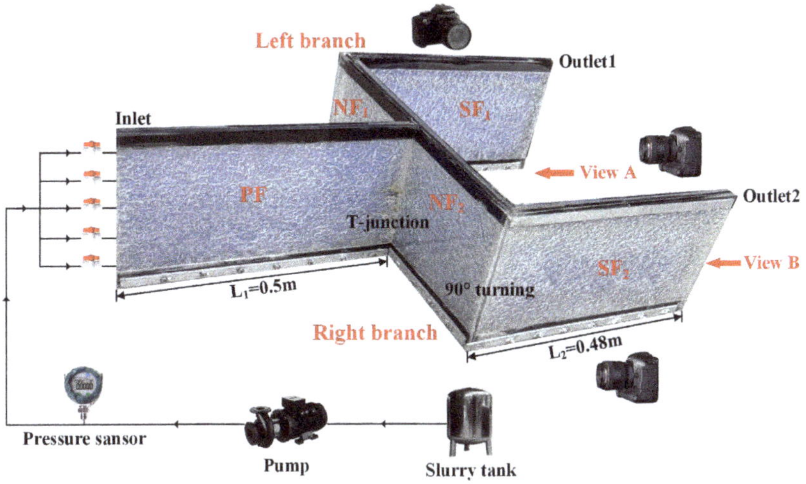

Figure 2. Experimental apparatus for proppant transport.

The perforations on the casing had diameters ranging from 8 to 12 mm, with a perforation density of 16–20 holes per meter. Based on this standard, five 10 mm-diameter perforations were arranged along the inlet, spaced 50 mm apart. The slurry is injected from five holes into the slot. The outlet, located at the top-right corner, had an opening height of 0.07 m, and the overall height of the right boundary was 0.2 m. The outlet remained open throughout the experiments, and no backflow pressure was applied. Injection pressure is monitored by a pressure sensor installed at the inlet and displayed in real time by a pressure gauge. Three cameras are mounted perpendicular to the slots.

Figure 3a presents the NF at View A. Two red dashed lines mark the PF aperture. Roughness characteristics of the plates were replicated from natural rock surfaces, as documented in [20]. Two sets of rough panels were combined to form the fracture channel. A white opaque plastic plate of defined dimensions is inserted into the NF_2 to generate an asymmetric configuration. Varying the height and thickness of the panels allows the left and right branches to differ in size, resulting in a non-symmetrical fracture configuration. In Figure 3a, the panel height is 0.135 m, and the pathway height is 0.135 m. Figure 3b shows a local profile of Outlet2. The aperture ranges from 3.19 mm to 5.22 mm. The real fracture width typically varies between 1 and 6 mm, and the average width is 4 mm. The joint roughness coefficient (JRC), which quantifies the roughness of rock fracture surfaces and ranges from 0 to 20, was calculated as 11.5 for the rough plate using the empirical equation provided in the study [21]. A JRC value exceeding 10 is generally considered representative of rock fractures.

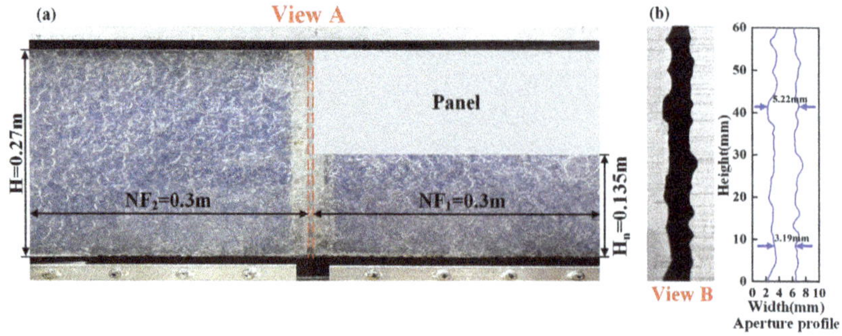

Figure 3. (**a**) NF slot at View A; (**b**) Rough profiles of rough pathway.

2.2. Experimental Materials

The slurry is prepared using 40/70 mesh sand and slickwater. In Figure 4, the average diameter of sands is 311 μm, and the size ranges from 0.106 mm to 0.43 mm. The bulk density of the sand is 1670 kg/m^3, and the slickwater viscosity is 1.5 mPa·s. The parameters are consistent with the field condition [1].

Figure 4. (**a**) 40/70 mesh sands; (**b**) The measurement of a laser particle sizer.

2.3. Experimental Scheme

Table 2 summarizes the experimental scheme. Case 1, with symmetric left and right branches, is the benchmark. In Cases 2–4 and 5–7, panels are inserted at the top and bottom of NF_1, respectively, creating flow pathways at the opposite sides. In Cases 8–9, a transparent panel is inserted into NF_1 to modify its width. The panel dimensions are 270 mm in height and 300 mm in length, with thicknesses of 3, 2, and 1 mm. The injection rate in the primary fracture (PF) is 2.93 m^3/h, corresponding to an average flow velocity of 0.42 m/s. The velocity is within the typical range of fluid velocities (0.1–0.5 m/s) observed in real hydraulic fractures [16,18]. The sand volume fraction is a constant value of 3%.

Table 2. Experimental scheme.

Case	Fracture Height (mm)	Panel Position	Fracture Weight (mm)	Height Ratio H_r	Width Ratio W_r
1	270	/	4	1	1
2	NF_1 = 202.5	Bottom	4	0.75	1
3	NF_1 = 135	Bottom	4	0.5	1
4	NF_1 = 67.5	Bottom	4	0.25	1
5	NF_1 = 202.5	Top	4	0.75	1
6	NF_1 = 135	Top	4	0.5	1
7	NF_1 = 67.5	Top	4	0.25	1
8	270	/	NF_1 = 3	1	0.75
9	270	/	NF_1 = 2	1	0.5
10	270	/	NF_1 = 1	1	0.25

2.4. Dimensionless Parameters

The proppant bed dimensions, including height and length, were nondimensionalized to facilitate analysis, as expressed in Equations (1) and (2).

$$r_L = \frac{L_s}{L} \tag{1}$$

$$r_H = \frac{H_s}{H} \tag{2}$$

where r_L is the normalized length; r_H is the normalized height; L_s is the sand bed length, m; H_s is the sand bed height, m; L is the fracture lenght, m; and H is the fracture height, m.

Equation (3) indicates the coverage of the proppant bed within the fracture.

$$CR = \frac{A_s}{A} \times 100\% \tag{3}$$

where CR is the area coverage ratio; A is the cross-sectional area of the fracture, calculated along the length and height directions, m^2; and A_s is the sand bed area, m^2.

3. Results and Analysis

3.1. Proppant Distribution Evolution

Due to lithological influences, the height of the NF_1 is constrained to half of NF_2. Figure 5 shows the bed geometries in the slot at four moments for Case 3 in Table 2. At the initial stage, the asymmetric structure is found to have little influence on the sand bed morphology. The 40/70 mesh sands can pass through the T-junction and negotiate two 90° turns, depositing along the entire slot, while some particles remain suspended and exit directly through the outlet. During this process, strong collisions occur at the junction and turns. Frequent collisions with convex surfaces deflect sands, while impacts with concave regions lead to particle trapping and localized deposition, which is identical to the literature [17]. The trapped sands are later dispersed by the flowing fluid, facilitating the dynamic redistribution of proppants within the slot.

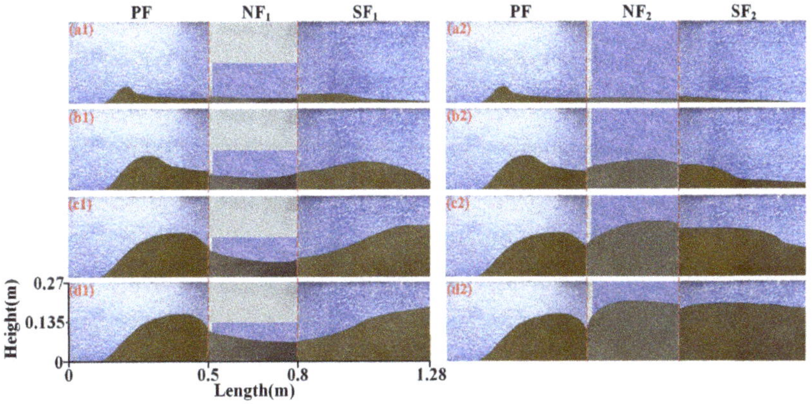

Figure 5. Snapshots of 40/70 mesh sand beds at four moments. (**a1**,**a2**) 250 s; (**b1**,**b2**) 500 s; (**c1**,**c2**) 750 s; (**d1**,**d2**) 1000 s.

At 500 s, the asymmetric structure of the NF begins to impact the sand distribution. With no obstruction in NF_2, the lower fluid velocity facilitates sand settling, resulting in a taller sand bed. Sands undergo frequent collisions near the turns in NF_2 and SF_2, where changes in flow direction and velocity gradients enhance particle–wall interactions. These collisions promote particle deposition, leading to locally elevated sand beds around the 90° turn. As the height of NF_1 is reduced by half, the fluid entering NF_1 from the PF contracts due to the decreased cross- sectional area, resulting in an increase in fluid velocity. This acceleration enhances downward-directed shear stress, generating erosion that forms a downward-sloping sand bed in NF_1. When the flow moves from NF_1 into SF_1, it expands, causing a sudden reduction in velocity and turbulence near the bed surface. This expansion induces surface erosion and leads to the formation of an upward-sloping sand bed in SF_1.

At 1000 s, the beds reach equilibrium. In the left branch, the sand bed further reduces the flow area of NF_1. The resulting high-velocity erosion produces a lower equilibrium height and a more irregular bed morphology, with an upward-sloping bed formed in SF_1. In contrast, in the right branch, sands settle more readily, resulting in a higher and more uniform equilibrium sand bed. It proves that the significant influence of geometric asymmetry on local flow dynamics, particle–wall interactions, and proppant distribution patterns within branched fractures. Local vortices induced by flow separation and recirculation near the inner corners of the turning sections erode the sand bed surface, resulting in the formation of small pits. This phenomenon is consistent with the observations reported by Wen and Tong [9,19]. Qu Hai et al. further demonstrated through numerical simulations that vortices generated at the turning sections of narrow fractures can erode the sand bed surface, leading to the formation of small pits [22].

Figure 6a shows normalized profiles of sand beds at four moments. At equilibrium, the bed profile becomes the most regular, while small pits appear near the turns. In Figure 6b, the CR exhibits a logarithmic correlation with time, indicating a rapid early-stage increase that gradually levels off as equilibrium is approached. High flow velocity initially promotes rapid particle transport and deposition, resulting in a sharp rise in coverage ratio. As the particle bed develops, the flow area becomes constricted, leading to an increase in local fluid velocity and enhanced particle washout from the slot. Consequently, the coverage ratio gradually approaches a steady value. After 500 s, the coverage ratio in the right branch exceeds that in the left branch. At equilibrium, the difference in coverage ratio between the left and right branches is the largest. The reduced height of NF_1 decreases the coverage area in the left branch. Furthermore, suppose the injection time is insufficient and the sand bed has not yet reached equilibrium. In that case, the effective proppant coverage will be further reduced, which may significantly compromise the stimulation performance.

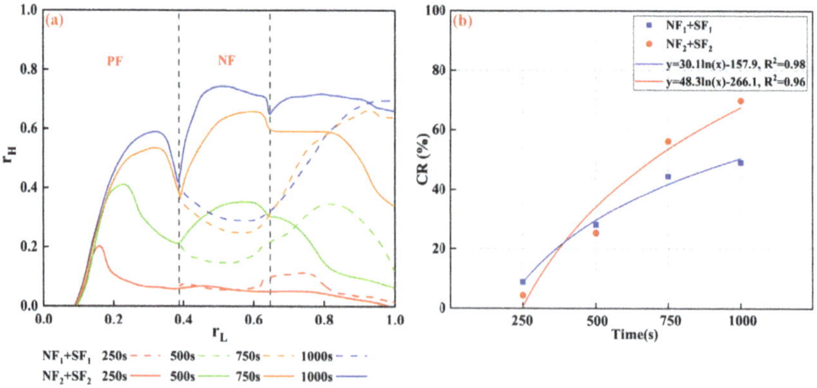

Figure 6. (**a**) Normalized profiles of sand beds at four moments; (**b**) Coverage ratios of sand beds in two branches.

3.2. Proppant Distribution with NF_1 at the Bottom

When the upper section of the left branch experiences higher rock stress, the fracture tends to extend downward. In Figure 7, the NF_1 is at the bottom for Cases 1, 2 to 4 (Table 2). In Figure 7a, the structures of the two branches are identical, and the bed morphologies are also similar, with only small pits observed at the two 90° turns. The sand bed evolution is similar to the bed in the straight slot [7,23]. In Figure 7b, the height of NF_1 is 75% of that of NF_2. Even though NF_1 has a smaller height, it reduces sand bed height and coverage ratio and alters the bed profile. It indicates that the bed morphology is highly sensitive to fracture height. The lower height also increases local fluid velocity, intensifying bed erosion and further limiting proppant placement. In Figure 7d, the height of NF_1 decreases to 67.5 mm. Since NF_1 is positioned lower, when the sand bed height in PF exceeds that of NF_1, sand bridging occurs at the PF–NF_1 turn, forcing the slurry to flow entirely into the left branch. The increased fluid velocity in the left branch subsequently leads to a slight reduction in the bed equilibrium height. Besides, the small NF_1 causes the violent fluid flow in the right branch, and the sand bed in SF_1 develops a steeper slope and exhibits a lower coverage ratio. The sand bed around the turn between NF_1 and SF_1 is discontinuous. Thus, the discontinuous and small sand beds in the right branch cannot support the fracture effectively. The right branch becomes a weak link, and the fracture network connected to SF_1 can probably not contribute to production.

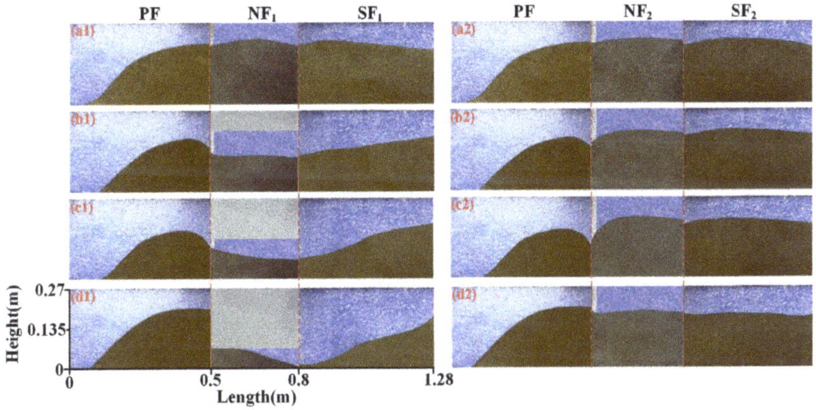

Figure 7. Snapshots of sand beds at the equilibrium with varied height ratios at the bottom of NF_1. (**a1**,**a2**) Two identical branches at H_r = 1; (**b1**,**b2**) H_r = 0.75; (**c1**,**c2**) H_r = 0.5; (**d1**,**d2**) H_r = 0.25.

In Figure 8a, as the height ratio of NF_1 decreases from 1 to 0.25, the bed profile in the left branch becomes increasingly irregular. When CR = 0.25, sand plugging occurs in NF_1, causing the bed in the PF to shift closer to the inlet. In Figure 8b, this change in height ratio causes the CR of NF_1 to decrease from 73.8% to 35.1%, whereas the CR of NF_2 increases from 68.8% to 76.1%. The reduced height of NF_1 directs more sand toward NF_2, leading to an increased bed height in NF_2.

Figure 9 shows the liquid pressure at the inlet. A reduction in NF_1 height increases the fluid flow resistance, leading to higher pumping pressure. For a height ratio of 1, the pathway height remains unchanged, flow resistance is low, and the initial pressure is 35 kPa. When the height ratio is reduced to 0.25, the flow area is constricted, and the initial pressure rises to 42 kPa. As the sand bed gradually accumulates, the effective flow region in the slot decreases, causing a steady increase in liquid pressure. When the coverage ratio reaches 0.25, the injection pressure exhibits a sudden spike from 43.1 kPa to 44.8 kPa at 520 s due to sand bridging at NF_1. Due to the constant pump speed, the reduction in effective flow area resulted in a sudden increase in pressure. The phenomenon is identical to the injection pressure at the field [24].

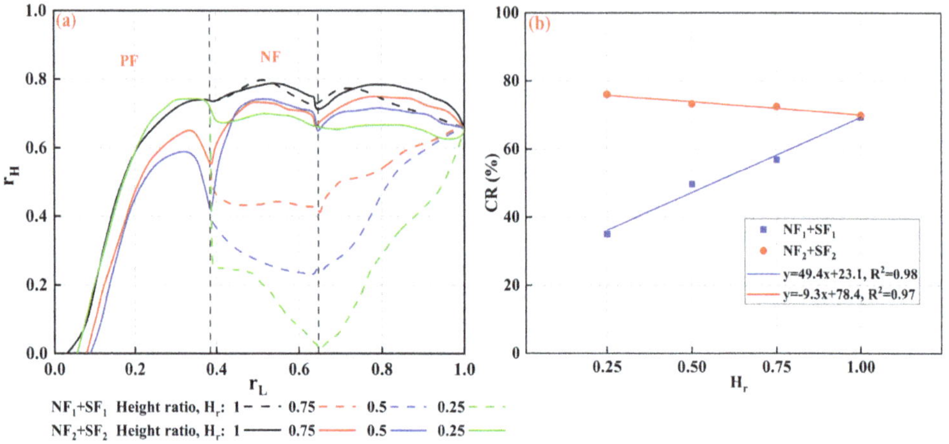

Figure 8. Quantitative characteristics of sand bed for height ratio at the bottom of NF_1. (**a**) Sand bed profiles at equilibrium; (**b**) Coverage ratios of sand beds in two branches.

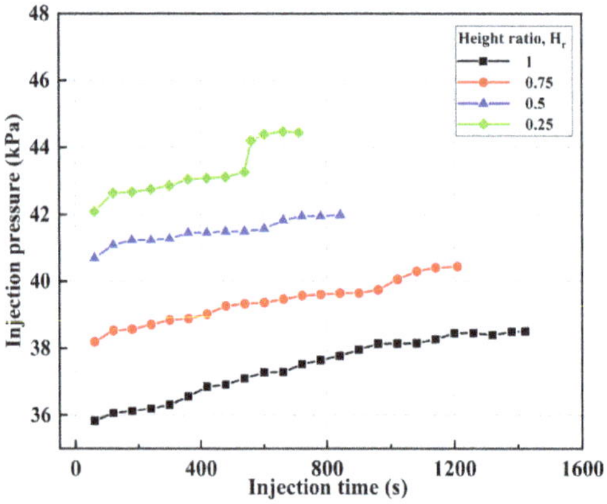

Figure 9. Injection pressure at the inlet with varied height ratios at the bottom of NF_1.

3.3. *Proppant Distribution with NF_1 at the Top*

When the lower section of the left branch experiences higher rock stress, the fracture tends to extend upward. In Figure 10, the NF_1 is at the top for Cases 1, 5 to 7 (Table 2). The asymmetric structure primarily affects the sand distribution in the left branch. The sand bed exhibits minimal morphological differences in the right branch. In Figure 10b, the sand bed in NF_1 becomes slightly thinner, producing a modest asymmetry, while the overall coverage remains relatively extensive. In Figure 10c, the sand bed morphology begins to exhibit changes. The sand bed heights in PF and SF_1 decrease moderately, whereas those in NF_1 and NF_2 remain largely comparable. In Figure 10d, significant changes in sand bed morphology occur within the left branch. Owing to the limited flow area in NF_1, the flow velocity increases markedly, inhibiting sand deposition. A thin sand layer forms mainly near the 90° turn, likely resulting from deposition obstructed by the 90° turn. When the slurry flows from NF_1 into SF_1, the expansion of the flow area induces oblique erosion of the sand bed, forming a concave surface. Also, the sand bed height in SF_1 is reduced.

Because of the lack of proppant support, NF_1 will rapidly close after the fracturing. In complex hydraulic fractures, branches like NF_1 serve as critical bridging links between upstream and downstream sections, governing local fracture connectivity. The result is identical to the literature [25]. Fracture coverage can be improved by reducing proppant transport capacity, for example, through lower injection rates, higher fluid viscosity, or larger proppant sizes.

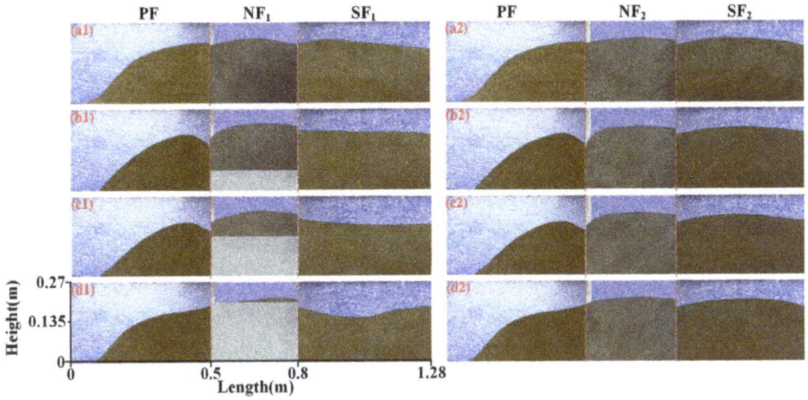

Figure 10. Snapshots of sand beds at the equilibrium with varied height ratios at the top of NF_1. (**a1**,**a2**) Two identical branches at $H_r = 1$; (**b1**,**b2**) $H_r = 0.75$; (**c1**,**c2**) $H_r = 0.5$; (**d1**,**d2**) $H_r = 0.25$.

In Figure 11a, when the H_r is less than 0.5, the bed height in SF_1 is significantly reduced. In Figure 11b, as the H_r of NF_1 decreases from 1 to 0.25, the CR of the left branch decreases from 74.5% to 53.8%, while the average CR of the right branch remains around 74.5%.

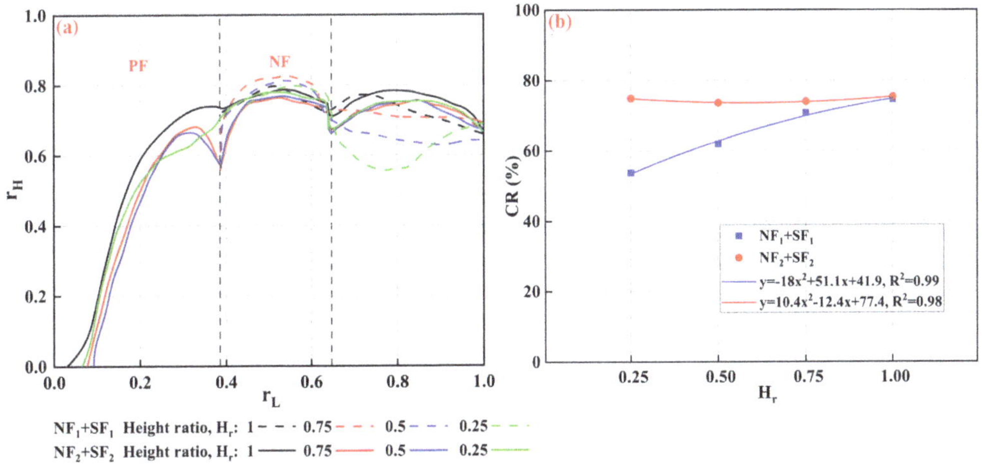

Figure 11. Quantitative characteristics of sand bed for height ratio at the top of NF_1. (**a**) Sand bed profiles at equilibrium; (**b**) Coverage ratios of sand beds in two branches.

In Figure 12, as the NF_1 height decreases, the reduced flow area increases hydraulic resistance, thereby requiring a higher injection pressure. For the height ratio of 0.25, the initial pressure is 41.9 kPa, which is higher than the initial pressure of 35.8 kPa for the height ratio of 1. When H_r is 0.25, the sand bed reaches equilibrium at 1254 s. It is because the flow is turbulent, making it difficult for the sand to settle into the slot, thus requiring a longer injection time.

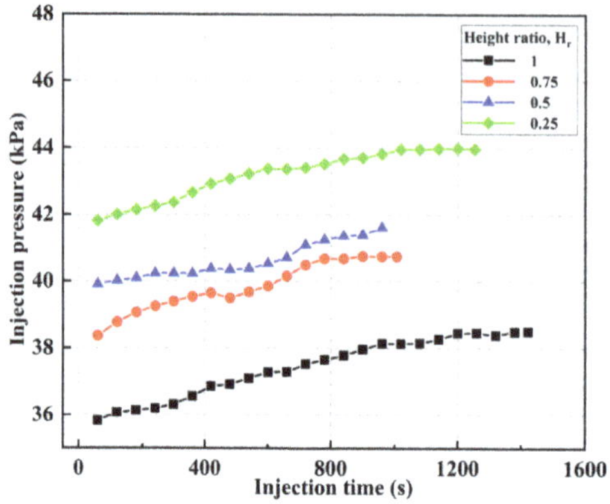

Figure 12. Injection pressure at the inlet with varied height ratios at the top of NF_1.

3.4. *Effect of Branch Width*

When the in situ stress is higher on one side, the fracture width becomes smaller, leading to asymmetric widths between the two branches. In Figure 13, the width of the NF decreases from 4 mm to 2 mm for Cases 1, 8 to 10 (Table 2). As the width of NF_1 decreases, the overall heights of the sand bed in the left branch and in PF both decrease, while the height of the sand bed in the right branch remains largely unchanged. Unlike height variations, changes in fracture width do not lead to irregular sand bed morphologies. It indicates that the effects of fluid contraction and expansion on sand bed morphology in the fracture width direction differ significantly from those in the height direction.

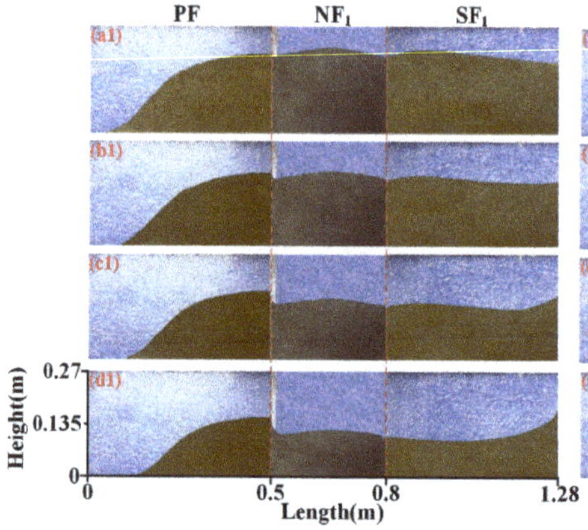

Figure 13. Apshots of sand beds at the equilibrium with varied width ratios. (**a1**,**a2**) Two identical branches at $W_r = 1$; (**b1**,**b2**) $W_r = 0.75$; (**c1**,**c2**) $W_r = 0.5$; (**d1**,**d2**) $W_r = 0.25$.

In Figure 14a, as the width of NF_1 decreases, the sand bed height in both the PF and the left branch is reduced, and when W_r is lower than 0.25, the sand bed height in the right branch decreases significantly. In Figure 14b, as the W_r decreases from 1 to 0.25, the CR of the left branch decreases from 74.5% to 40.8%, and the CR of the right branch decreases from 75.4% to 70.3%.

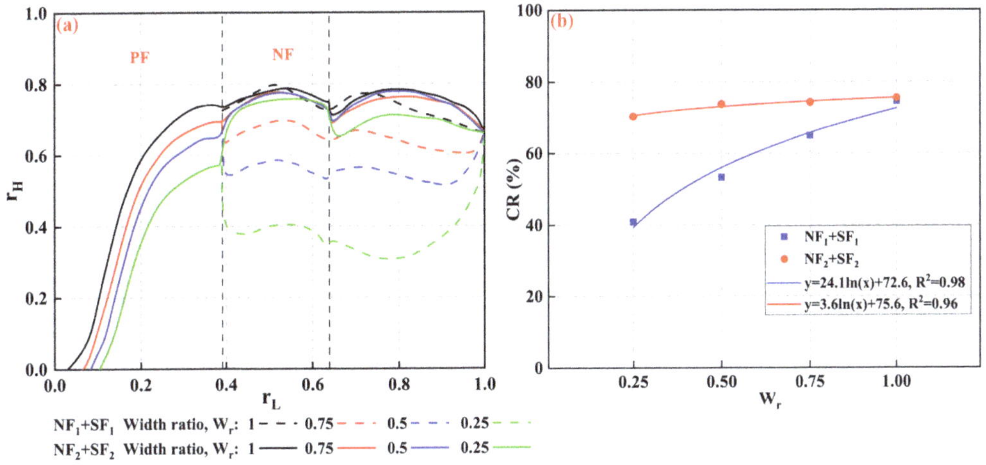

Figure 14. Quantitative characteristics of sand bed for the width ratio. (**a**) Sand bed profiles at equilibrium; (**b**) Coverage ratios of sand beds in two branches.

In Figure 15, as the NF width decreases, flow resistance increases due to the reduced flow area, resulting in higher fluid pressure at the injection point. For a width ratio of 1, the pressure rises from 35.9 kPa to 38.4 kPa, whereas for a width ratio of 0.25, it increases from 40.8 kPa to 42.2 kPa. Notably, the sand bed height in the model with a width ratio of 1 is higher than that in the model with a width ratio of 0.25. These results indicate that fracture width exerts a greater influence on injection pressure than sand bed height. When the width ratio is 0.25, the sand bed reaches equilibrium in the shortest time of 780 s due to the reduced coverage.

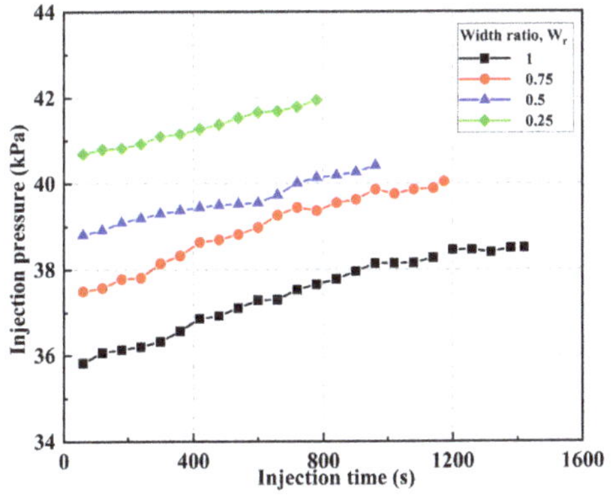

Figure 15. Injection pressure at the inlet with varied branch width.

4. Limitations

Several limitations should be noted. The model is based on a laboratory scale, which differs significantly in size from actual fractures that can extend to hundreds of meters. The slot walls do not possess leak off capability, unlike fracture surfaces in the formation. In addition, the sand concentration was assumed to be constant. Therefore, the particle bed configuration in the model may differ from that in real fractures. In future work, numerical simulations will be combined to account for variations in fracture dimensions and injection parameters, enabling a more in-depth understanding of proppant transport mechanisms in asymmetric fractures.

5. Conclusions

Hydraulic fracturing in naturally fractured reservoirs often generates complex fracture networks, in which asymmetric structures are commonly present. This study is the first to experimentally investigate proppant transport in asymmetric, branched fractures. The effects of fracture width, height, and branch position on proppant distribution are systematically analyzed. Several new insights have been obtained for engineering applications.

1. At a height ratio of 0.25, the branch is located in the lower section, and sand bridging is likely to occur once the sand bed in the primary fracture reaches a certain height. In naturally fractured reservoirs, it is recommended to increase the proportion of small-sized proppants (e.g., 70/140 or 200 mesh) to ensure smooth transport through narrow fractures and minimize the risk of premature bridging.
2. When the branch is located at the upper section, proppants hardly settle to form a bed, leading to closure of the natural fracture. In the later stage of fracturing, lowering the injection rate could enhance particle accumulation in the natural fracture.
3. A reduction in fracture width significantly increases the injection pressure while decreasing the sand bed area. In naturally fractured reservoirs, this leads to higher operational pressures during fracturing. Increasing fluid viscosity or reducing the injection rate decreases injection pressure and particle transport.
4. The bed morphology within asymmetric branch fractures is more irregular than in regular slots, leading to a lower sand bed coverage ratio. The more complex slurry flow in asymmetric fractures can transport particles deeper into the fracture, further reducing coverage. In the later stage of fracturing, lowering the injection rate or using larger-sized proppant can enhance particle settling, improving overall coverage.

Author Contributions: Conceptualization, Z.L. (Zhitian Lu) and H.Q.; methodology, Z.L. (Zhitian Lu); software, Y.L.; validation, Y.L. and P.Z.; formal analysis, Z.L. (Zhonghua Liu); investigation, P.Z. and K.Y.; resources, Y.L.; data curation, S.L.; writing—original draft preparation, Z.L. (Zhitian Lu); writing—review and editing, H.Q.; visualization, S.L.; supervision, H.Q.; funding acquisition, Z.L. (Zhitian Lu), H.Q., Y.L. and Z.L. (Zhonghua Liu). All authors have read and agreed to the published version of the manuscript.

Funding: This research is supported by Chongqing University of Science and Technology Graduate Student Innovation Program Project (YKJCX2420152), and the National Natural Science Foundation of China (Grant No. 52274035), and the Chongqing Natural Science Foundation General Project (Grant No. CSTB2025NSCQ-GPX0914, CSTB2022NSCQ-MSX0917).

Data Availability Statement: The data presented in this study are available on request from the corresponding author.

Conflicts of Interest: The authors declare no conflicts of interest.

Abbreviations

The following abbreviations are used in this manuscript:

A	fracture area, m^2
A_s	sand bed area, m^2
CR	coverage ratio, %
H	fracture height, m
H_r	height ratio
H_s	sand bed height, m
JRC	Joint Roughness Coefficient

L	fracture length, m
L_s	sand bed length, m
NF	natural fractures
NF_1	left branch of natural fractures
NF_2	right branch of natural fractures
PF	primary fracture
r_H	normalized height
r_L	normalized length
SF	secondary fractures
SF_1	left branch of secondary fractures
SF_2	right branch of secondary fractures
W_r	width ratio

References

1. Miskimins, J. *Hydraulic Fracturing: Fundamentals and Advancements*; Society of Petroleum Engineers: The Woodlands, TX, USA, 2020. [CrossRef]
2. Hou, B.; Chang, Z.; Fu, W.; Muhadasi, Y.; Chen, M. Fracture Initiation and Propagation in a Deep Shale Gas Reservoir Subject to an Alternating-Fluid-Injection Hydraulic-Fracturing Treatment. *SPE J.* **2019**, *24*, 1839–1855. [CrossRef]
3. Bahorich, B.; Olson, J.E.; Holder, J. Examining the Effect of Cemented Natural Fractures on Hydraulic Fracture Propagation in Hydrostone Block Experiments. In *SPE Annual Technical Conference and Exhibition*; Society of Petroleum Engineers: San Antonio, TX, USA, 2012; p. 21. [CrossRef]
4. Guo, T.; Tang, S.; Liu, S.; Liu, X.; Xu, J.; Qi, N.; Rui, Z. Physical Simulation of Hydraulic Fracturing of Large-Sized Tight Sandstone Outcrops. *SPE J.* **2021**, *26*, 372–393. [CrossRef]
5. Lv, M.; Qu, Z.; Guo, T.; Chen, M.; Zhang, Y.; Jia, X.; Liu, X.; Liu, Y. Experimental Study on the Transport Law of Different Fracturing Granular Materials in Fracture. *Energy Fuels* **2022**, *36*, 10886–10898. [CrossRef]
6. Liu, Y.; Qu, H.; Lu, Z.; Fei, Y. Experimental investigation of particle transport behaviors and distribution characteristics in a dual-branch fracture with rough surfaces. *Powder Technol.* **2025**, *452*, 120602. [CrossRef]
7. Guo, J.; Gou, H.; Zhang, T.; Zeng, X.; Li, M.; Tang, T. Experiment on proppant transport in near-well area of hydraulic fractures based on PIV/PTV. *Powder Technol.* **2022**, *410*, 117833. [CrossRef]
8. Qu, H.; Wang, R.; Ao, X.; Xue, L.; Liu, Z.; Lin, H. The Investigation of Proppant Particle-Fluid Flow in the Vertical Fracture with a Contracted Aperture. *SPE J.* **2022**, *27*, 274–291. [CrossRef]
9. Tong, S.; Mohanty, K.K. Proppant transport study in fractures with intersections. *Fuel* **2016**, *181*, 463–477. [CrossRef]
10. Isah, A.; Hiba, M.; Al-Azani, K.; Aljawad, M.S.; Mahmoud, M. A comprehensive review of proppant transport in fractured reservoirs: Experimental, numerical, and field aspects. *J. Nat. Gas Sci. Eng.* **2021**, *88*, 103832. [CrossRef]
11. Lv, M.; Guo, T.; Qu, Z.; Chen, M.; Dai, C.; Liu, X. Experimental Study on Proppant Transport within Complex Fractures. *SPE J.* **2022**, *27*, 2960–2979. [CrossRef]
12. Sahai, R.; Moghanloo, R.G. Proppant transport in complex fracture networks—A review. *J. Petrol. Sci. Eng.* **2019**, *182*, 106199. [CrossRef]
13. Kern, L.; Perkins, T.; Wyant, R. The mechanics of sand movement in fracturing. *J. Pet. Technol.* **1959**, *11*, 55–57. [CrossRef]
14. Patankar, N.A.; Joseph, D.; Wang, J.; Barree, R.; Conway, M.; Asadi, M. Power law correlations for sediment transport in pressure driven channel flows. *Int. J. Multiph. Flow* **2002**, *28*, 1269–1292. [CrossRef]
15. Woodworth, T.R.; Miskimins, J.L. Extrapolation of Laboratory Proppant Placement Behavior to the Field in Slickwater Fracturing Applications. In *SPE Hydraulic Fracturing Technology Conference*; Society of Petroleum Engineers: College Station, TX, USA, 2007; p. 12. [CrossRef]
16. Alotaibi, M.A.; Miskimins, J.L. Slickwater Proppant Transport in Hydraulic Fractures: New Experimental Findings and Scalable Correlation. *SPE Prod. Oper.* **2018**, *33*, 164–178. [CrossRef]
17. Huang, H.; Babadagli, T.; Li, H.A.; Develi, K.; Wei, G. Effect of injection parameters on proppant transport in rough vertical fractures: An experimental analysis on visual models. *J. Petrol. Sci. Eng.* **2019**, *180*, 380–395. [CrossRef]
18. Sahai, R.; Miskimins, J.L.; Olson, K.E. Laboratory Results of Proppant Transport in Complex Fracture Systems. In *SPE Hydraulic Fracturing Technology Conference*; Society of Petroleum Engineers: The Woodlands, TX, USA, 2014; p. 26. [CrossRef]
19. Wen, Q.; Wang, S.; Duan, X.; Li, Y.; Wang, F.; Jin, X. Experimental investigation of proppant settling in complex hydraulic-natural fracture system in shale reservoirs. *J. Nat. Gas Sci. Eng.* **2016**, *33*, 70–80. [CrossRef]
20. Qu, H.; Xu, Y.; Liu, Y.; Li, Z.; Liu, X.; Zeng, Z.; Guo, R. Experimental study of fluid-particle flow characteristics in a rough fracture. *Energy* **2023**, *285*, 129380. [CrossRef]

21. Tse, R. Estimating joint roughness coefficients. *Int. J. Rock Mech. Min. Sci. Geomech. Abstr.* **1979**, *16*, 303–307. [CrossRef]
22. Qu, H.; Liu, Y.; Lin, H.; Tang, S.; Wang, R.; Xue, L.; Hu, Y. 3D CFD-DEM simulation and experiment on proppant particle-fluid flow in a vertical, nonplanar fracture with bends. *Int. J. Multiph. Flow* **2022**, *146*, 103873. [CrossRef]
23. Chun, T.; Zhu, D.; Zhang, Z.; Mao, S.; Wu, K. Experimental Study of Proppant Transport in Complex Fractures with Horizontal Bedding Planes for Slickwater Fracturing. *SPE Prod. Oper.* **2020**, *Preprint*, 14. [CrossRef]
24. Nolte, K.G. Application of Fracture Design Based on Pressure Analysis. *SPE Prod. Eng.* **1988**, *3*, 31–42. [CrossRef]
25. Wang, X.; Yao, J.; Gong, L.; Sun, H.; Liu, W. Numerical simulations of proppant deposition and transport characteristics in hydraulic fractures and fracture networks. *J. Petrol. Sci. Eng.* **2019**, *183*, 106401. [CrossRef]

processes

MDPI

Article

An Integrated Tubing String for Synergistic Acidizing-Flowback: Simulation and Optimization Targeting Offshore Dongying Formation

Liangliang Wang [1,2], Minghua Shi [2], Yi Chen [3], Tengfei Wang [4,*] and Jiexiang Wang [2,*]

1 Bailie School of Petroleum Engineering, Lanzhou City University, Lanzhou 730070, China; llwang2017@163.com
2 School of Petroleum Engineering, China University of Petroleum (East China), Qingdao 266580, China; 18419366657@163.com
3 PipeChina Engineering Technology Innovation Co., Ltd., Tianjin 300450, China; chyi2020@163.com
4 College of Chemistry & Environmental Engineering, Yangtze University, Jingzhou 434023, China
* Correspondence: wangtengfeiforever@126.com (T.W.); wangjxupc@126.com (J.W.)

Abstract

The oil layers in the Dongying Formation offshore oilfield are severely contaminated. The near-wellbore reservoir pore throats are blocked, which seriously affects the development effect. It has become urgent to implement acidizing stimulation measures. However, the target reservoir is deeply buried, has high reservoir pressure, and is highly sensitive. These factors result in high pressure during acidizing operations, a long single-trip time for raising and lowering the tubing string, and high costs. Moreover, acid that is not promptly returned to the surface after acidizing can cause secondary pollution to the reservoir. This work proposes an integrated tubing string to perform reverse displacement and reverse squeeze. With this, acid can be injected into the formation through the annulus between the casing and tubing. The residual acid and its post-acidizing derivative residues are rapidly lifted to the surface by the reciprocating suction action of the return pump. Based on this, the structure and specifications of the acidizing-flowback tubing string are designed through the flow rate analysis method. The tubing string is mainly affected by mechanical effects, including buoyancy, piston effect, flow viscosity effect, helical bending effect, temperature difference effect, and expansion effect. The maximum deformations are 1.4 m, 1.9 m, 0.18 m, 2.7 m, 1.8 m, and 2.5 m, respectively. The total deformation is less than 3 m. Simulation results from three groups of oil wells at different depths indicate that the axial force of the tubing string ranges from 400 to 600 kN. The stress ranges from 260 to 350 MPa, deformation is 1.1–2.4 mm, and the safety factor exceeds 3.0. This can effectively ensure the safety of on-site operations. Based on the actual field conditions, the acidizing-flowback tubing string is evaluated. This verifies the effectiveness of the acidizing-flowback tubing string. This research provides an economical and efficient operation process for acidizing operations in the Dongying Formation offshore oilfield. It achieves the goal of removing reservoir contamination and provides guidance for the unblocking and stimulation of low-permeability and sensitive reservoirs in the middle and deep layers of offshore oilfields.

Keywords: Dongying Formation offshore oilfield; synergistic acidification-flowback; dynamic simulation; integrated tubing string

1. Introduction

Well acidizing is an important measure to eliminate the contamination of pore throats in the near-well reservoir of sandstone oil reservoirs, restore the original formation permeability, and improve the development effect [1]. After a century of development, acidizing technology has become an important technique for increasing production in oil and gas fields and was successfully applied in offshore oilfields in the 1990s, becoming an important measure to ensure the stable exploitation of offshore oil fields. With the successive breakthroughs of technical bottlenecks, the development of acidizing pipe columns, the configuration of acidizing fluids, and other supporting technologies have provided important support for the construction of the acidizing technology system. To meet the needs of safety, efficiency, speed, and economy in offshore oilfield development operations, it is necessary to analyze the development problems faced by offshore oilfields, propose corresponding countermeasures, improve the acidizing process technology, and ensure the smoothness of acidizing construction [2].

At present, the offshore Dongying formation Oilfield suffers from severe oil layer contamination and blockage of pore throats in the near-well area, which seriously affects the development efficiency. Therefore, it is urgent to implement acidification measures to restore the formation permeability and ensure the efficient development of the offshore Dongying formation Oilfield. However, the offshore Dongying formation reservoir is deeply buried, under high pressure and low permeability, and highly sensitive. The construction operation is quite difficult. Moreover, during the acidification operation, there are engineering problems such as a long single trip circulation time of the tubing string, large span of some well sections, uneven layer pressure, high acidification construction pressure, and secondary pollution of the reservoir due to failure to drain the acid after acidification.

Current research on acidification in offshore oilfields mainly focuses on optimizing and combining acid fluid systems [3,4]. However, there is little research on optimizing tubing string technology. Researching acidizing pipe column technology suitable for offshore oilfields can enhance the existing acidizing process. In addition, the current integrated tubing string technology has a cumbersome production process, high requirements for reservoir conditions, and a complex downhole structure, and is mostly for stratified production integration. The main purpose is to promptly discharge the acid fluid that may contaminate the formation within the wellbore in a short period of time, thereby ensuring the smooth progress of actual production [5]. However, the research on acidification and drainage integration technology is relatively scarce. It is worth noting that the current theoretical research on the mechanics of pipe columns has reached a relatively mature stage. By applying the classical mechanical analysis models subsequently and conducting checks on the safety of the pipe columns, a theoretical basis can be provided for ensuring the safety of the pipe columns [6,7].

In response to the above development requirements and technical difficulties, an acidizing and flowback integrated technology suitable for acidizing operations in sensitive oil reservoirs of the Dongying formation was proposed. The technical principle is to use the integrated tubing string to implement reverse displacement and reverse pumping to displace the blocking acid fluid into the formation through the annulus between the tubing and the casing [8,9]. Then, through the reciprocating suction effect of the flow-back pump, the residual substances after acid treatment are rapidly lifted to the surface, achieving large-volume pumping and flowback, significantly compressing the time required for traditional fluid discharge operations after conventional well completion, minimizing the secondary sedimentation pollution of the oil layer after acid treatment, and achieving the purpose of discharging and increasing production in the medium and deep reservoirs with low permeability and sensitive properties [10–12].

This work analyzes the reservoir characteristics and development status of the offshore Dongying formation Oilfield and constructs a new type of acidizing-flowback tubing string—by injecting acid fluid into the annular space of the casing and performing reaming and pumping to achieve the goal of acidizing and negative pressure reaming of the target reservoir with a single pipe column. Based on this, the adaptability of this acidizing-flowback tubing string technology to the offshore Dongying formation Oilfield is analyzed. Using the pipe column mechanical analysis model, the safety of the tubing string's working conditions is analyzed through numerical simulation to ensure the safety and effectiveness of the construction pipe column. It is also compared with the field test to evaluate the effect of the acidizing-flowback tubing string, fully considering its rationality and reliability, to make it more suitable for the field conditions and technical requirements.

2. Adaptive Analysis of the Integrated Acidizing and Flowback Tubing System Technology

2.1. Analysis of the Characteristics of the Offshore Dongying Formation Oilfields

The east slope of the Shengli offshore Chengdao Oilfield is located to the northeast of the main body of the Chengdao Oilfield, with a water depth of 15–20 m, and an exploration area of 200 km^2. The east slope area has discovered four oil-bearing formations: the Paleozoic, the lower tertiary Sheshigou formation, the Dongying formation, and the Upper Tertiary Guantao formation. Among them, the Dongying formation is the main oil-bearing formation. The burial depth of the target reservoir is 2584–3603 m.

The reservoir rock is mainly feldspar-charged sandstone with low mineral maturity. The quartz content ranges from 30% to 40%, with moderate sorting. The average median grain size is 0.355 mm, which classifies it as medium sandstone. The mud content is relatively low, ranging from 0.56% to 9.37%, with an average of 2.8%. The clay minerals are mainly kaolinite, with a content ranging from 68.5% to 79.2%, averaging 76.8%. The interbed of Yili and Mongol clays has a content of 10.0–11.5%, averaging 10.8%; the illite content ranges from 1.0% to 8.0%, averaging 4.2%; and the chlorite content ranges from 4.8% to 13.5%, averaging 8.5%.

The upper reservoir layer of the offshore Dongying formation Oilfield is a medium-porosity and medium-permeability reservoir. The reservoir porosity is mainly concentrated in the range of 9–30.8%, and the average permeability is 33–1151 $\times 10^{-3}$ μm^2; the 8th and 9th sandstone groups of the Dongying Formation, which are mainly oil-bearing, are low-porosity and low-permeability reservoirs, with an average porosity of 13.3–17.4% and an average permeability of 44 $\times 10^{-3}$ μm^2.

The target reservoir has weak water sensitivity, medium-to-weak hydrochloric acid sensitivity, and strong soil acid sensitivity. After acidizing operations, it requires the rapid return of the well fluid and acid–rock reaction derivatives to reduce formation pollution [13–15]. The conditions for the water sensitivity and acid sensitivity test are shown in Tables 1 and 2.

Table 1. Experimental conditions for water sensitive injury evaluation.

Experimental Condition	Experiment I	Experiment II
Temperature/°C	80	80
Core number	CB6	CB14
Permeability/10^{-3} μm^2	116.81	5.30
Pore volume/cm^3	4.69	3.41
Porosity/%	17.07	11.21
Length of rock sample/cm	5.60	6.20
Diameter of rock sample/cm	2.50	2.50
Volume of rock sample/cm^3	27.48	30.42

Table 2. Experimental conditions for acid sensitive injury evaluation.

Experimental Condition	Experiment I	Experiment II
Temperature/°C	19	19
Well section/m	3370.00–3377.66	3370.00–3377.66
Permeability/10^{-3} μm^2	16.70	101.20
Initial permeability/10^{-3} μm^2	4.51	43.80
Pore volume/cm^3	4.00	4.45
Porosity/%	14.10	16.70
Length of rock sample/cm	5.69	5.28
Diameter of rock sample/cm	2.52	2.33
Petrographic description	Oil-saturated sandstone	

The water sensitivity damage evaluation experiments of cores with different permeabilities are shown in Figure 1.

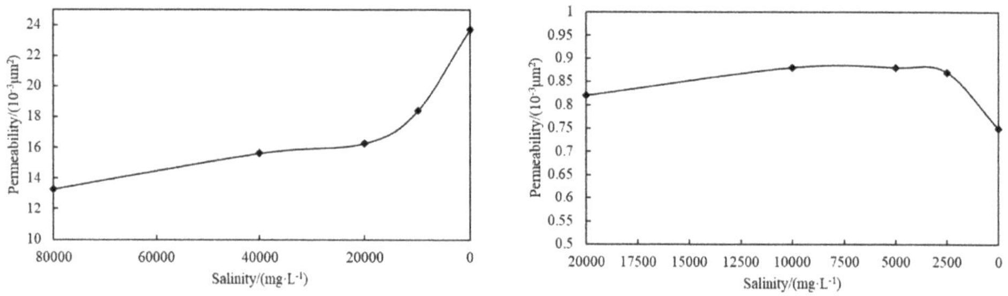

(a) (b)

Figure 1. Experimental results of water sensitivity damage evaluation of cores with different permeabilities. (**a**) Core No. 1 (116.81 mD). (**b**) Core No. 2 (5.30 mD).

The experimental results of evaluating the acid sensitivity damage of cores with different permeabilities are shown in Figure 2.

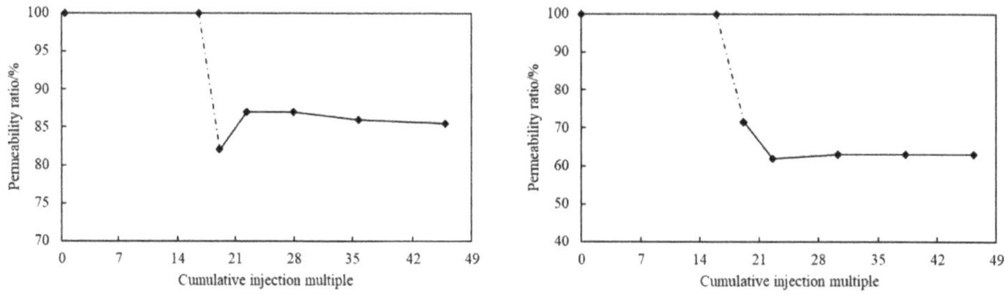

(a) (b)

Figure 2. Experimental results of the acid sensitivity damage evaluation of cores with different permeabilities. (**a**) Core No. 1 (16.70 mD). (**b**) Core No. 2 (101.20 mD).

2.2. *Adaptive Analysis of Integrated Tubing String*

The offshore Dongying formation Oilfield suffers from problems such as oil layer contamination and blockage of pore throats in the near-well area, which seriously affect the development effect and urgently require the implementation of acidification production enhancement measures [16–21]. The well-completion quality of the target oilfield is good, and the integrated technology of acidizing and fluid return is relatively time- and space-cost-saving, which can meet the requirements of the acidizing operation [22,23].

The main challenges encountered during the construction are as follows: the oilfield reservoirs have issues such as water sensitivity and acid sensitivity, which can lead to phenomena such as water lock, clay expansion, and reservoir blockage; the reservoirs are deep, and it is difficult to return the acid fluid and its acid–rock reaction derivatives into the well, especially when the drainage time is long [24–28].

3. Design and Mechanical Analysis of the Integrated Acidizing and Flowback Tubing String Model

3.1. *Integrated Acidizing and Flowback Process Design*

The offshore Dongying Formation reservoir exhibits sensitivity to velocity, water, and acid, leading to issues such as water blocking, clay swelling, and formation damage. Conventional completion strings further complicate post-acidizing flowback, resulting in prolonged cleanup times and significant wellhead backpressure [29]. For certain gas wells requiring post-perforation acidizing, additional steps such as well control and tubing replacement are necessary before pumping and flowback can proceed [30,31]. These repeated well interventions exacerbate formation damage and extend operational cycles, thereby increasing costs.

While acidizing can remove near-wellbore solids and soluble minerals to enhance permeability, it may also compromise rock integrity. Therefore, it is critical to optimize both the acidizing fluid system and operational procedures. A less-aggressive fluid with slower reaction kinetics should be selected to minimize damage to rock cementation. Furthermore, the efficient and thorough flowback of spent acid is essential to mitigate secondary formation damage [32,33].

3.1.1. Plan Concept and Expected Goals

(1) Plan concept

The simplified diagram of the acidizing flowback integration process flow is shown in Figure 3 [34,35].

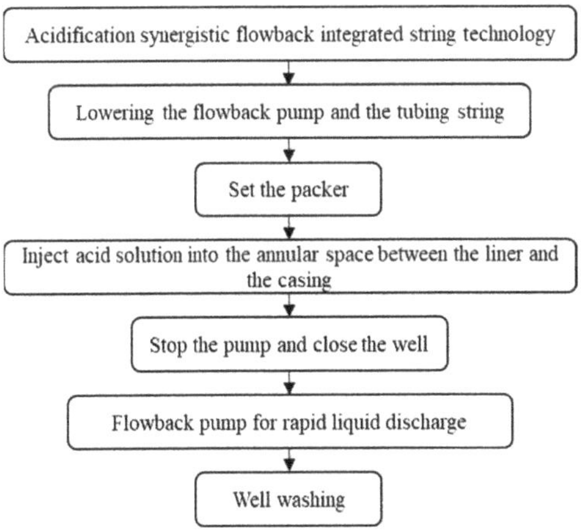

Figure 3. Schematic diagram of the acidizing flowback integration process.

(2) Expected goals

① A single tubing string can achieve acid fracturing for blockage removal, and rapid reaming, and enhance the effectiveness of acidizing operations.

② By using an integrated tubing string, costs can be reduced and efficiency increased, and economic and time costs can be lowered.

③ Optimizing the construction process can improve the safety of the operation and reduce operational risks.

3.1.2. Integrated Acidizing and Flowback Operation Construction Procedures

The detailed working process of acidizing and flowback tubing string is shown in Figure 4 [36–39].

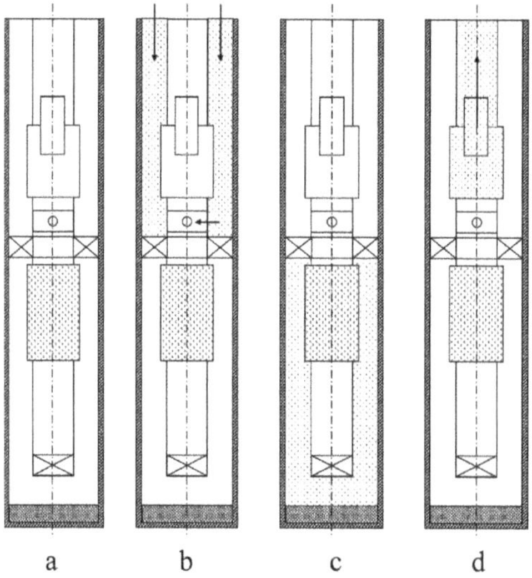

Figure 4. Detailed operation flowchart of the integrated acidizing and flowback tubing string. (**a**) Lowering the tubing string and setting it in place. (**b**) Injecting acid into the well. (**c**) Acidic fluid dissolving blockages. (**d**) Returning the acid fluid for re-extraction.

3.2. Integrated Tubing Column Structure Model for Acidizing and Flowback Recovery

3.2.1. Casing Structure Design

The injection structure is realized by devices such as packers and backwash valves. The return and displacement structure mainly consists of sucker rods and return pumps, and the return process is accomplished by the sucker rods driving the return pumps to move up and down [25–27,30–33,39–42]. The overall structure of the acidizing and flowback integrated construction pipe column designed for this work is shown in Figure 5.

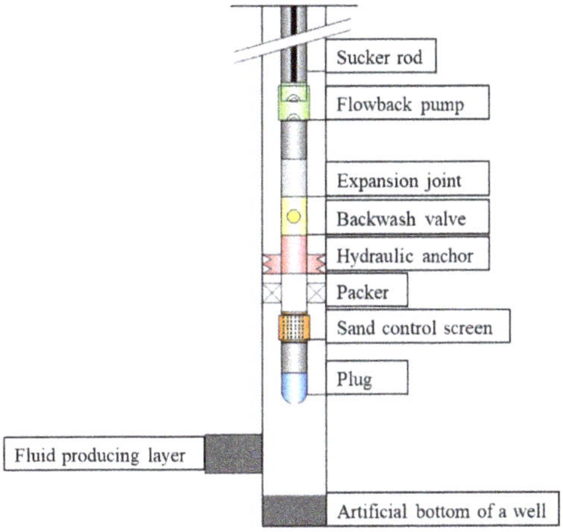

Figure 5. Integrated tubing column structure model for acidizing and flowback operations.

3.2.2. Casing Selection Design and Calculation

Taking the 7-inch casing well size of the offshore Dongying formation Oilfield as the design basis, a set of methods and processes suitable for the matching of the downhole tubing string was designed, as shown in Figure 6 [37,43–45].

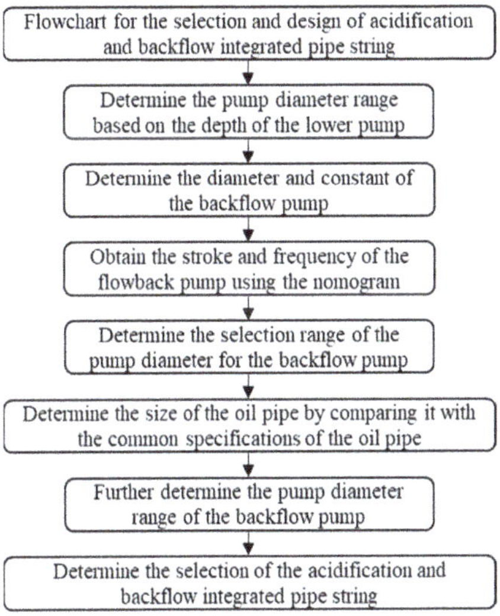

Figure 6. Design flow chart for the selection of acidizing flowback integrated tubing string.

To meet the needs of acidizing tubing string backflow, as important equipment for backflow injected fluid, the first task is to clarify the available range of pump diameters for backflow pumps, calculate the theoretical discharge of the backflow pump, and the mathematical model used is as follows (1):

$$Q_T = 1440 \times nSA_p = KnS \tag{1}$$

In the formula: Ap is the plunger area, m^2; Q_T is the pump theoretical displacement, $m^3 \cdot d^{-1}$; K is the flowback pump constant, m^2; N is the stroke rate of the pump, min^{-1}; and S is stroke length, m.

The matching relationship among the pump diameter of the return pump, the piston area of the return pump, and the pump constant of the return pump are shown in Table 3. The nomogram for selecting the stroke and frequency of the return pump under different working conditions is shown in Figure 7.

Table 3. Data sheet for pump diameter, plunger area, and pump constants of the return pump.

Pump Diameter (mm)	Plunger Area A_p (10^{-3}m^2)	Pump Constant K (m^2)
70	3.848	5.542
83	5.410	7.790
95	7.088	10.202
105	8.659	12.462
120	11.310	16.278

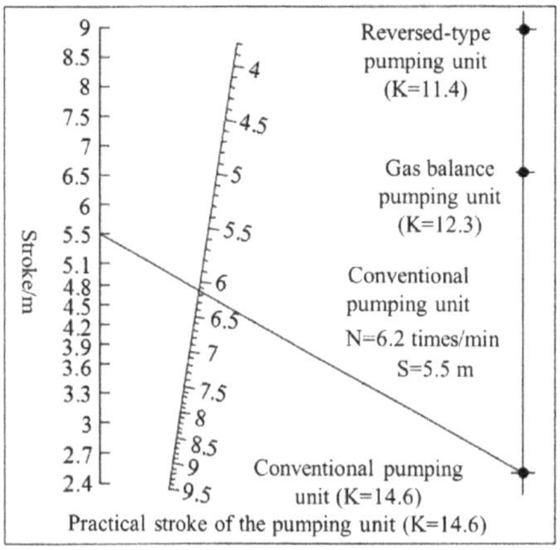

Figure 7. Nomogram for selecting the stroke and frequency of the return pump under different operating conditions.

The No model diagram that matches the pumping stroke and stroke frequency commonly used in oilfield production operations is selected, and the stroke and stroke frequency values required for the return operation conditions are determined [46].

The relationship between the pump diameter of the return pump and the size of the sucker rod is shown in Table 4.

Table 4. Matching table of pump diameter and sucker rod specification.

Dimension Symbol	Pump Diameter/mm	Pumping Rod Specification
30–275	69.85	GYG16(5/8)
30–325	82.55	GYG19(3/4)
35–375	95.25	GYG22(7/8)
40–425	107.95	GYG22(7/8)
45–475	120.65	GYG25(1)

Based on the matching results of the pump diameter of the return pump and the specifications of the sucker rod, combined with the commonly used tubing specifications and sizes in actual production, the matching was carried out as shown in Table 5.

Table 5. Specification and size of the oil pipe.

Gauge Size (in)	Diameter D/mm	Wall Thickness S/mm	Inner Diameter d/mm
2–7/8	73.0	6.0	62.0
3–1/2	88.9	6.5	75.9
4	101.6	6.5	88.6
4–1/2	114.3	7.0	100.3

3.2.3. Main Components of the Acidification Pipe Column

(1) Extension and retraction joint

The extension and retraction joint is mainly designed with a key slot structure, and its size is slightly larger than the inner diameter of the tubing. The extension and retraction joint generally compensates for the movement of the tubing during oil well acidizing and fracturing operations and is also convenient for adjusting the length at the wellhead [30].

(2) Packer

As shown in Table 6, by comparing the technical parameters of the Y211 packer and RTTS packer, according to the requirements of the integrated tubing string, the RTTS packer can be applied in acidizing operations and can meet the reverse flushing conditions. Therefore, the RTTS packer is selected for the tubing string structure.

Table 6. Technical parameters of packer (Y211 and RTTS).

Model	Adaptive Casing /mm	Total Length /mm	External Diameter /mm	Inner Diameter /mm	Working Pressure /MPa	Working Temperature /°C
Y211-102	112	2095	102	36	50	130
Y211-110	118–121	2080	110	48	50	130
Y211-112	121–124	2115	112	48	50	130
Y211-114	121–124	2080	114	48	50	130
Y211-136	143–150	2080	136	60	50	130
Y211-148	154–164	2080	148	60	50	130
Y211-150	154–164	2080	149	60	50	130
RTTSY221-148	157–161	1545	150	50	35–50	130
RTTSY221-146	154–157	1545	146	50	35–50	130
RTTSY221-114	121–124	1370	114	48	35–50	130
RTTSY221-110	118–121	1370	110	48	35–50	130

(3) Reverse flushing valve

The annular pressure reverse flushing valve is a reverse circulation valve that opens by relying on the annular pressure and can be spring-reset closed. It is commonly used in 7-inch casing wells and is mainly used for reverse circulation operations or for returning after acid fracturing.

(4) Hydraulic anchor

The insertion of the sealing outer diameter is 101.6 mm, and the inner diameter of the plug-in opening tool is 47 mm. One hydraulic anchor is installed in the tubing string to anchor the tubing string and prevent the tubing string from moving upward during acidizing.

3.2.4. Basic Mechanical Effects of Downhole Pipe Column

During the acidizing and reverse circulation operation, the tubing string is mainly subjected to various loads such as buoyancy force, flow viscous effect, piston effect, bulging effect, temperature difference effect and helical bending effect [47,48].

(1) Floating weight effect

The direction of the gravitational force acting on the acidizing operation tubing string is vertically downward. The gravitational force acting on each unit length of the operation tubing string is

$$q_{oi} = A_{oi}\rho_o g \tag{2}$$

In the formula, q_{oi} is the gravitational distributed force acting on the target tubing string, N·m^{-1}; A_o is the cross-sectional area of the target tubing string, m^2; ρ_o is the density of the tubing string, kg·m^{-3}; and g is the gravitational acceleration, m·s^{-2}.

The buoyancy of the well fluid acting on the acidizing operation tubing column is perpendicular to the axis of the tubing column in the direction. The force distribution per unit length of the buoyancy is

$$q_{oli} = A_{oi}\rho_l g \cos\theta \tag{3}$$

In the formula, q_{oli} represents the buoyancy distribution force of the target tubing string, N·m^{-1}; ρ_l is the density of the liquid in the wellbore, kg·m^{-3}; and θ is the angle between the axis of the tubing string and the horizontal plane, rad.

(2) Viscous flow effect

During the process of being lowered into the well, the operation string will encounter frictional resistance from the fluids in the well. The direction of this resistance is parallel to the axis of the operation string and is opposite to the direction of the string's movement during the lowering process. The distribution force of the frictional resistance is calculated using the following formula.

The distribution force of frictional resistance is

$$q_{fr} = 2\pi\mu_L\left[\frac{m^2-1}{(m^2+1)\ln m - (m^2-1)}\right]v_{fp} \tag{4}$$

In the formula, q_{fr} is the frictional distribution force between the operation tubing and the wellbore fluid, N·m^{-1}; μ_L is the dynamic viscosity of the fluid in the annular space, mPa·s; m is the ratio of the inner diameter of the casing to the outer diameter of the tubing; D_r is the inner diameter of the casing, m; D_{ri} is the inner diameter of the tubing, m; and v_{fp} is the relative velocity of the wellbore fluid and the operation tubing, m·s^{-1}.

The Reynolds number of the annular flow is

$$R_{eo} = \frac{4\rho_{oi}Q_o}{\pi\mu_o(d_w - d_o)} \tag{5}$$

In the formula, R_{eo} is the Reynolds number of the fluid in the annular space; v_o is the average flow velocity of the fluid in the annular space, m·s^{-1}; d_w is the inner diameter of the casing, m; d_o is the outer diameter of the tubing, m; ρ_{fo} is the density of the fluid in the oil-tubing annular space, kg·m^{-3}; μ_o is the viscosity of the fluid in the oil-tubing annular space, mPa·s; and Q_o is the flow rate of the fluid in the oil-tubing annular space, m^3·s^{-1}.

If $R_{eo} \leq 2000$, the fluid is in a laminar flow state, and the total pressure of the fluid inside the annular space between the oil and casing is

$$P_o = P_{o0} - \int_0^l \frac{192\mu_o Q_o}{\pi(d_w+d_o)(d_w-d_o)^3}dl + \int_0^l \rho_{fo} g \cos\alpha dl \tag{6}$$

In the formula, P_o represents the total pressure of the fluids inside the annulus of the oil and casing in the well, MPa; and P_{o0} represents the total pressure of the fluids inside the annulus of the oil and casing at the wellhead, MPa.

If $R_{eo} > 2000$, the fluid is in a turbulent state, and the total pressure of the fluid inside the annular space between the oil and casing is

$$P_o = P_{o0} - \int_0^l \frac{2.5168\rho_{fo}Q_o^2}{\pi^2(d_w - d_o)(d_w^2 - d_o^2)^2 R_{ei}^{0.25}} dl + \int_0^l \rho_{fo} g \cos\alpha dl \quad (7)$$

(3) Plunger effect

During the process of raising and lowering the pipe column, the packer connected to the pipe column moves in the wellbore fluid, resulting in inconsistent pressure on the upper and lower ends of the packer and generating axial loads. At this time, the working pipe column experiences the plunger effect. The schematic diagram of the piston effect is shown in Figure 8.

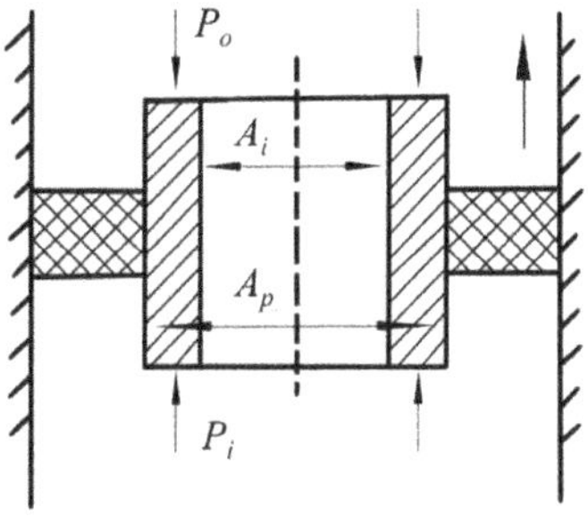

Figure 8. The schematic diagram of the piston effect.

The forces acting on the piston from below upwards and from above downwards are, respectively,

$$F_1 = P_i(A_{p1} - A_i) \quad (8)$$

$$F_2 = P_o(A_{p1} - A_o) \quad (9)$$

Assuming that the packer exerts a negative force when subjected to tension and a positive force when subjected to pressure, then the piston force is

$$F_p = F_1 + F_2 = P_i(A_{p1} - A_i) - P_o(A_{p1} - A_o) \quad (10)$$

In the formula, A_i represents the inner cross-sectional area of the pipe column, m^2; A_o represents the outer cross-sectional area of the pipe column, m^2; P_o represents the internal pressure of the pipe column, MPa; P_i represents the external pressure of the pipe column, MPa; and A_{p1} represents the cross-sectional area of the sealing chamber of the packer, m^2.

Based on Hooke's law, calculate the extension and contraction amount of the pipe column caused by the piston effect:

$$L_1 = -\frac{F_1 L}{EA_S} \quad (11)$$

In the formula, L represents the length of the pipe column, m; E is the elastic modulus, MPa; and A_{p2} is the cross-sectional area of the pipe column, m^2.

Due to the changes in the environment where the operation tubing is located, there are fluctuations in the internal and external pressures. When the internal and external pressures of the operation tubing change, it will cause a change in the piston force:

$$\Delta F_1 = (A_{P2} - A_i)\Delta P_i - (A_{P2} - A_o)\Delta P_o \tag{12}$$

The deformation of the pipe column caused by the change in piston force is

$$\Delta L_1 = -\frac{\Delta F_1 L}{E A_S} \tag{13}$$

(4) Bulging effect

When the fluid pressures inside and outside the pipe column change, it will cause the pipe column to undergo deformation towards the inside or outside, which is known as the bulging effect. The schematic diagram of the bulging effect is shown in Figure 9.

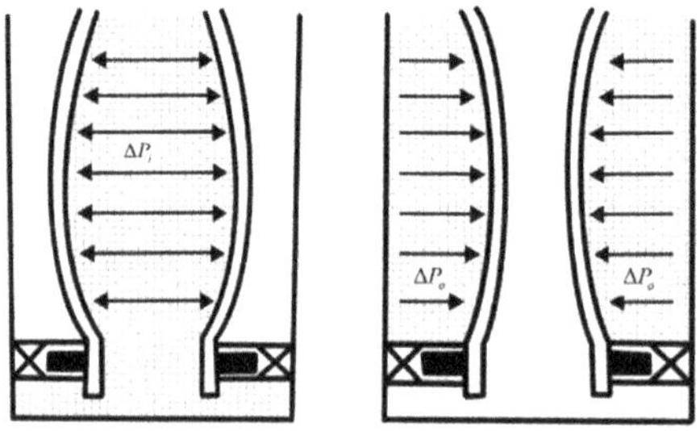

Figure 9. The schematic diagram of the expansion effect.

The force generated by the bulging effect is

$$\Delta F_e = 0.6 A_i \Delta P_{ia} - 0.6 A_o \Delta P_{oa} \tag{14}$$

In the formula, ΔF_e is the force caused by the bulging effect, kN; ΔP_{ia} is the average pressure change within the tubing string, MPa; and ΔP_{oa} is the average pressure change within the annular space between the oil and casing, MPa.

The deformation caused by the bulging effect is

$$\Delta L_e = -\frac{\mu\left[\Delta\rho_i - R^2\Delta\rho_o - \frac{1+2\mu}{2\mu}\delta\right]L^2}{E(R^2-1)} - \frac{2\mu(\Delta P_{is} - R^2\Delta P_{os})}{E(R^2-1)}L \tag{15}$$

In the formula, μ represents the Poisson's ratio of the material; R is the ratio of the outer diameter to the inner diameter of the tubing; $\Delta\rho_i$ is the change in the density of the fluid inside the tubing, kg·m^{-3}; $\Delta\rho_o$ is the change in the density of the fluid inside the annular space, kg·m^{-3}; ΔP_{is} is the change in the pressure inside the tubing at the wellhead, MPa; ΔP_{os} is the change in the pressure in the annular space at the wellhead, MPa; and δ is the pressure drop per unit length of the workover string caused by fluid flow, MPa·m^{-1}.

(5) Spiral bending effect

When the axial force at the position where the packer is located exceeds the critical force that causes the operation tubing string to undergo spiral bending deformation, and when the pressure change inside the tubing is less than the pressure change inside the annular space between the tubing and the casing, the tubing string will exhibit a spiral bending effect. The schematic diagram is shown in Figure 10.

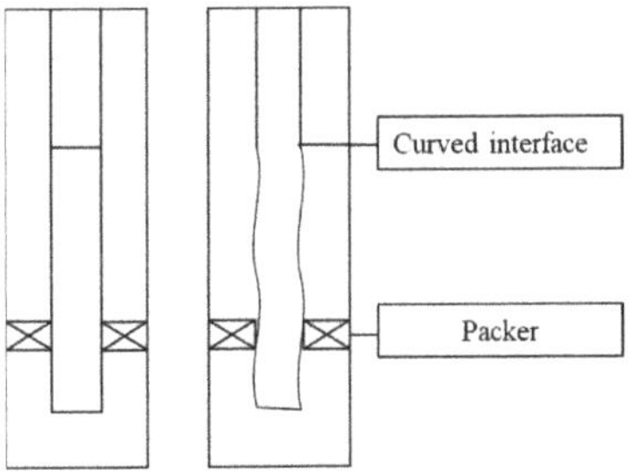

Figure 10. The schematic diagram of the spiral bending effect.

The helical bending stress is

$$F = F_f + F_g = A_p(P_i - P_o) + F_g \tag{16}$$

The spiral bending elongation and shortening amounts of the operation tubing column are as follows:

$$\Delta L_1 = \frac{r^2}{8EIW} F_f^2 \tag{17}$$

$$\Delta L_2 = -\frac{F_f L}{EA_s} \tag{18}$$

In the formula, ΔL_1 represents the elongation of the tubing string due to bending, m; F_f is the effective bending force at the end of the tubing string, kN; EI is the bending stiffness of the tubing string; and r is the radius gap of the annular space between the tubing and casing, m.

By combining Equations (17) and (18), the axial deformation of the pipe column caused by the helical bending effect can be obtained as

$$\Delta L_{hb} = \Delta L_1 + \Delta L_2 = \frac{r^2}{8EIW} F_f^2 - \frac{F_f L}{EA_s} \tag{19}$$

(6) Temperature effect

When the temperature changes, the working tubing string undergoes thermal expansion and contraction, and at this point, a temperature difference effect occurs in the tubing string.

The temperature effect results in the following force on the pipe column:

$$F_w = -\Delta T \frac{\alpha E \pi (D^2 - d^2) L}{4} \times 10^3 \tag{20}$$

The temperature effect results in the column shape variable being

$$\Delta L_w = \Delta T \alpha L \tag{21}$$

In the formula, α represents the temperature linear expansion coefficient of the tubing string, in units of $^{\circ}C^{-1}$; L denotes the length of the tubing string, m; ΔT represents the temperature gradient within the wellbore, $^{\circ}C \cdot m^{-1}$; E is the elastic modulus of the tubing material, MPa; D is the outer diameter of the tubing string, m; and d is the inner diameter of the tubing string, m.

In particular, if the column-shaped variable is restricted by the packer, the following two phenomena will occur. When $\Delta T > 0$, $\Delta Lw > 0$, the tubing string elongates and is subjected to axial compressive load; when $\Delta T < 0$, $\Delta Lw < 0$, the tubing string shortens and is subjected to axial tensile load.

3.3. Study on the Force and Deformation of the Acidizing Tubing String Under Different Mechanical Effects

The acidizing operation tubing string is affected by factors such as buoyancy, gravity, and viscosity. At the same time, the existence of the piston effect, helical bending effect, bulging effect, temperature effect, and other basic effects, play a significant role in the force acting on the acidizing operation tubing string [36,41,42]. The force and deformation analysis of the acidizing tubing string is carried out, taking $2^7/_8$ in and $3^1/_2$ in as examples. The materials selected are common N80 and P105 steel grades, and the basic data are shown in Table 7.

Table 7. Basic data of tubing.

Nominal Diameter/in	Steel Grade	Unit Weight/(N·m^{-1})	Wall Thickness/mm	Yield Strength/MPa	Thread Tensile Strength Limit/kN	Resistance to Internal Pressure/MPa
$2^7/_8$	N80	92.2	5.51	552	657.4	72
		125	7.82	552	901.1	102
	P105	92.2	5.51	724	863.0	95
		125	7.82	724	1113	135
$3^1/_2$	N80	132	6.45	552	939.9	70
		186	9.52	552	1336	102
	P105	132	6.45	724	1233.4	92
		186	9.52	724	1753.4	136

(1) Evaluation of the influence of buoyancy force on the acidizing operation tubing string

Taking the tubing string for the acidizing operation in a 3000 m oil well as an example, the $2^7/_8$ in and $3^1/_2$ in oil tubes were analyzed. The axial force and axial stress caused by the buoyancy force are shown in Figure 11. The axial force of the large-sized operation tubing string is smaller than that of the small-sized tubing string, and the tubing string with a larger wall thickness for acidizing operation experiences a greater axial force due to the buoyancy force. The deformation amount is related to the length of the tubing string. As the well depth increases, the deformation amount gradually increases, and the deformation trend gradually accelerates with the increase in tubing string length. At the same time, the deformation amount caused by the buoyancy force due to the floating weight of different diameters and different wall thickness tubing strings is very similar. The deformation amount caused by the buoyancy force of the 3000 m length tubing string is about 1.4 m.

(2) Evaluation of the impact of the piston effect on the acidizing operation tubing string

The axial force and axial stress caused by the piston effect are shown in Figure 12. They are mainly related to the pressure difference between the inside and outside of the tubing string. Analysis reveals that as the pressure difference between the inside and outside of the tubing string increases, the axial force caused by the piston effect also increases. Moreover, when the pressure difference between the inside and outside of the tubing string is small, the axial force caused by the piston effect on the large-sized tubing string is greater than that on the small-sized tubing string. When the pressure difference is larger, the opposite effect occurs. Particularly, when the pressure difference between the inside and outside of the tubing string exceeds 15 MPa, the axial force on the small-sized tubing string exceeds that on the large-sized tubing string. It is analyzed that when the pressure difference between the inside and outside of the acidizing tubing string is small, the axial force caused by the piston effect has a more significant effect on the tubing string with a larger outer diameter. However, when the pressure difference exceeds a certain range, the tubing string with a smaller outer diameter experiences a more significant force [33,47]. The axial stress caused by the piston effect is related to the pressure difference in the acidizing tubing string. As the pressure difference increases, the axial stress also increases. At the same pressure difference, the stress on the larger-sized tubing string is relatively smaller, and the tubing string with a thicker wall thickness experiences a smaller stress.

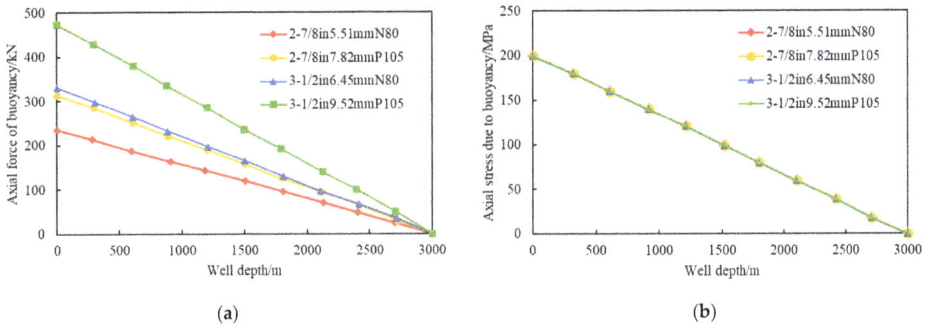

Figure 11. Force diagram of the floating weight tubing string. (**a**) Axial force. (**b**) Axial stress.

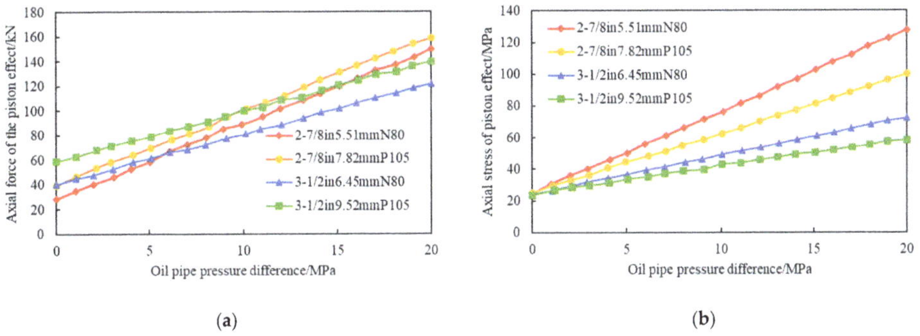

Figure 12. Force diagram of the piston effect tube column. (**a**) Axial force. (**b**) Axial stress.

In addition, due to the tendency of the tubing string deformation caused by the piston effect to contract, the contraction deformation increases with the increase in tubing string length [31]. And when the well depth is the same, the larger the tubing string size, the smaller the deformation. The larger the wall thickness of the tubing string, the smaller the deformation.

(3) Evaluation of the effect of bulging on the acidizing operation tubing string

The axial force caused by bulging is shown in Figure 13. The axial force has both positive and negative values. When the axial force is negative, it indicates that the internal pressure of the tubing string is less than the external pressure, and an internal bulging effect occurs, resulting in a reduction in the diameter of the tubing string. Conversely, when the axial force is positive, it indicates that the internal pressure of the tubing string is greater than the external pressure, and an external bulging effect occurs, resulting in an increase in the diameter of the tubing string. When the outer diameter of the tubing is the same, under certain conditions—when the pressure difference is small between the inside and outside of the tubing string, an internal bulging effect occurs, and the thicker-walled tubing string receives a relatively larger axial force. When the pressure difference is large, an external bulging effect occurs, and the thinner-walled tubing string receives a relatively larger axial force. Overall, the axial force on the large-sized tubing string is smaller than that on the small-sized tubing string. The stress on the tubing string of the same steel grade is relatively close. The influence is mainly related to the internal and external pressure differences. When the internal and external pressure differences in the tubing string are greater than 10 MPa, the stress received by the tubing string with a larger wall thickness is relatively smaller. The change in the tubing string shape due to the bulging effect is a contraction deformation. The deformation amount increases with the increase in well depth, and the overall deformation is less than 0.18 m. Therefore, the axial deformation of the tubing string caused by the bulging effect is relatively small.

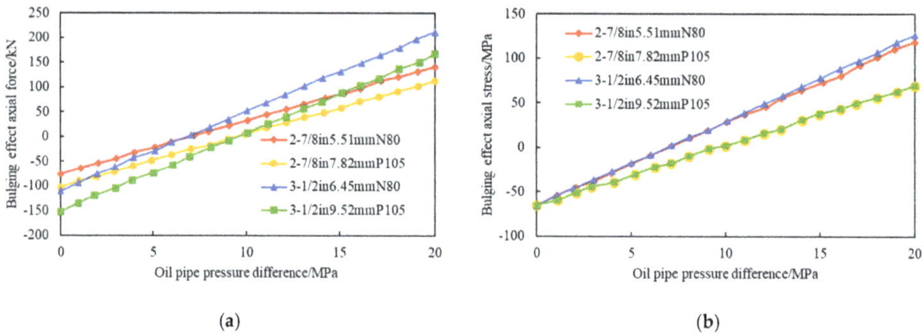

Figure 13. Force diagram of the bulging effect pipe column. (**a**) Axial force. (**b**) Axial stress.

(4) Evaluation of the effect of helical bending on the acidizing operation tubing string

When the axial force of the tubing string at the packer location exceeds the critical force that causes the tubing string to spiral bend, the tubing string will undergo spiral bending. The axial force that effectively acts on the end of the tubing string and causes the spiral bending effect is independent of the tubing string itself and is only related to the packer and the pressure difference between the inside and outside of the tubing string [46]. As shown in Figure 14, the axial stress caused by the spiral bending effect of different tubing strings due to the spiral bending effect and the change in the tubing string shape caused by bulging are calculated under different pressure differences. Analysis shows that under the same pressure difference conditions, the axial stress on the small-sized tubing string is relatively larger, and the axial stress on the tubing string with a smaller wall thickness is also larger.

The tubing string shape variable of 3000 m length increases with the increase in the end load. The negative values in the figure indicate that under the axial load at the end, the tubing string has a tendency to shorten in shape. Due to the fact that in actual production, the tubing string is not only affected by the end load but also by other factors such as self-weight and friction, the end axial load will not be so large, and the actual axial shape variable will be relatively lower than the calculated result.

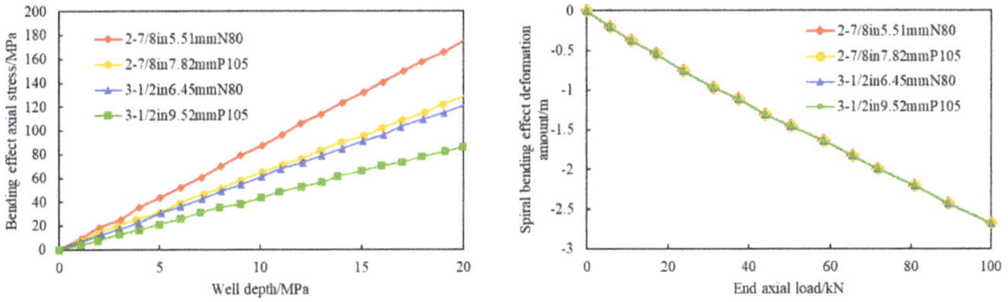

(**a**) (**b**)

Figure 14. Diagram of force and deformation of the spiral bending effect tube column. (**a**) Axial stress. (**b**) Deformation quantity.

(5) Evaluation of the influence of the temperature effect on the acidizing operation tubing string

Due to the difference between the wellhead temperature and the temperature inside the well, the operation tubing string will be affected by the temperature, resulting in axial deformation. The magnitude of the axial stress caused by the temperature effect is only related to the formation temperature, the depth of the oil well, and the elastic modulus of the tubing string, and has no relation to the size of the tubing string [48]. The axial force variation in the tubing string caused by the temperature effect is shown in Figure 15. The axial force gradually increases with the increase in well depth, and larger-sized tubing strings are more affected by temperature. The tubing string deformation is mainly related to factors such as temperature and the expansion coefficient of steel and has little relation to the size of the tubing string. As the well depth increases, the deformation of the tubing string due to the temperature effect becomes more obvious and the trend becomes greater. The 3000 m operation tubing string deforms under the influence of the temperature effect, and its deformation is close to 1.8 m. The temperature has a very significant impact on the tubing string.

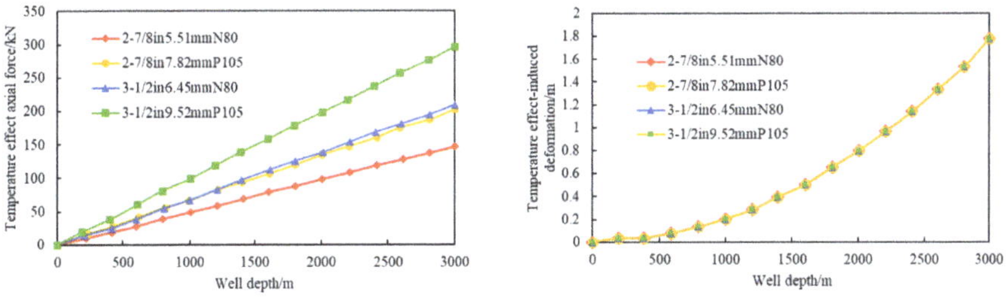

(**a**) (**b**)

Figure 15. Diagram of the force and deformation of the temperature effect tube column. (**a**) Axial stress. (**b**) Deformation quantity.

(6) Evaluation of the influence of the flow viscosity effect on the tubing string in acidizing operations

The axial force and deformation of the tubing string caused by the flow viscosity effect are shown in Figure 16. As the well depth increases, the axial force gradually increases, and the larger the outer diameter of the tubing string, the smaller the axial force it receives. For the same-sized tubing string, the smaller the inner diameter, the greater the axial force it receives. For the same-sized tubing string, the stress is relatively close, but the stress received by the tubing string with a smaller inner diameter is relatively smaller, while the stress received by the tubing string with a smaller outer diameter is relatively larger. The flow viscosity effect has a significant impact on the deformation of the tubing string. Among them, the larger-sized tubing string undergoes deformation less than the smaller-sized tubing string. For the same-sized tubing string, the deformation is relatively close, and the tubing string with a larger inner diameter receives a greater influence than the tubing string with a smaller inner diameter.

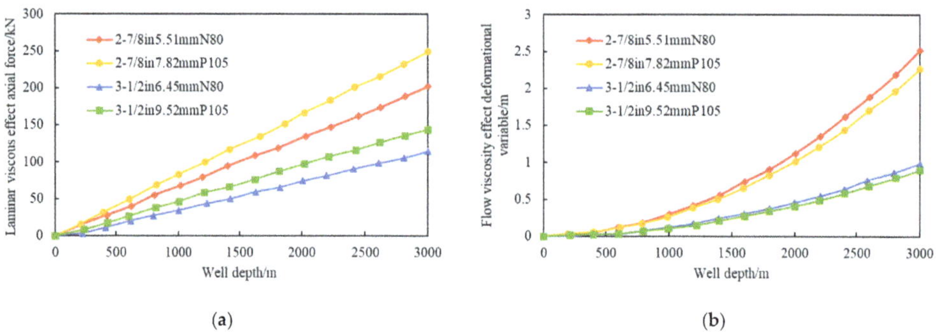

Figure 16. Diagram of force and deformation of the flowing viscosity effect pipe column. (**a**) Axial stress. (**b**) Deformation quantity.

(7) Analysis of total deformation of the tubing string caused by different effects

As shown in Figure 17, the deformation of the tubing string under the influence of different mechanical effects is calculated by superimposing, and the total deformation of the working tubing string is obtained. There are differences for pipes of different sizes. The deformation of the tubing string in an oil well at a depth of 3000 m is approximately 1.45 m.

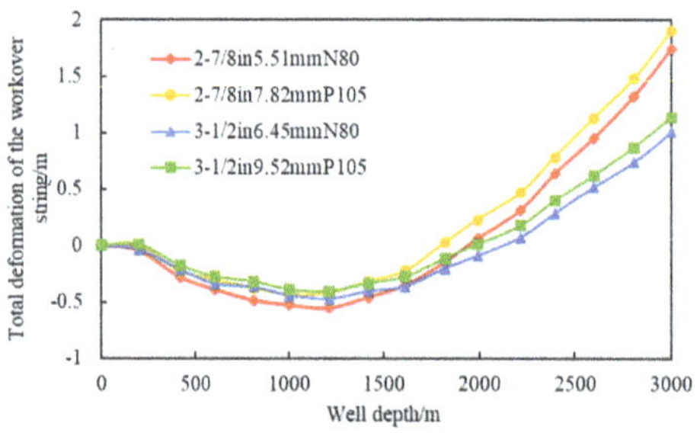

Figure 17. Total deformation of the workover string.

4. Analysis of Key Equipment and Calculation of Force Distribution Simulation

4.1. Mechanical Analysis of Expansion Joints

Using the Fluent dynamic simulation software, the common working states of the expansion joint were simulated. Based on its safety, the simulation mainly focused on the tensile state when the pins were not broken and the entire tensile state when the pins were broken. The maximum stress that the expansion joint was subjected to was obtained through numerical simulation, and the maximum stress under different working conditions was calculated using the Lame formula, as shown in Table 8. By comparing the theoretical solution with the numerical solution, the results were similar, and the error between the two was controlled within 15%.

Table 8. Theoretical solution of expansion joint strength under different working conditions.

Operating Condition	Axial Force/kN	Internal Pressure/MPa	External Pressure/MPa	Maximum Stress/MPa		Error/%
				Numerical Analysis	Theoretical Calculation	
Normal well setting	170	70	80	521.18	446.25	14.3
Acidic conditions	190	70	100	575.37	495.26	13.9
Reverse operation condition	230	110	80	687.08	589.64	14.2

4.2. Mechanical Analysis of Anti-Fouling Valve and Causes of Failure

The anti-washing valve, as a key tool for connecting the oil and casing, has its failure mainly caused by two aspects: sealing failure and mechanical faults.

(1) Sealing Failure

The main seals of the anti-washing valve are O-ring seals and T-type seals. Affected by factors such as the fluid flow state through the valve, fluid pressure, and friction loss, the stress on the sealing rings will fluctuate continuously as the anti-washing valve operates [20,35]. Under this influence, failure is more likely to occur. Especially near the inlet, the T-type seals are subjected to fluid erosion for a long time, making the damage probability even greater.

(2) Mechanical Faults

The fluid in the operation well has corrosiveness, and the equipment is in a high-temperature environment. The strength of the pins in the equipment has weakened. Moreover, in complex well conditions, the anti-washing valve equipment is constantly exposed to the fluid in the well with frequent pressure fluctuations. The frequent and complex changes in pressure may cause premature failure of the pins of the anti-washing valve. At the same time, affected by the environment, the spring used to close the anti-washing valve will have weakened elasticity, which may lead to a situation where the anti-washing valve cannot be closed tightly.

The design precision between the inner cylinder, outer cylinder, and the anti-washing valve is very high. Any problem in the cooperation, materials, sealing, and assembly quality of these parts may cause jamming and affect the normal operation of the anti-washing valve. Moreover, the downhole environment of the operation well is complex, and there is tool debris such as formation sand, pipe column fragments, pins, and other solid particles. During the operation of the anti-washing valve, solid particles entering the mechanical gap will also cause mechanical wear and jamming, and even damage the mechanical structure of the anti-washing valve.

4.3. Mechanical Analysis of Reverse Discharge Pump and Calculation of Displacement

Based on the geological characteristics of the offshore Dongying formation Oilfield and the requirements of the integrated acidizing and flowback tubing string technology, with a 7-inch wellhead size as the design basis, the theoretical displacement of the flow-back pump was calculated and analyzed. The basic parameter table for offshore flowback construction shown in Table 9 indicates that in actual production, to enhance the flowback effect, the stroke can be raised to 4–6 m.

Table 9. Basic parameters of offshore flowback construction.

Pump Type/mm	Outer Diameter/mm	Pump Constant/($m^3 \cdot d^{-1}$)	Stroke/m	Stroke Rate/min^{-1}	Theoretical Displacement/($m^3 \cdot d^{-1}$)	Work Pressure/MPa
95	108	10.202	3 3	2 3	61.212 91.818	15
105	120	12.462	3 3	2 3	74.772 112.158	
120	136	16.278	3 3	2 3	97.668 146.502	
130	148	19.104	3 3	2 3	114.624 171.936	

4.4. Dynamic Simulation of Stress and Deformation

A dynamic simulation and optimization software for pipe column mechanics was developed using the C#. By inputting basic data, the wellbore trajectory, the types of casing and pipe column structures, the pipe column dimensions, the relative density of the well entry fluid, and the downhole depths of the pipe column-related equipment can be constructed. Based on this, the software can conduct mechanical analysis of the pipe column under different working conditions. The data well for the target block is established based on the actual well inclination data of the target block. Numerical simulation is carried out using this software to construct a three-dimensional wellbore trajectory and pipe column structure model diagram. According to the information of the offshore Dongying formation reservoir, it is known that the burial depth of the oil layer in this area is within the range of 2596–3602 m. The acidification operation simulation of the target block is carried out, with the pipe columns at 2700 m, 3000 m, and 3300 m as the research objects. Three data wells that conform to the target block reservoir are established and numerical simulation is conducted using this software for mechanical analysis. The data for the wellbore trajectory is shown in Table 10. The results of the mechanical analysis and calculation for the oil well when the depths are 2700 m, 3000 m and 3300 m are shown in Figures 18–20.

Table 10. X oil well deviation part data.

Depth Measurement/m	Segment Length/m	Well Deflection/°	Orientation/°	Vertical Depth/m
0.00	0.00	0.00	0.00	0.00
288.00	28.63	3.20	320.00	287.78
604.09	28.73	0.20	38.04	603.43
900.65	28.83	1.20	137.73	899.97
1215.74	28.62	0.82	217.93	1215.02
1502.14	28.65	1.48	215.83	1501.37
1817.25	28.76	0.99	212.63	1816.40
2104.30	9.65	1.92	195.83	2103.42
2390.85	28.54	21.00	237.43	2382.14
2706.89	28.58	20.74	230.33	2677.11

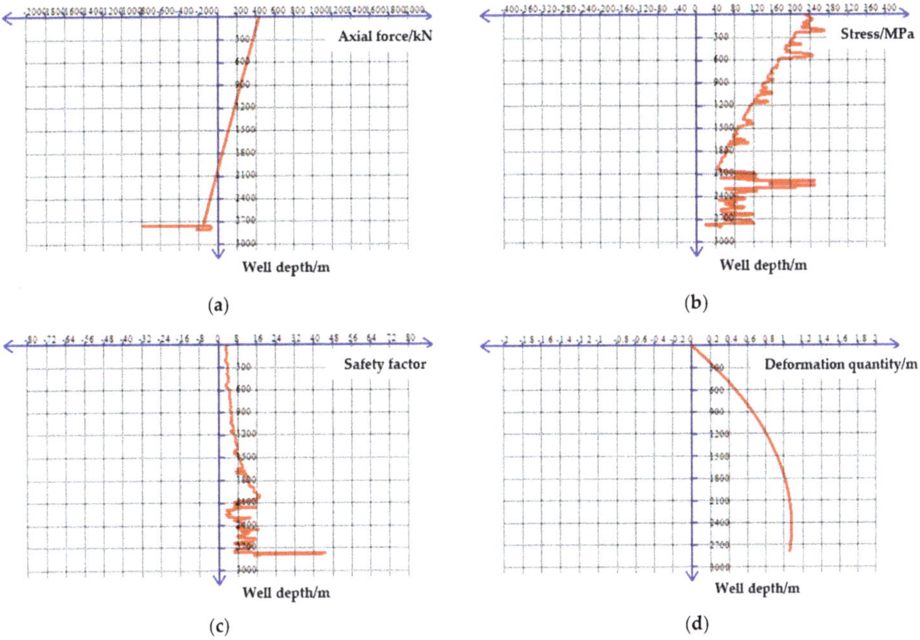

Figure 18. Mechanical analysis of the acidizing operation process for a 2700 m tubing string. (**a**) Axial stress. (**b**) Stress. (**c**) Safety factor. (**d**) Deformation quantity.

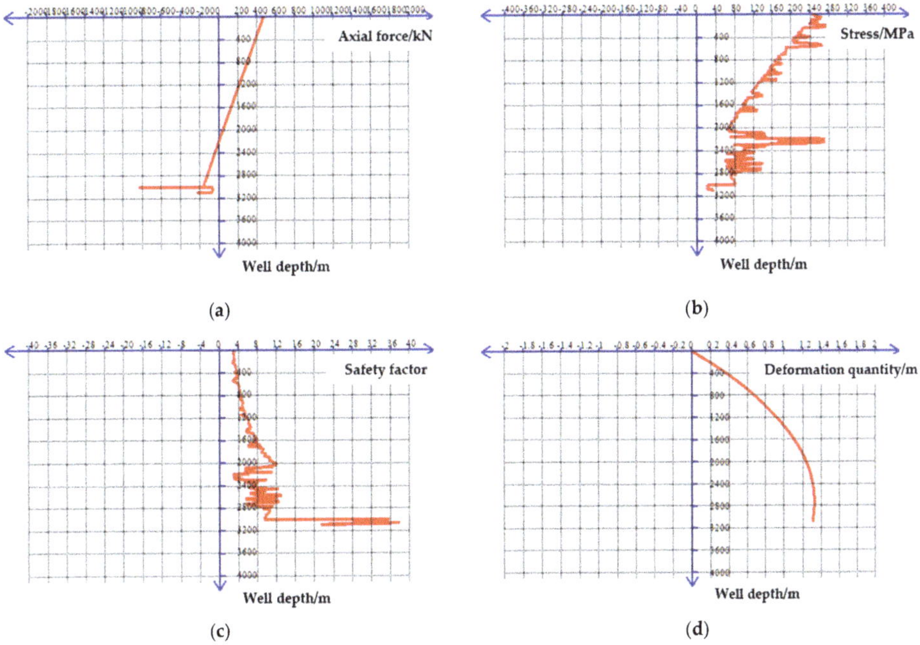

Figure 19. Mechanical analysis of the acidizing operation process for a 3000 m tubing string. (**a**) Axial stress. (**b**) Stress. (**c**) Safety factor. (**d**) Deformation quantity.

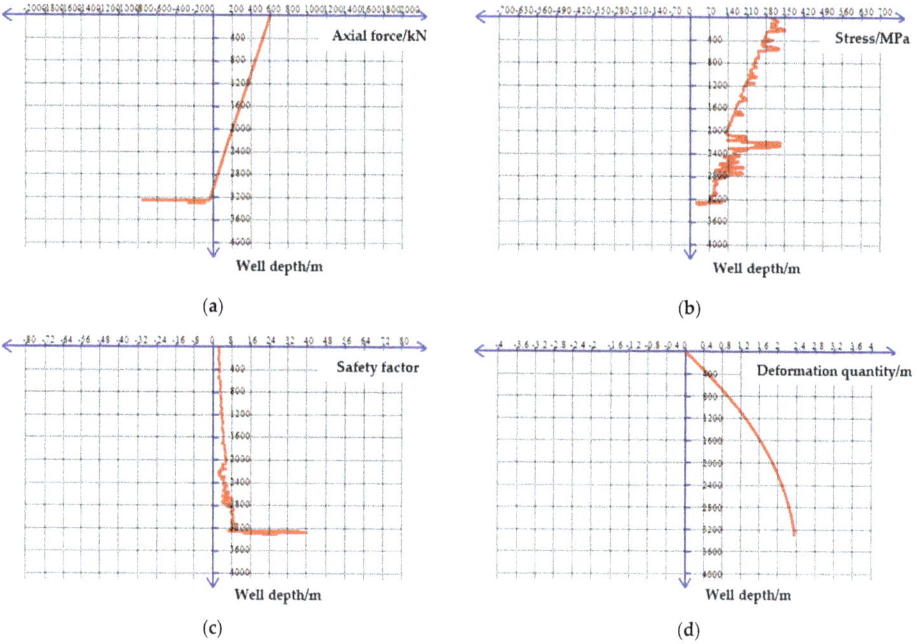

Figure 20. Mechanical analysis of the acidizing operation process for a 3300 m tubing string. (**a**) Axial stress. (**b**) Stress. (**c**) Safety factor. (**d**) Deformation quantity.

It is known that the expansion joint compensation distance is 3 m. The individual maxima arise under different boundary conditions and do not occur at the same time. Through the dynamic simulation and calculation of the integrated tubing string for acidizing and flow-back in Section 3.3, the deformation of different operation tubing strings under different conditions was measured [32]. As shown in Figure 21, the deformation of the tubing string is within the safe range within the expansion tube compensation range. The maximum axial force of the 2700 m operation tubing string is approximately 400 kN, the maximum axial stress is approximately 260 MPa, the safety factor is greater than 3.0, and the maximum deformation of the tubing string is approximately 1.1 m. The maximum axial force of the 3000 m operation tubing string is approximately 500 kN, the maximum axial stress is 300 MPa, the safety factor is greater than 3.0, and the maximum deformation of the tubing string is approximately 1.4 m. Meanwhile, the maximum axial force of the 3300 m operation tubing string is approximately 600 kN, the maximum axial stress is 350 MPa, the safety factor is greater than 3.0, and the maximum deformation of the tubing string is approximately 2.4 m. During the operation, the tubing string is relatively safe.

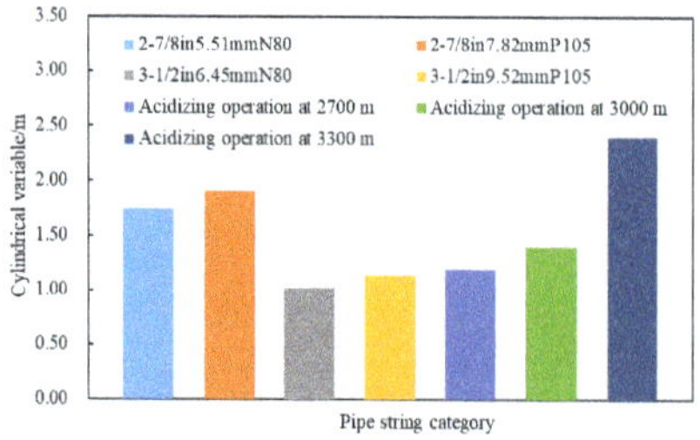

Figure 21. Comparison chart of different operation pipe column shapes.

5. Evaluation of Production Increase Effect

5.1. Application Example

Based on on-site applications, the acidizing and flowback integrated process technology was tested, and its effect was analyzed [49–52]. Through comparative analysis, the economic and social effects of negative pressure flowback pump flowback compared with other flowback methods were compared. The comprehensive evaluation of the acidizing and flowback integrated process was improved. The application example well was the CBG-10 well in the offshore Dongying formation Oilfield. The total well depth was 3260.68 m, the maximum well inclination depth was 2448.46 m, the maximum well inclination angle was 23.30°, and the maximum well inclination azimuth was 238.43°. The casing of the operation well was 762 mm in specification, with a depth of 95.93 m. The surface casing specification was 339.7 mm, with a depth of 701.54 m. The technical casing specification was 177.8 mm, with a depth of 3226.68 m. The cementing quality was excellent. The pump injection volumes of the pre-acid pretreatment solution, the main acid plugging solution, and the post-active flushing acid are 15–20 m^3. The construction displacement is $\leq$0.4 m^3/min, and the injection method is to inject while closing the casing. The detailed casing data and cementing data are shown in Table 11.

Table 11. Casing data and cementing data of the operation well.

Packing Tube Name	Specification/mm	Pipe Steel Grade	Wall Thickness/mm	Deep Down/m	Well Cementing Quality
Separator pipe	762	D32	28.00	95.93	High-quality well
Surface casing	339.7	J55	9.65	701.54	
Technical casing	177.8	P110	9.19	3226.68	

The structure of the construction tubing string is as follows: plug + 2 strands of spiral screen pipe + 22 pieces of 89 mm EU oil tubes + safety joint + hydraulic anchor + RTTS packer + 1 piece of 89 mm EU oil tube + acid-resistant flowback pump + pin-type oil release device + 115 pieces of 89 mm EU oil tubes + stop valve + 1 piece of 89 mm EU oil tube + 2 pieces of 89 mm EU oil tubes + sanding movable elbow.

The construction effects before and after the acidizing operation of the target well were compared. Before the acidizing operation, the daily liquid production of the operation well was 50.5 $t{\cdot}d^{-1}$, and the daily oil production was 6.3 $t{\cdot}d^{-1}$. After the acidizing operation, the daily liquid production of the operation well was 150.9 $t{\cdot}d^{-1}$, and the daily oil production was 15.7 $t{\cdot}d^{-1}$. After the implementation of the acidizing operation, the daily liquid production capacity and the daily oil production capacity have been significantly improved, with the liquid production increasing by approximately 100 $t{\cdot}d^{-1}$ and the oil production increasing by approximately 10 $t{\cdot}d^{-1}$. The construction effects are shown in Table 12.

Table 12. Construction effect after operation.

Comparison	Oil Mouth/mm	Voltage/V	Current/A	Tubing Pressure/MPa	Back-Pressure/MPa	Daily Fluid Output/($t{\cdot}d^{-1}$)	Daily Oil Output/($t{\cdot}d^{-1}$)	Water Volume Fraction/%
Before the assignment	5	1230	32	1.8	0.9	50.5	6.3	87.5
After the assignment	5	1251	33	6.3	0.9	150.9	15.7	89.6

Before the acidizing and fluid return operation was carried out in other wells of the target block, the maximum daily liquid production was 89.0 $t{\cdot}d^{-1}$, the maximum daily oil production was 13.4 t·d^{-1}, and the average daily liquid production of the 9 wells was 42.1 t·d^{-1}, with an average daily oil production of 11.3 t·d^{-1}. The comparison of liquid production before and after acidizing operations in other wells of the research area is shown in Figure 22. After the acidizing and fluid return operation, the maximum daily liquid production was 164.2 t·d^{-1}, the maximum daily oil production was 54.1 t·d^{-1}, the average daily liquid production of the 9 wells was 89.4 t·d^{-1}, and the average daily oil production was 29.4 t·d^{-1}. Among them, the maximum increase in daily liquid production per well exceeded 150.0 t·d^{-1}, the maximum increase in daily oil production exceeded 45.0 t·d^{-1}, the average daily liquid production increased by 47.3 t·d^{-1}, and the daily oil production increased by 18.07 t·d^{-1}.

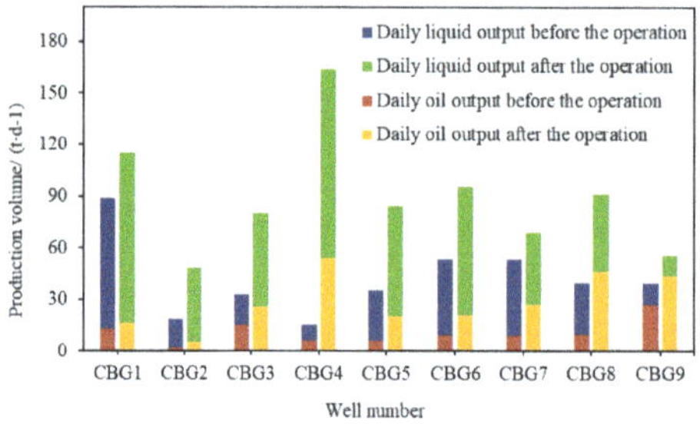

Figure 22. Comparison of fluid production before and after acidizing operations in other wells.

5.2. Economic and Social Impact Evaluation

Using the negative pressure return flow of the return pump to reduce costs can save more than CNY 500,000 compared to the other two processes; it can also save more than 40% of construction time. The comparison for the construction of different backflow operation methods is shown in Table 13. The negative pressure return flow of the return pump can eliminate the need for ship support, and the acidification return fluid can be incorporated into the pipeline, significantly reducing the operational risks of the construction and alleviating the environmental pressure of offshore operations. Through multiple safety auxiliary devices, the safety risk points are relatively low.

Table 13. Comparison table for the construction of different backflow operation methods.

Platform Type	Construction Project and Cost/CNY Ten Thousand		
	The Backflow Pump Performs Negative Pressure Backflow	**Nitrogen Foam Is Discharged Under Negative Pressure**	**Three-in-One Test for Liquid Discharge**
Mobile unit	80	180	380
Integrated platform for recovery and maintain	30	80	280

6. Conclusions

(1) An innovative integrated process of acidizing-flowback was achieved by injecting acid through the tubing-casing annulus and conducting acidizing via the reverse flushing valve in the target zone, combined with the rapid flowback of spent acid using a return pump. This approach reduces time and space costs while effectively preventing secondary formation damage.

(2) The tubing string is primarily subjected to mechanical effects, including buoyancy, piston effect, flow viscosity effect, helical buckling, temperature differential effect, and

ballooning effect. The maximum deformations are 1.4 m, 1.9 m, 0.18 m, 2.7 m, 1.8 m, and 2.5 m, respectively, all remaining below 3 m, indicating overall string safety.

(3) Dynamic simulation and optimization software for tubing string mechanics, developed in C#, was used to simulate three groups of wells at different depths. Results showed axial forces of 400–600 kN, stress of 260–350 MPa, deformation of 1.1–2.4 m, and safety factors exceeding 3.0, confirming operational safety.

(4) Field application demonstrated significant production enhancement, with example wells showing an approximately 100 $t \cdot d^{-1}$ increase in total fluid production and about a 10 $t \cdot d^{-1}$ increase in oil production. The integrated acidizing flowback technology achieved economic savings exceeding CNY 500,000, reduced time costs by over 40%, and effectively ensured operational safety.

Author Contributions: Conceptualization, L.W. and J.W.; methodology, Y.C. and M.S.; software, Y.C.; validation, T.W. and J.W.; formal analysis, T.W.; investigation, Y.C.; data curation, T.W.; writing—original draft preparation, L.W. and M.S.; writing—review and editing, M.S. and T.W. All authors have read and agreed to the published version of the manuscript.

Funding: This research was funded by the Lanzhou City University Doctoral Research Fund Project (LZCU-BS2025-10), and the National Natural Science Foundation of China (42202182).

Data Availability Statement: Data are contained within the article.

Acknowledgments: The authors would like to extend their heartfelt thanks to Lanzhou City University, China University of Petroleum (East China) and Yangtze University for the invaluable educational foundation and academic support that have significantly contributed to the development of this work, and express their sincere gratitude to PipeChina Engineering Technology Innovation Co., Ltd. for materials support and for providing access to laboratory facilities that were essential for the successful completion of this study. Additionally, the authors also express their sincere appreciation to the anonymous reviewers for their valuable and constructive comments.

Conflicts of Interest: Author Yi Chen was employed by the company PipeChina Engineering Technology Innovation Co., Ltd. The remaining authors declare that the research was conducted in the absence of any commercial or financial relationships that could be construed as a potential conflict of interest.

References

1. Al-Shargabi, M.; Davoodi, S.; Wood, D.A.; Ali, M.; Rukavishnikov, V.S.; Minaev, K.M. A critical review of self-diverting acid treatments applied to carbonate oil and gas reservoirs. *Petrol. Sci.* **2023**, *2*, 922–950. [CrossRef]
2. Cai, J.; Yang, J.; Huang, Z.; Xu, S.; Zhang, L.; Wang, H. Quantitative evaluation of residual acid invasion and flowback in fractured-vuggy carbonate reservoirs using microfluidics. *Energies* **2025**, *18*, 1162. [CrossRef]
3. Zhang, J.; He, J.; Zhang, R.; Zhang, L.; Xu, W. Optimization and performance evaluation of an atomized acid system for the expansion of carbonate gas injection. *Processes* **2023**, *11*, 3080. [CrossRef]
4. Jafarpour, H.; Moghadasi, J.; Khormali, A.; Petrakov, D.; Ashena, R. Increasing the stimulation efficiency of heterogeneous carbonate reservoirs by developing a multi-bached acid system. *J. Pet. Sci. Eng.* **2019**, *172*, 50–59. [CrossRef]
5. Hu, J.; Liu, H.; Wu, D.; Zhang, J. Acidizing flowback optimization for tight sandstone gas reservoirs. *J. Nat. Gas. Sci. Eng.* **2015**, *24*, 311–316. [CrossRef]
6. Nasriani, H.R.; Jamiolahmady, M. Optimising flowback strategies in unconventional reservoirs: The critical role of capillary forces and fluid dynamics. *Energies* **2024**, *17*, 5822. [CrossRef]
7. Zhang, W.; Gao, D.; Liu, Y.; Bai, J.; Wang, C. Study of the failure mechanism of an integrated injection-production string in thermal recovery wells for heavy oil. *Energies* **2023**, *16*, 3246. [CrossRef]
8. Saini, R.K.; Sayed, M.A.; Cairns, A.J. Flowback additive for acidizing fluid to stimulate carbonate gas reservoirs. *Soc. Pet. Eng. J.* **2024**, *29*, 4841–4855. [CrossRef]
9. Wang, Y.; Zhou, F.; Su, H.; Li, Y.; Yu, F.; Dong, R.; Wang, Q.; Li, J. Performance evaluation of microemulsion acid for integrated acid fracturing in Middle Eastern carbonate reservoirs. *Petrol. Explor. Dev* **2023**, *50*, 1196–1205. [CrossRef]

10. Ren, J.; Liu, Z.; Deng, Y.; Guo, J.; Lai, J.; Wang, J.; He, S. The quantitative study of residual acid damage of low permeability carbonate rock based on NMR test. *Geoenergy. Sci. Eng.* **2024**, *241*, 213109. [CrossRef]
11. Li, P.; Wang, L.; Lai, X.; Gao, J.; Dang, Z.; Wang, R.; Mao, F.; Li, Y.; Jia, G. Two-level self-thickening mechanism of a novel acid thickener with a hydrophobic-associated structure during high-temperature acidification processes. *Polymers* **2024**, *16*, 679. [CrossRef]
12. Cai, W.; Huang, Z.; Mo, X.; Zhang, H. Velocity string drainage technology for horizontal gas wells in Changbei. *Processes* **2022**, *10*, 2640. [CrossRef]
13. Zhang, L.P.; Zhang, L.; Lan, X.T.; Liu, C.L.; Zou, J.; Gao, S. Application of non-acid block removing technology in Bohai Oilfield. *Drill. Fluid Complet. Fluid* **2018**, *35*, 116–122.
14. Yuan, H.; Sun, Y.; Xiong, W.; Niu, W.; Tang, Z.; Li, Y. Advances in evaluation methods for artificial fracture networks in shale gas horizontal wells. *Appl. Sci.* **2025**, *15*, 9008. [CrossRef]
15. Chen, D.; Sun, Z. Gas–liquid flow pattern and hydrate risk in wellbore during the deep-water gas-well cleanup process. *ACS Omega* **2023**, *8*, 12911–12921. [CrossRef] [PubMed]
16. Liu, J.W.; Xian, G.Q.; Li, C.Y.; Qu, H.M.; Tan, Y.X.; Cui, Y.H.; Ding, Z.F.; Xiao, H.; Zhao, T.H. Acidizing, Negative-Pressure and Flowback Integrated Equipment and Technology. CN Patent CN105672976, 15 June 2016. Available online: https://ss.zhizhen.com/detail_38502727e7500f261a7b5f441dec441545103c5d99a9155e1921b0a3ea25510133e7f87ae1903bd62163d3a8b505c2ddfc7d188a48f9f2e7424194cada808d01a613e5dd08e158c2f826a6a8f4e512e2? (accessed on 1 November 2025).
17. Lin, C.; Kang, Y.; He, H.; Han, L.; Xu, Q.; Li, G.; Zhou, H. Geoscience-engineering integration-based formation damage control drill-in fluid (GEI-FDC-DIF): An essential technology for successful extraction of deep mid-high permeability sandstone reservoirs. *Geoenergy Sci. Eng.* **2024**, *237*, 212828. [CrossRef]
18. Wang, C.; Zhang, J.; Huang, B.; Du, H.; Meng, X.; Li, X.; Xue, X.; Su, Y.; Li, C.; Guo, H. Polymer flooding injectivity maintaining and enhancement strategies: A field case study of chinese offshore EOR project. *Processes* **2025**, *13*, 903. [CrossRef]
19. Wu, S.W.; Zhou, H.Y.; Lin, K.X.; Yuan, H.; Luo, G.; Hu, Y.L. An integrated working fluid for blocking removal and sand control in offshore wells blocked by particle migration. *Drill. Fluid Complet. Fluid* **2021**, *38*, 391–396.
20. Ming, E.; Li, C.; Lan, H.; Yu, J.; Zheng, L.; Pei, X. Modelling wave transmission for transient flow and amplitude-frequency characteristics of tubular string in a water injection well. *Appl. Sci.* **2023**, *13*, 3917. [CrossRef]
21. Li, D.; Gao, B.; Cui, X. Emulsifying behavior between oil and water of flowback fluids from acidizing oil wells based on high-voltage electric field oil–water separation technology. *New J. Chem.* **2022**, *46*, 16858–16873. [CrossRef]
22. Li, S.N.; Yan, W. Commonly used treatment technology of acidizing fluid flowback. *Open J. Nat. Sci.* **2017**, *5*, 302–305. [CrossRef]
23. Zhai, Z.B.; Qi, S.W.; Kartoatmodjo, G. Pressure survey, analysis and diagnostic of small diameter velocity string to improve well monitoring and surveillance. *Pet. Sci. Technol.* **2022**, *41*, 1019–1037. [CrossRef]
24. Yin, D.; Luo, P.; Zhang, J.; Yao, X.; Wang, R.; Wang, L.; Wang, S. Synthesis of oligomeric silicone surfactant and its interfacial properties. *Appl. Sci.* **2019**, *9*, 497. [CrossRef]
25. Kong, X.; Liu, B.; Xu, H.; Shen, J.; Li, S. Optimization and performance evaluation of foam acid systems for plugging removal in low pressure oil and gas reservoirs. *Processes* **2023**, *11*, 649. [CrossRef]
26. Li, J.T.; Han, J.J.; Xu, Q.H.; Xue, X.C.; Zhang, S.A. Third-Section Acid Pickling Integrated Well Completion Pipe String and Method. CN Patent CN105041267, 11 November 2015.
27. Yu, Z.S.L. Perforating, Detecting, Acidating and Liquid-Discharging Integrated Testing Pipe String. CN Patent CNU2005201062073, 18 July 2007. Available online: https://ss.zhizhen.com/detail_38502727e7500f263168c25d27cffd2178ea8fa7e004e9451921b0a3ea25510133e7f87ae1903bd60cbc8308fe832728c7f667a6b29a344ca9480c637d9474d22eb13cf9a5c1052d6fd5fc3f93bf62bf? (accessed on 1 November 2025).
28. Li, Q.Z.; Xu, D.K.; Wu, E.C. Deep Gas Well Subsection Acidizing and Completion Integral Tool String. CN Patent CNU2007201516785, 23 April 2008. Available online: https://ss.zhizhen.com/detail_38502727e7500f261a3b92c20e49cd472bc66bc1e2315ea81921b0a3ea25510133e7f87ae1903bd674cd55a89d1ef7a51c317007611b90ac4abc1a683a53ee43d75c28f4a098b629af4f8e85cdc5fa72? (accessed on 1 November 2025).
29. Deng, S.; Li, X.; Yang, Z.; Yi, L.; Rao, D. Flow and heat transfer for non-Newtonian CO_2 mixed fluid injection in the wellbore. *Fuel* **2025**, *381*, 133647. [CrossRef]
30. Zhang, F.; Yang, X.; Peng, J.; Li, N.; Lv, S.; Zeng, N.; Zhang, R. Well Integrality technical practice of ultra deep ultrahigh pressure well in Tarim Oilfield. In Proceedings of the 6th International Petroleum Technology Conference, Beijing, China, 26–28 March 2013.
31. Shishlyannikov, D.; Zverev, V.; Ivanchenko, A.; Zvonarev, I. Increasing the time between failures of electric submersible pumps for oil production with high content of mechanical impurities. *Appl. Sci.* **2022**, *12*, 64. [CrossRef]
32. Li, J.; Ding, Y.Y.; Wang, L.R.; Shi, S.Z.; Liao, R.Q.; Zhang, M.L.; Zhang, Q. The mechanic analysis of integrated tubing string for acid fracturing and production. *J. Oil Gas Technol.* **2017**, *39*, 79–88. [CrossRef]

33. Awotiku, O.; Hackney, D.; Ledergerber, A.; Ugah, I.; Femi-Saliu, B. Well clean out and acidization of a deepwater intelligent completion well—A case history. In Proceedings of the Nigeria Annual International Conference and Exhibition, Lagos, Nigeria, 1 July–2 August 2017.
34. Wang, J.; Li, D.; Du, X.; Li, H.; Yang, S. Development of a design method for casing and tubing strings under complex alternating loads. *Processes* **2023**, *11*, 2582. [CrossRef]
35. Al Ramadan, M.; Salehi, S.; Ezeakacha, C.; Teodoriu, C. Analytical and experimental investigation of the critical length in casing–liner overlap. *Sustainability* **2019**, *11*, 6861. [CrossRef]
36. Feng, Z.; Guo, B.; Cai, Z.; Yuan, H. Eccentric wear mechanism and centralizer layout design in 3D curved wellbores. *Appl. Sci.* **2025**, *15*, 1494. [CrossRef]
37. Li, Y.K.; Jia, Y.Y.; Zhang, G.Z.; Wang, H.W.; Cui, Y.H. Research progress and development suggestion of stratified acidizing strings in water injection wells of Shengli Oilfield. *Pet. Drill. Tech.* **2021**, *49*, 129–134.
38. Zhao, X.; Geng, Q.; Qiu, Z.S.; Geng, T.; Zhou, G.W.; Xing, X.J. Gel-breaHng free drill-in nuid technology for deepwater high-porosity and high-permeability reservoirs. *Nat. Gas Ind.* **2021**, *41*, 107–114.
39. Wang, Q.; Yang, H.; Han, X.; Huang, X.; Zeng, J.; Chen, W.; Zhang, H.; Li, L.; Zhou, C.; Wang, H. Method for Acid Fracturing Comprises e.g. Sandblasting and Perforating to Form Perforation Channel in Formation, Running Acid Fracturing String Into Wellbore, Injecting Acid Liquid into Perforation Channel, and Liquid Passes Through Perforation Channel to form Fractures in Formation. CN Patent CN10431573, 21 October 2022. Available online: https://ss.zhizhen.com/detail_38502727e7500f26a94cc9e8088d1781fbd65685d76678fe1921b0a3ea25510133e7f87ae1903bd6b466ec2028a731074210bf2a3c9923bf3bab857c045c8b9b3070a6704eede4aafe61327bf5125744? (accessed on 1 November 2025).
40. Dore, F.P.; Pereira, A.Z.I.; Germino, F.D.S.M.; Ferreira, D.S.M.G.; Ignacio, P.A.Z.; Fernandes, P.D. Uniformization of Flow for Stimulation Operations Involves Using the Wash Pipe Tool to Stimulate the Reservoir, and Performing Reservoir Stimulation with the Wash Pipe Tool with Divergent Perforation During the Well Completion. US Patent US983606, 11 May 2023. Available online: https://ss.zhizhen.com/d-tail_38502727e7500f269f46c8a6849ffd8f01cd3ee1b951b0441921b0a3ea25510133e7f87ae1903bd6653e84ab8524ea2a70e511d2dca4593b77b7c8986150a9f2a384695c6c9a9b7f36834484f53973e0? (accessed on 1 November 2025).
41. Pan, W.H.; Qian, Y.C.; Tao, J.Q.; Pan, Y.; Gu, S.B.; Wang, H. Application of original production string acidizing technology in oil well. *Shiyou Huagong Yingyong* **2013**, *32*, 7–11. [CrossRef]
42. Yu, Y.; Cao, Y.; Qu, Z.; Dou, Y.; Wang, Z. Finite-element analysis on energy dissipation and sealability of premium connections under dynamic loads. *Processes* **2023**, *11*, 1927. [CrossRef]
43. Martin, F.; Miquilena, E.; Molero, N.J.; Acevo, A.K.; Franco, E.; Garcia, M.G.; Castillo, M. Corrosion inhibition workflow allows injection of high-volume acid treatments through coiled tubing Strings in Offshore Mexico. In Proceedings of the International Conference and Exhibition on Oilfield Corrosion, Aberdeen, UK, 9–10 May 2012.
44. Wang, Y.; Yang, J.; Zhang, Y.; Huo, L.; Song, Z.; Cao, X. Development and application of integrated injection system for injection and profile control in Dagang Oilfield. *J. Phys. Conf. Ser.* **2024**, *2834*, 012044. [CrossRef]
45. Wang, J.; Xiong, Y.; Lu, Z.; Shi, J.; Wu, J. Influence of volume fracturing on casing stress in horizontal wells. *Energies* **2021**, *14*, 2057. [CrossRef]
46. Cheatham, J.B.; Patillo, P.D. Helical postbuckling configuration of a weightless column under the action of m axial load. *Soc. Pet. Eng. J.* **1984**, *24*, 467–472. [CrossRef]
47. Kyriakides, S.; Corona, E. *Mechanics of Offshore Pipelines, Vol. 1: Buckling and Collapse*; Gulf Professional Publishing: Houston, TX, USA, 2023.
48. Lu, Y.P.; Wu, X.D.; Guo, S.S.; Fu, H.; Zhang, G.Y. Tubing string stress and axial deformation in gas wells. *Nat. Gas Ind.* **2008**, *28*, 100–102.
49. Wang, C. Integrated technology for layered sand control and layered water injection. *J. Phys. Conf. Ser.* **2024**, *2683*, 012044. [CrossRef]
50. Zou, J.; Huang, L.P.; Zhao, P.; Yang, M.; Gao, S. Analysis of the content of solid suspended matter in acid fracturing flowback fluids from oil wells in Bohai Oilfield. *Drill. Fluid Complet. Fluid* **2020**, *37*, 536–540. [CrossRef]
51. Xu, B.; Chen, H.; Yin, G.; Qin, R.; Gao, J.; He, X. Acoustic transmission characteristics and model prediction of upper and lower completion pipe strings for test production of natural gas hydrate. *Appl. Sci.* **2025**, *15*, 9174. [CrossRef]
52. Bai, Y.J.; Huang, B.; Zou, C.; Xu, X.N.; Liu, W.B.; Wang, Z.; Zhang, N.; Li, D.; Cao, G.S.; Fu, C. Mechanism of enhanced oil recovery by nanoparticles extracted from flowback fluids of acidizing oil wells. *Colloid Surf. A* **2024**, *690*, 133782.

processes

Article

A New Kind of Thermosensitive Screen Used for Wellbore Stability

Yanlong Zhao [1,*], Yuheng Wei [1], Xing Qin [2] and Yifei Ran [1]

[1] School of Petroleum, China University of Petroleum-Beijing at Karamay, Karamay 834000, China; 2024216901@st.cupk.edu.cn (Y.W.); 2025216924@st.cupk.edu.cn (Y.R.)
[2] Sinopec Research Institute of Petroleum Engineering, Beijing 100101, China; qinxing.sripe@sinopec.com
* Correspondence: zhaoyanlong@cupk.edu.cn; Tel.: +86-18010092639

Abstract

In light of frequently occurring wellbore instability such as wellbore collapse and sand production that often occur in drilling and the completion of shale oil and gas development, we propose one-run shape memory thermosensitive screen technology that can expand spontaneously at a specific temperature to help strengthen the formation. Based on the theory of thermal expansion and large deformation of shape memory materials, the expansion process of the thermosensitive screen is calculated by the finite element method. After expanding to the wellbore wall, the effects of the screen squeezing force on the formation production parameters are evaluated theoretically. The analysis shows that the radial compressive stress of the thermosensitive screen decreases with the increase in the radial distance, but as the original outer diameter of the thermosensitive screen is greater than the wellbore diameter, it can provide extrusion force for the wellbore wall. According to the in situ stress model, the extrusion force after the screen contacts the wellbore can effectively improve the stress distribution near the wellbore and reduce the impact of sand production caused by formation instability. Moreover, in shale oil and gas completion, it can effectively increase the bottom hole flowing pressure and drawdown pressure.

Keywords: thermosensitive screen; large deformation; expansion process; wellbore stability; mechanical analysis

1. Introduction

The wellbore stability of deep and ultra-deep wells is a complex problem often encountered in drilling and completion engineering [1–4]. During the drilling stage, borehole collapse will cause accidents such as sticking and obstruction, affecting the drilling rate and cementing quality. In the completion stage, wellbore collapse often occurs due to the decrease in wellbore pressure or excessive drawdown pressure. Wellbore collapse will plug the borehole and affect subsequent production or operations. In oil fields, perforated tubing or slotted screens are often run to the collapsed layer to support the wellbore. However, due to the limitation of fixed size, the running ability of common screens in the wellbore is limited, especially in horizontal wells or extended reach wells. It is difficult to run them in a well section with serious dogleg. As the screen string of designed size cannot be run to the target layer, the operators usually have to replace it with a screen string of a smaller size to run it into the well, which will weaken the support effect and lead to the re-accumulation of cuttings in the wellbore, thereby affecting oil production.

Therefore, we introduce a kind of expansion string, namely the thermosensitive screen (hereafter abbreviated as TSS) [5–7]. Its advantage is that the screen can maintain a smaller size (i.e., compressed state) than the conventional oil screen in the lowering process. After the screen is run to the target layer, the formation temperature stimulates the expansion of the screen. As the original size of the thermosensitive screen is greater than the wellbore size, it can provide extrusion force for the wellbore wall when expanded and pressed against the wellbore wall. Compared with traditional mechanical expansion screen technology [8,9], this technology can greatly shorten the construction period and features a simple expansion process. It can be used for the completion of oil and gas wells with limited wellbore conditions such as long horizontal section wells, complex structure wells, and offshore wells [10,11].

2. Principle of Thermosensitive Expansion

The thermosensitive screen proposed in this paper is composed of a kind of shape memory polymer. Shape memory polymer refers to a kind of environment-responsive material that can change their initial shape through external environmental stimulation and can be restored to their initial shape through reverse stimulation [12,13]. Under the stimulation of external temperature, the TSS has two different states, namely a glassy state at low temperatures and a rubbery state at high temperatures. The transition temperature from the glassy state to the rubbery state is called the glass-transition temperature Tg. As shown in Figure 1, the TSS in the initial state is first heated to the glass-transition temperature Tg and then compressed to the designed shape in the mold by external force. After reducing the temperature below the transition temperature Tg and removing the external force, the TSS will fix its compressed shape at a lower temperature. Once the TSS in the low-temperature glassy state is heated to the transition temperature (i.e., when the ambient temperature rises to the transition temperature), the TSS will regain its initial state [14–17].

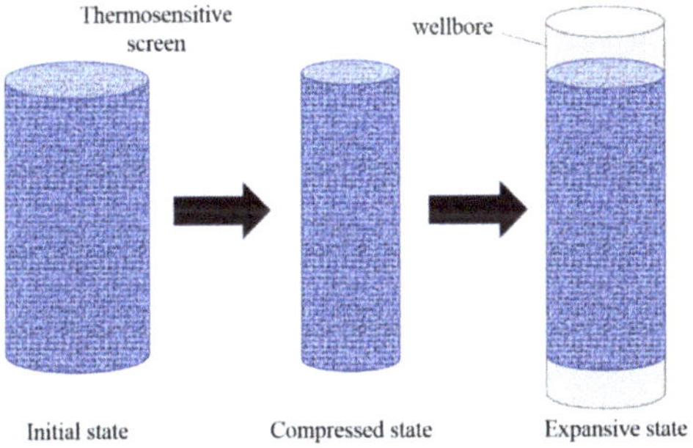

Figure 1. Transformation progress of thermosensitive screen.

In the field of aerospace, the Japanese NEC company has used shape memory materials combined with 4D printing technology to manufacture self-deformation pipes and self-repair pipes. The former can expand or shrink according to service conditions, and the latter has the ability of environmental perception and can repair itself after injury. In biomedical science, a vascular stent made of shape memory alloy can support the blood vessels and restore the normal function of the stenotic blood vessels [18,19].

After the thermosensitive screen composed of the above shape memory material is sent to the predetermined underground stratum, the formation temperature is used to stimulate the expansion of the screen, which begins to recover its initial shape. As its initial shape is larger than the wellbore size, after the screen expands and is pressed against the wellbore wall, a positive extrusion force will be generated on the wellbore wall to stabilize the formation and facilitate later production.

3. Expansion Process of Thermosensitive Screen

According to the above shape memory characteristics, we establish the large deformation model of shape memory polymers by simplifying the research problems. This method can simulate and predict the stress and strain variation in the thermosensitive screen after expansion in place. In order to simplify the problem, the assumptions adopted in this model are as follows:

1. When the temperature rises to the critical value, the memory shape of the material can be completely restored.
2. The time effect is ignored, and the recovery time of expansion is long enough that the deformation of the material is stable.
3. The constitutive model of the thermosensitive screen is described by the Odgen model [20,21].

The expansion process of the thermosensitive screen can be simplified to an elastic problem including geometric nonlinearity and constitutive nonlinearity based on Assumptions (1)–(3). Because the relevant mechanical parameters of the thermosensitive screen are unknown in advance, it needs to be solved by inversion according to the experimental data.

Therefore, firstly, according to the uniaxial compression experiment, the axial pressure under different axial compressions is measured, and the values of the unknown parameters in the Odgen model can be inverted by using the experimental results. The inverse constitutive model is applied to simulate the deformation and stress distribution of the thermosensitive screen in the process of real well completion.

3.1. Mechanical Parameter Inversion

The deformation potential energy of the thermosensitive screen can be expressed as

$$E = \sum_{i=1}^{N} \frac{2\mu_i}{{\alpha_i}^2} [{\lambda_z}^{\alpha_i} + {\lambda_r}^{\alpha_i} + {\lambda_\theta}^{\alpha_i} - 3 - \alpha_i \ln(\lambda_z \lambda_r \lambda_\theta)] \tag{1}$$

where λ_z, λ_r, and λ_θ are the principal eigenvalues of deformation, λ_z is the value in the axial direction, λ_r is the value in the radial direction, λ_θ is the value in the circumferential direction, μ_i and α_i are dimensionless real numbers, and N is a positive integer. These three values need to be inversed and calculated by using the experimental data.

In the uniaxial compression test, the sample has an outer diameter of 25 mm and height of 30 mm. The universal mechanical testing machine for materials is set to displacement control mode, the compression speed is set to 2 mm/min, and the compression distance is set to 15 mm. The compression properties were measured in 70 °C water bath and 25 °C air environment, respectively. The sample is soaked in a 70 °C water bath for one hour to be fully heated. The measurement results of stress ($\sigma = \frac{P}{A}$) and strain ($\varepsilon = \frac{l-l_0}{l_0}$) during unidirectional compression are shown in Figure 2. According to the experimental results, the axial length of the cylindrical sample becomes shorter after compression, while the radial deformation is negligible (see Figure 2), so it can be approximately assumed that $\lambda_2 = 1$, $\lambda_3 = 1$.

The formula of axial stress can be obtained according to the potential energy as follows:

$$t_z = \sum_{i=1}^{N} \frac{2\mu_i}{\alpha_i} \left({\lambda_z}^{\alpha_i - 1} - {\lambda_z}^{-1} \right) \tag{2}$$

where t_z is the Biot stress in the axial direction, N/m^2.

The results of axial deformation and measured stress in the experiment are substituted into the above calculation formula, and the results of μ_i, α_i, and N are determined by the nonlinear fitting method. The values of the results are

$$\begin{cases} N = 2 \\ \alpha_1 = 0.11335 \quad \mu_1 = 1.42299 \\ \alpha_2 = 28.2891 \quad \mu_2 = 5.44801 \end{cases} \tag{3}$$

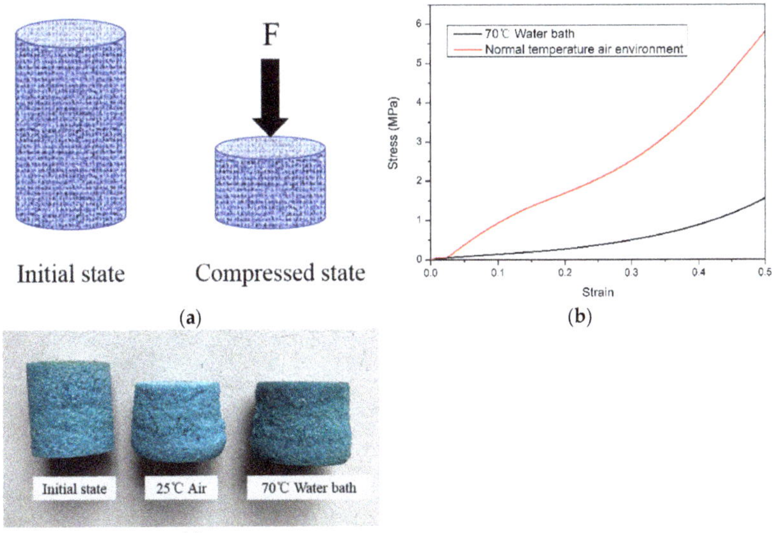

Figure 2. Compression experiment under constant temperature. (**a**) Compression progress; (**b**) compression stress and strain curve. (**c**) Experimental samples.

The fitting calculation results and experimental results are shown in the figure below (Figure 3), which shows that the above fitting results are very accurate and can be used for case analysis in the next step.

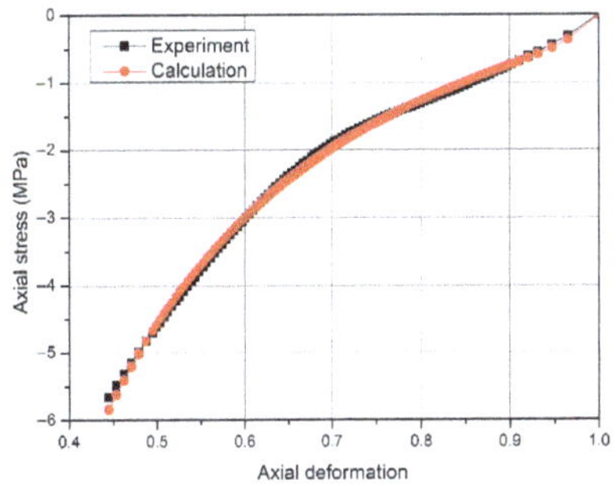

Figure 3. Comparison of parameter inversion results.

3.2. *Load Distribution After Expansion*

When the shape of the thermosensitive screen is restored and becomes stable, it experiences significant deformation. The outer diameter of the thermosensitive screen is the same as the inner diameter of the wellbore wall, and the inner diameter is the same as the outer diameter of the base tube (the base tube is sleeved in the thermosensitive screen; see Figure 4). As the radial compression deformation has little effect on the axial deformation, the calculation problem can be simplified to a plane stress problem.

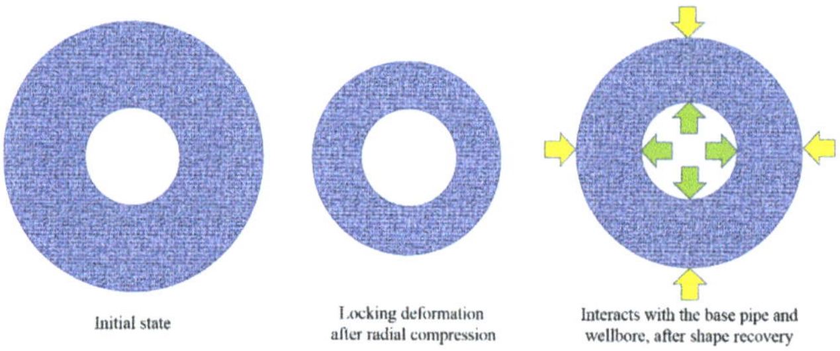

Figure 4. Schematic diagram of TSS expansion process. The green arrow represents the contact force between the base pipe and the material, while the yellow arrow represents the contact force between the wellbore and the material.

3.2.1. Basic Model

According to the characteristics of the axisymmetric problem, the deformation of the thermosensitive screen is

$$r = r(R) = u(R) + R \tag{4}$$

where R is the radial coordinate of the TSS under the initial configuration, u is the radial displacement, and r is the new radial coordinate after radial deformation.

The deformation gradient and strain are expressed as follows, respectively:

$$\mathbf{F} = \frac{dr}{dR}\mathbf{e}_r\mathbf{e}_r + \frac{r}{R}\mathbf{e}_\theta\mathbf{e}_\theta \tag{5}$$

$$\mathbf{K} = \mathbf{F} - \mathbf{I} = \frac{du}{dR}\mathbf{e}_r\mathbf{e}_r + \frac{u}{R}\mathbf{e}_\theta\mathbf{e}_\theta \tag{6}$$

The principal eigenvalues of the radial and circumferential deformation are

$$\lambda_r = \tfrac{dr}{dR} \quad \lambda_\theta = \tfrac{r}{R} \tag{7}$$

According to the function of deformation potential energy (1), each component of Biot stress $\mathbf{t} = t_r\mathbf{e}_r + t_\theta\mathbf{e}_\theta$ is as follows:

$$t_r = \tfrac{dE}{d\lambda_r} \quad t_\theta = \tfrac{dE}{d\lambda_\theta} \tag{8}$$

Since the expansion problem is axisymmetric, Biot stress is equivalent to nominal stress. According to the principle of virtual work, the following equilibrium conditions are obtained:

$$\iint_{\Omega^0} \frac{\partial E}{\partial \mathbf{K}} : \delta\mathbf{K} \cdot d\Omega = \iint_{\Omega^0} \delta\mathbf{u}^T\mathbf{f}_b \cdot d\Omega + \int_{\Gamma_s^0} \delta\mathbf{u}^T\mathbf{t} \cdot d\Gamma \tag{9}$$

where $d\Omega$ represents the area element and $d\Gamma$ represents the boundary element, both of which are integrated on the initial configuration. $\mathbf{u}$ is the displacement field and needs to satisfy the displacement boundary conditions. δ is an imaginary variable, which refers to the difference between any two variables that meet the constraints of the displacement boundary conditions. $\mathbf{K}$ is a strain tensor, $\mathbf{f}_b$ is a volume force vector, and $\mathbf{t}$ is a nominal stress vector.

The solution of the expansion problem can be obtained by using the above equation. However, as the problem involves significant deformation and the material constitutive relation is nonlinear, the analytical solution cannot be obtained directly, so it needs to be solved by the finite element method.

3.2.2. Numerical Solution Method

The circular screen is decomposed into several small rings, and each small ring is an axisymmetric problem. Due to the small difference between the internal and external diameters of the micro ring, the internal displacement field can adopt the following interpolation mode:

$$u = N_1 U_1 + N_2 U_2 \tag{10}$$

where U_1 and U_2 are the displacement values of the internal and external nodes, and N_1 and N_2 are the shape function.

The interpolation of the radial coordinates of the initial configuration of the ring also meets the following requirements:

$$R = N_1 R_1 + N_2 R_2 \tag{11}$$

where R_1 and R_2 are the radial coordinates of the internal and external nodes.

The calculation formula of shape function is as follows:

$$N_1 = \frac{1-\xi}{2} \quad N_2 = \frac{1+\xi}{2} \tag{12}$$

where ξ represents isoparametric deformation, and its value range is $[-1, 1]$.

Figure 5 shows the established finite element discrete model.

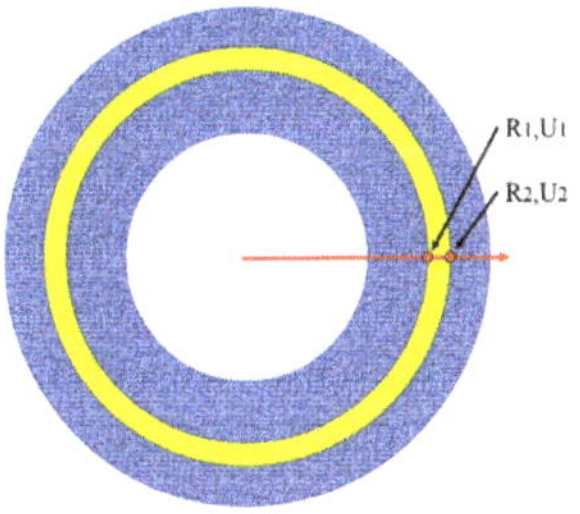

Figure 5. Finite element discrete model. The yellow line represents the material element.

The transformation matrix B_N of strain and displacement is defined, and its calculation formula is

$$B_N = \begin{bmatrix} \frac{\partial N_1}{\partial R} & \frac{\partial N_2}{\partial R} \\ \frac{N_1}{R} & \frac{N_2}{R} \end{bmatrix} \tag{13}$$

The left side of the equilibrium equation for this element is converted to

$$A(U, \delta U) = \delta U^T \cdot \left\{ \iint_{\Omega^0} B_N{}^T T \cdot d\Omega \right\} = \delta U^T P^{\text{int}} \tag{14}$$

where U is the node displacement vector, and the definition formula is $U = \begin{bmatrix} U_1 & U_2 \end{bmatrix}^T$. T is the nominal stress vector, and its definition formula is $T = \begin{bmatrix} t_r & t_\theta \end{bmatrix}^T$; P^{int} is the equivalent internal force at the node, which can be solved by the Gaussian numerical integration method.

Here, according to the Odgen model, the calculation formula of the nominal stress vector is

$$T = \begin{bmatrix} t_r \\ t_\theta \end{bmatrix} = \begin{bmatrix} \frac{\partial U}{\partial \lambda_r} \\ \frac{\partial U}{\partial \lambda_\theta} \end{bmatrix} = \begin{bmatrix} \sum_{i=1}^{N} \frac{2\mu_i}{\alpha_i}\left(\lambda_r^{\alpha_i - 1} - \lambda_r^{-1}\right) \\ \sum_{i=1}^{N} \frac{2\mu_i}{\alpha_i}\left(\lambda_\theta^{\alpha_i - 1} - \lambda_\theta^{-1}\right) \end{bmatrix} \tag{15}$$

In this problem, the effect of volume force is ignored, so the first term on the right is zero. For the second item on the right, it can be directly simplified as

$$L(\delta U) = \delta U^T \cdot 2\pi \begin{bmatrix} -t_{r,1}R_1 \\ t_{r,2}R_2 \end{bmatrix} = \delta U^T \cdot p^{ext} \tag{16}$$

where $t_{r,1}$ and $t_{r,2}$ are the radial stress at the inner and outer nodes, respectively. P^{ext} is the equivalent external force at the node.

According to the equilibrium equation,

$$R(U, \delta U) = A(U, \delta U) - L(\delta U) = \delta U^T \cdot \left\{ p^{\text{int}} - p^{ext} \right\} = 0 \tag{17}$$

The above equation gives the nonlinear equation of an element displacement; P^{ext} is independent of displacement and P^{int} is a nonlinear function of displacement. The above equation needs to be solved by an iterative method, such as the Newton–Raphson method.

In the Newton–Raphson method, the linear increment $R(U, \delta U)$ needs to be solved in advance, i.e.,

$$\begin{aligned} &\Delta R(U, \delta U) = \tfrac{d}{d\omega} R(U + \omega \cdot \Delta U, \delta U) \\ &= \tfrac{d}{d\omega} A(U + \omega \cdot \Delta U, \delta U) \\ &= \delta U^T \cdot \left\{ \iint_{\Omega^0} B_N{}^T D B_N \cdot d\Omega \right\} \cdot \Delta U \\ &= \delta U^T \cdot K_T \cdot \Delta U \end{aligned} \tag{18}$$

where D is the transformation matrix of strain and displacement, and its calculation formula is

$$D = \begin{bmatrix} \frac{\partial t_r}{\partial \lambda_r} & \frac{\partial t_r}{\partial \lambda_\theta} \\ \frac{\partial t_\theta}{\partial \lambda_r} & \frac{\partial t_\theta}{\partial \lambda_\theta} \end{bmatrix} = \begin{bmatrix} \sum_{i=1}^{N} \frac{2\mu_i}{\alpha_i}\left((\alpha_i - 1)\lambda_r^{\alpha_i - 2} + \lambda_r^{-2}\right) & \\ & \sum_{i=1}^{N} \frac{2\mu_i}{\alpha_i}\left((\alpha_i - 1)\lambda_\theta^{\alpha_i - 2} + \lambda_\theta^{-2}\right) \end{bmatrix} \tag{19}$$

K_T is the tangent stiffness matrix, which can be solved by the Gaussian numerical integration method.

At this time, the iterative equation in the Newton–Raphson method is expressed as

$$\delta U^T \cdot K_T \cdot \Delta U = \delta U^T \cdot \left(p^{ext} - p^{\text{int}} \right) \tag{20}$$

In order to solve the global displacement field, the elements need to be assembled to form the solution of the global problem. Considering the arbitrariness of virtual displacements, the non-displacement constraint points δU^T can be removed directly and the constraint of the displacement field can be added. The final Newton–Raphson iterative equation is as follows:

$$\begin{cases} M\left(U^k\right) \cdot \Delta U^k = R\left(U^k\right) \\ U^{k+1} = U^k + \Delta U^k \end{cases} \tag{21}$$

where U^k and U^{k+1} are the vectors composed of the radial displacement of each node in the K and K + 1 iterative steps, and they can be expressed as

$$U^k = \begin{bmatrix} U_1{}^k & \cdots & U_i^k & \cdots & U_{N+1}{}^k \end{bmatrix} \quad (22)$$

$M\left(U^k\right)$ is the stiffness matrix related to U^k, and $R\left(U^k\right)$ is the residual vector related to U^k. Both quantities can be solved according to the k-th iteration step.

3.2.3. Sensitivity Analysis

The parameters in the case analysis are as follows: the outer diameter of the base tube is 89 mm, the inner diameter of the thermosensitive screen is 90 mm, the outer diameter is 132 mm (before expansion, namely the compressed state), the original outer diameter is 190 mm, and the length is 2 m. The inner diameter of the wellbore is 152 mm.

After the shape of the thermosensitive screen is restored, its inner diameter is equal to the outer diameter of the base tube, and its outer diameter is equal to the inner diameter of the wellbore. It is a typical large deformation problem with known displacement boundary conditions. The stress and strain distribution of the TSS during the expansion process can be obtained with the proposed method based on the experimental results.

Figure 6 shows the radial displacement of a point along different radial positions. The radial compression displacement at the outermost part is the largest.

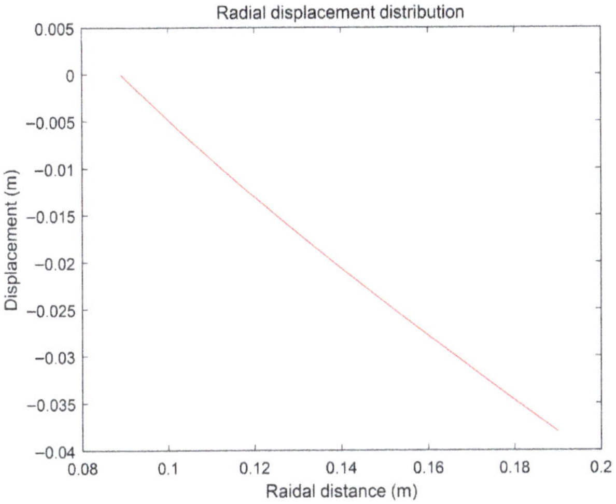

Figure 6. Radial displacement distribution.

Figure 7 shows the radial strain and circumferential strain along different radial positions. The radial compressive strain at the outermost part is the largest and the circumferential compressive strain is the smallest.

Figure 8 shows the change in nominal and Cauchy radial stress and circumferential stress with radial distance, in which the radial compressive stress decreases with the increase in radial distance, and the circumferential compressive stress increases with the increase in radial distance. Figure 9 shows the distribution of radial and circumferential stress clearly. The radial stress near the base tube is the largest, while that near the outermost layer is the smallest. Therefore, after the shape of the memory material is restored, the extrusion force on the base pipe is greater than that on the wellbore.

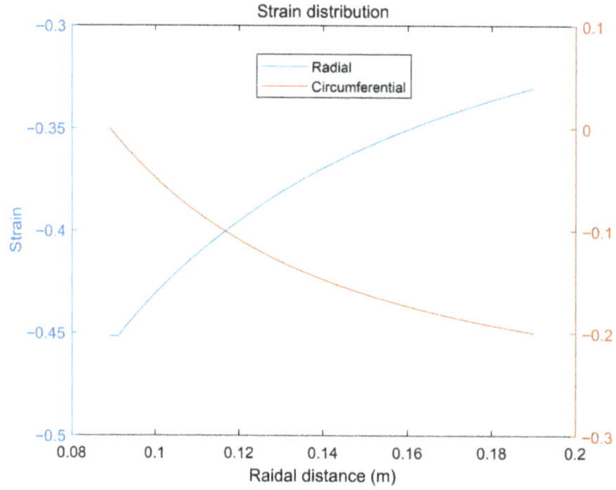

Figure 7. Radial and circumferential strain distribution.

Figure 8 shows the change in nominal and Cauchy radial stress and circumferential stress with radial distance, in which the radial compressive stress decreases with the increase in radial distance, and the circumferential compressive stress increases with the increase in radial distance. Figure 9 shows the distribution of radial and circumferential stress clearly. The radial stress near the base tube is the largest, while that near the outermost layer is the smallest. Therefore, after the shape of the memory material is restored, the extrusion force on the base pipe is greater than that on the wellbore.

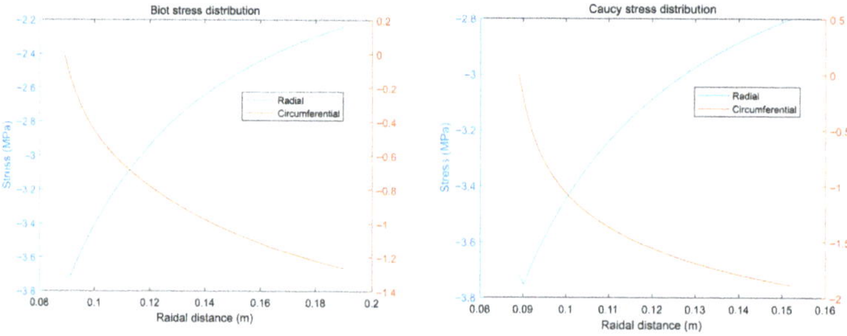

Figure 8. Biot stress distribution and Cauchy stress distribution.

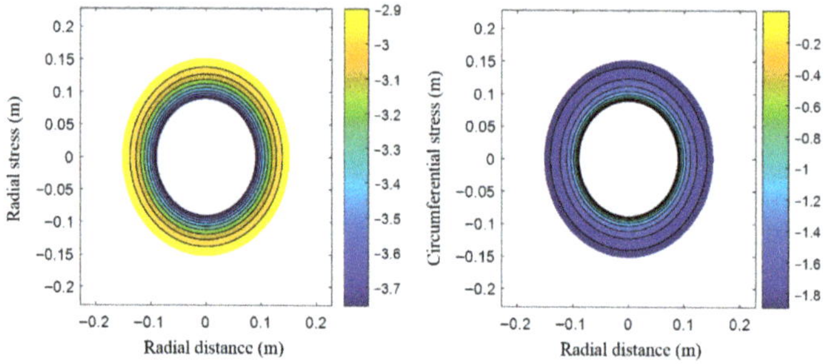

Figure 9. Radial and circumferential stress distribution.

Figure 10 shows the calculation results of radial stress and circumferential stress of three hypothetical materials after shape memory restoration. The parameters of material

2 are the same as the experimental results, while the parameters of material 1 are half of that of material 2, representing low elastic modulus. The parameter of material 3 is twice that of material 2, which represents high elastic modulus. It can be seen from the above results that with the increase in the elastic modulus of the material, the contact force of the material on the base pipe and the wellbore wall increases after shape recovery. When the contact force between the material and the wellbore is so small that the support effect of the material on the wellbore wall is not effective, the wellbore may experience yield failure. When the contact force between the material and the wellbore wall is so large that the extrusion force of the material on the base pipe is above the yield strength of steel, the base pipe may experience collapse failure. Therefore, the optimization of material properties should be between the two failure cases.

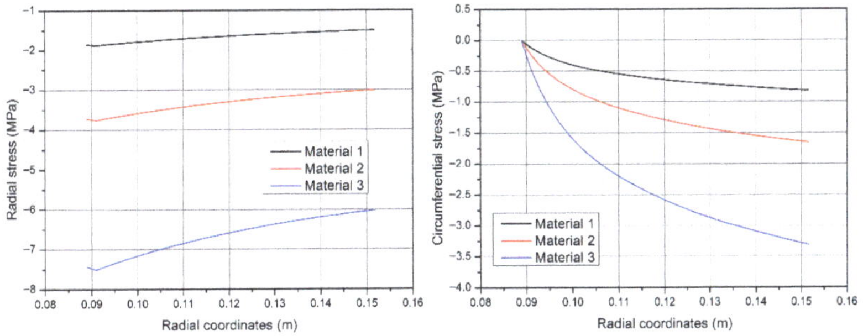

Figure 10. Comparison of radial stress and circumferential stress of three materials.

Figure 11 shows the calculation results of radial stress and circumferential stress after shape recovery under three kinds of initial outer diameters. The parameters of outer diameter 2 are the same as the experimental results. Outer diameter 1 is 0.9 times that of outer diameter 2, representing a small initial outer diameter. Outer diameter 3 is 1.1 times that of outer diameter 2, representing a large initial outer diameter. It can be seen from the above results that with the increase in the initial outer diameter, the contact force of the material on the base pipe and the wellbore wall increases after the shape is restored. The effects of the initial outer diameter on the failure of the base tubes and wellbores are consistent with the above elastic modulus analysis. Similarly, the optimization of the initial outer diameter of the material should be between the two failure cases.

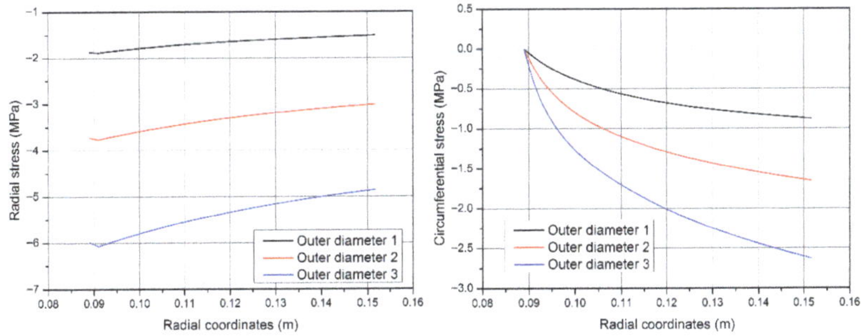

Figure 11. Comparison of radial stress and circumferential stress of different outer diameters.

4. Wellbore Strengthening of Thermosensitive Screen

Whether the wellbore is stable or not ultimately depends on the stress state of the surrounding rock, which will produce plastic deformation once the stress level exceeds the elastic range under the combined action of in situ stress, pore pressure, and wellbore fluid pressure.

4.1. Effects of Thermosensitive Screen on Elastic Solution

The formation of stress distribution around the wellbore refers to the calculation of the vertical stress, tangential stress, and radial stress at any point around the wellbore under specific conditions according to the original principal stress of the formation, that is, the vertical principal stress and the horizontal maximum and minimum principal stress. Considering the influence of the thermosensitive screen on the formation, the assumptions are as follows [22]:

1. The formation rock is a homogeneous, isotropic, and linear elastic porous medium.
2. The thermosensitive screen has good permeability (the screen and the base pipe can be perforated or slotted to form a flow channel).
3. The influence of the tectonic stress field and temperature field on the formation is ignored.
4. The surrounding rock of the wellbore wall is in the plane strain state.

According to the force balance of the element, the distribution of effective stress at any position of the surrounding rock under the influence of the thermosensitive screen in the vertical well is

$$
\begin{aligned}
\sigma_r(r,\theta) &= \tfrac{\sigma_H+\sigma_h}{2}\left(1-\tfrac{r_i^2}{r^2}\right)+\tfrac{\sigma_H-\sigma_h}{2}\left(1-\tfrac{4r_i^2}{r^2}+\tfrac{3r_i^4}{r^4}\right)\cos 2\theta+\tfrac{r_i^2}{r^2}(P_i+\sigma_{\mathrm{TSS}})\\
\sigma_\theta(r,\theta) &= \tfrac{\sigma_H+\sigma_h}{2}\left(1+\tfrac{r_i^2}{r^2}\right)-\tfrac{\sigma_H-\sigma_h}{2}\left(1+\tfrac{3r_i^4}{r^4}\right)\cos 2\theta-\tfrac{r_i^2}{r^2}(P_i+\sigma_{\mathrm{TSS}})\\
\sigma_z(r,\theta) &= \sigma_v-2\mu(\sigma_H-\sigma_h)\tfrac{r_i^2}{r^2}\cos 2\theta\\
\tau_{r\theta}(r,\theta) &= -\tfrac{\sigma_H-\sigma_h}{2}\left(1+\tfrac{2r_i^2}{r^2}-\tfrac{3r_i^4}{r^4}\right)\sin 2\theta
\end{aligned}
\tag{23}
$$

where r is the radial distance from the borehole center, m. θ is the included angle with the maximum horizontal principal stress, rad. μ is the Poisson's ratio of rock, dimensionless. r_i is the boundary radius in the elastic region, m. P_i is the inner boundary pore pressure in the elastic region, MPa. σ_H and σ_h are the original maximum and minimum horizontal principal stresses after considering pore pressure, MPa. σ_v is the original vertical in situ stress, MPa. σ_r, σ_θ, and σ_z are the effective radial, tangential, and vertical stress in the elastic zone, MPa. $\tau_{r\theta}$ is the shear stress of the borehole surrounding rock, MPa. σ_{TSS} refers to the interaction stress between the thermosensitive screen and the wellbore after being extruded against the wellbore.

Considering the fluid pressure in the formation pores, the Mohr Coulomb criterion expressed by principal stress can be expressed as

$$\sigma_1-\beta\cdot P=2C_0\tan\alpha+(\sigma_3-\beta\cdot P)\cdot\tan^2\alpha \tag{24}$$

where σ_1 and σ_3 are the maximum and minimum stresses, MPa. β is the Biot constant. α is the failure angle, (°).

Let us assume that the rock stress state just conforms to the Mohr Coulomb failure criterion at the characteristic position $r=r_x$; then, the pore pressure is expressed as a function of bottom flowing pressure P_{wf}:

$$P(r_x)=P_{wf}+\frac{P_o}{\ln\frac{r_o}{r_w}}\cdot\ln\frac{r_x}{r_w} \tag{25}$$

where P is the pore pressure; P_o is the outer boundary pore pressure in the elastic region, MPa; and r_o is the boundary radius in the elastic region, m. r_w is the wellbore radius, m.

The maximum stress σ_1 and minimum stress σ_3 at the characteristic position are also a function of bottom flowing pressure P_{wf}. By substituting the two stresses into the Mohr Coulomb failure criterion, we obtain the balance equation of bottom flowing pressure P_{wf}:

$$\sigma_1\left(P_{wf}\right) - \beta \cdot P\left(P_{wf}\right) - 2C_0 \tan\alpha - \left[\sigma_3\left(P_{wf}\right) - \beta \cdot P\left(P_{wf}\right)\right] \cdot \tan^2\alpha \geq 0 \quad (26)$$

The bottom hole flowing pressure obtained by solving Equation (26) is the critical bottom flowing pressure for sand production, and the critical drawdown pressure can be obtained by

$$\Delta P_c = P_{wfc} - P_r \quad (27)$$

where P_{wfc} is the critical bottom flowing pressure for sand production, MPa. ΔP_c is the critical drawdown pressure of sand production, MPa. P_r is the formation static pressure, MPa.

4.2. Case Analysis

Suppose a well has a borehole diameter of 152 mm, outer boundary radius of 200 m, boundary pressure of 12 MPa, and formation static pressure of 9.5 MPa. The original horizontal principal stress is 19.73 MPa and 14.35 MPa, and the original vertical principal stress is 17.68 MPa. The Poisson's ratio of formation rock is 0.25 and the elastic modulus is 2017 MPa. According to the elastic solution, the stress distribution near the wellbore and the critical drawdown pressure can be analyzed. The shear failure of the formation mainly depends on the difference between the maximum and minimum principal stresses. The greater the stress difference, the greater the shear effect on the formation, and the easier the wellbore yields and loses stability.

The elastic stress distribution near the open hole after considering the influence of the thermosensitive screen is shown in Figure 12 (In the figure, "ext" refers to the extrusion stress of TSS. "Radial" refers to the stress in the radial direction. The abbreviation also applies to Figures 13 and 14). Under these formation parameters, the radial stress is the smallest and the circumferential stress is the largest. The formation stress gradually tends to be stable from near the wellbore zone to far from wellbore zone. After considering the influence of the thermosensitive screen, the stress distribution of the formation near the wellbore is significantly improved. The radial stress increases with the increase in the contact stress of the thermosensitive screen, and the circumferential stress decreases with the increase in the contact stress of the thermosensitive screen. Therefore, the stress difference borne by the formation near the wellbore is reduced due to the expansion of the thermosensitive screen.

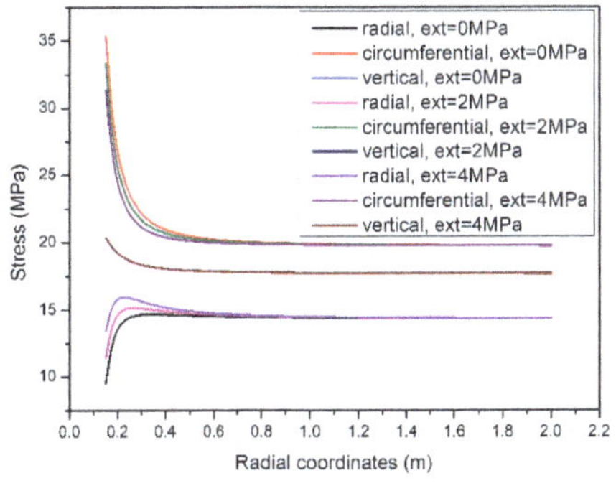

Figure 12. Elastic stress along single direction (α = 90°).

Figure 13 shows the variation trend in the formation stress difference in the diameter direction in detail. The stress difference around the wellbore reaches the maximum on the open hole wall and gradually stabilizes with the radial extension of the formation. However, the stress difference will continue to decrease with the increase in extrusion stress. When the extrusion stress in the figure is 8 MPa, the minimum stress difference is about 2 MPa, which is smaller than that in the far well zone. Therefore, as long as the effective stress on the formation does not exceed the fracture pressure, the expansion extrusion force of the thermosensitive screen can be improved as much as possible. Figure 14 shows the formation stress distribution along 0.4 m of the well diameter within 360° of the well circumference. At 90° and 270° azimuth angles, the largest stress difference can be found. Under the influence of the TSS, it shows the same influence trend as that in a single direction.

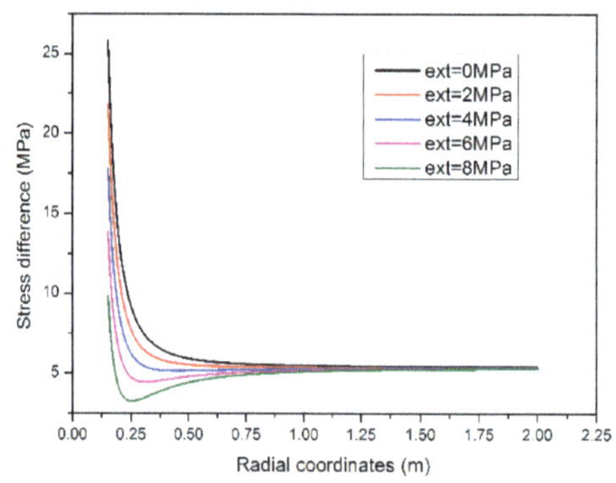

Figure 13. Maximum and minimum principal stress difference.

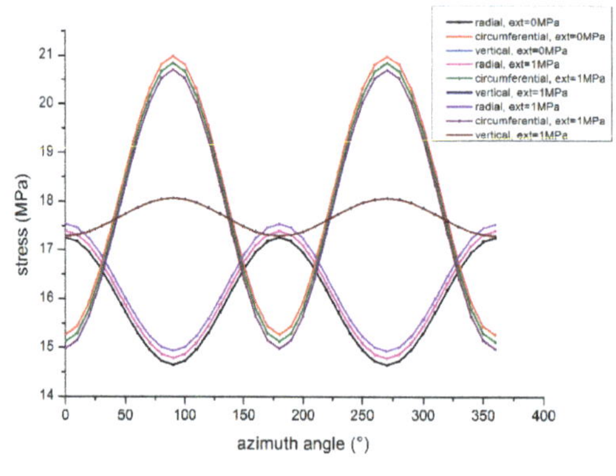

Figure 14. Elastic stress distribution around borehole.

Figure 15 shows the critical bottom flowing pressure under the influence of the thermosensitive screen. Obviously, in the near-wellbore zone (0.4 m around the well), the critical bottom flowing pressure increases with the increase in the extrusion stress of the thermosensitive screen. The critical bottom flowing pressure when there is no extrusion stress is 41.4 MPa, while the critical pressure increases to 71.7 MPa when the extrusion stress of the thermal screen is 10 MPa. According to the definition of critical drawdown pressure, the critical drawdown pressure can be increased by 95% under the current formation conditions. Therefore, the study of a high-expansion-rate and high-stiffness thermosensitive screen will play a significant role in strengthening the stability of wellbores.

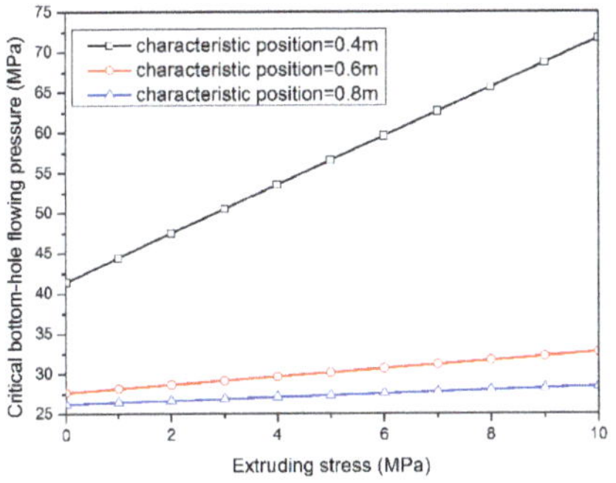

Figure 15. Critical bottom flowing pressure.

4.3. Effects of Thermosensitive Screen on Plastic Solution

According to the Coulomb failure criterion, once the stress exceeds the critical stress, the formation will change from the elastic state to plastic state. The elastic solution shows that there will be a plastic region around the wellbore in loose formations. According to the plastic flow law of formation and the rock failure criterion, the plastic region of formation can be determined and the influence of the TSS on the plastic radius can be further determined by introducing the influence of TSS boundary conditions.

The so-called sand production radius refers to the radius of the plastic zone near the wellbore. It is the maximum theoretical radius around a wellbore within which the in situ stresses exceed the formation's strength, leading to rock failure and the potential for sand production. It represents the boundary between the failed, yielded zone (where sand is dislodged) and the intact, stable formation. Before sand production, elastic deformation occurs in the whole formation. After sand production, plastic failure occurs in the region around the wellbore. The whole formation is divided into the plastic area and elastic area shown in Figure 16.

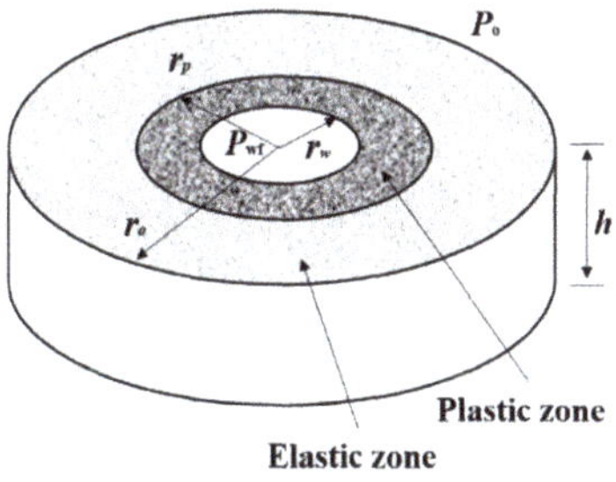

Figure 16. Elastic and plastic formation stress distribution model.

In this paper, we mainly discussed the effects of TSS extrusion force on the plastic region, so a detailed discussion of elastic and plastic solutions of formation is beyond the scope of this article. The effects of TSS extrusion force are simplified to an external force applied to the wellbore wall. Given the internal and external boundary conditions, namely the wellbore radius r_w, bottom effective internal pressure $P_i + \sigma_{TSS}$, external boundary radius r_o, and external boundary pressure P_o, the elastic–plastic stress distribution of the formation near the well can be calculated.

According to the Mohr Coulomb failure criterion, when the critical stress expressed by the Mohr Coulomb equation is reached, the rock material will change from elastic deformation to plastic deformation. When solving stress distribution in the plastic zone, it is necessary to confirm the maximum and minimum principal stresses first. According to the elastic stress analysis, the radial stress around the wellbore is the smallest, so we assume that $\sigma_r < \sigma_z < \sigma_\theta$ to discuss the extrusion effects [22,23].

If the vertical stress and tangential stress in the outer plastic zone are equal on the boundary of the outer plastic zone, we obtain

$$b_1\left(\frac{r_b}{r_w}\right)^t + b_2 \ln\frac{r_b}{r_w} + b_3 = 0 \tag{28}$$

where

$$\begin{array}{l} b_1 = \frac{1}{t}\left(2S_o \tan\alpha - \frac{\mu q}{2\pi h k_c}\right)[(t+1) - v(t+2)],\ b_2 = (1-2v)(1-\beta)\frac{\mu q}{2\pi h k_c} \\ b_3 = (1-2v)(1-\beta)P_i + (1-v)\frac{\mu q}{2\pi h k_c} - \frac{(1+v)(1-2v)}{1-v}(\sigma_{zo} - \beta P_o) \\ \quad -\frac{1}{t}\left(2S_o \tan\alpha - \frac{\mu q}{2\pi h k_c}\right)(1-2v) \end{array} \tag{29}$$

where r_b is the radius of the inner plastic zone. t is an intermediate variable, $t = \tan^2\alpha - 1$. q is the fluid flow rate, m^3/s. μ is the fluid viscosity, Pa·s. h is the reservoir thickness, m. v is Poisson's ratio, dimensionless. k_c is the permeability, m^2. S_o is the cohesive strength.

The continuity of radial stress must be satisfied at the elastic–plastic stress boundary. By substituting the radial stress in the plastic region and the tangential stress in the elastic region into the Coulomb criterion, the equation for the plastic radius can be obtained as [23]

$$\begin{array}{l} c_1 r_p^{t+2} \ln\frac{r_o}{R_c} + c_2 r_p^t \ln\frac{r_o}{R_c} + c_3 r_p^2 + c_4 r_p^2 \ln\frac{r_p}{r_w} + c_5 r_p^2 \ln\frac{r_o}{r_p} \\ + c_6 \ln\frac{r_o}{r_p} \ln\frac{r_p}{r_w} + c_7 \ln\frac{r_p}{r_w} + c_8 \ln\frac{r_o}{r_p} + c_9 = 0 \end{array} \tag{30}$$

where r_p is the plastic radius, m. Other intermediate variables are as follows:

$$\begin{array}{l} c_1 = \left(2S_o \tan\alpha - \frac{\mu q}{2\pi h k_c}\right) r_w^{-t},\ c_2 = -\frac{t+2}{t} c_1 r_o^2,\ c_3 = -\frac{1-2v}{2(1-v)}\beta(P_o - P_i) \\ c_4 = \frac{1-2v}{2(1-v)}\beta\frac{\mu q}{2\pi h k_c},\ c_5 = \frac{\mu q}{2\pi h k_c},\ c_6 = -\left(2 - \frac{1-2v}{1-v}\beta\right) c_5 r_o^2,\ c_7 = -c_4 r_o^2 \\ c_8 = \left[\frac{2v}{1-v}\sigma_{zo} + \frac{1-2v}{1-v}\beta(P_o + P_i) - 2P_i + \frac{2}{t} 2S_o \tan\alpha - \frac{t+2}{t}\frac{\mu q}{2\pi h k_c}\right] r_o^2,\ c_9 = -c_3 r_o^2 \end{array} \tag{31}$$

4.4. Case Study

We refer to the data of reference [1] for the stress solution with no fluid flow. This corresponds to the situation after the well has been drilled and the fluid pressure in the well is in balance with the pore pressure in the formation. The relevant parameters of a well with formation instability are considered as follows: the Poisson ratio of the rock is 0.45, cohesion strength is 101.4 kPa, the internal friction angle is 60°, the elastic modulus is 5765 MPa, the Biot number is 1.0, and the formation pressure at the initial stage of production is 32.065 MPa. The outer boundary radius of the single well control area is 10 m, the outer boundary pressure is 32.065 MPa, and the borehole diameter is 100 mm. The original vertical principal stress is 65.5 MPa, and the original horizontal principal stress is 60.255 MPa. By substituting the relevant parameters of the example well into the plastic radius equation derived above, the elastic–plastic stress distribution of the formation near the wellbore can be calculated by a numerical method.

Figures 17 and 18 compare and analyze the plastic stress distribution under the influence of the TSS. The formation stress increases along the radial direction from the wellbore wall, the vertical stress and circumferential stress coincide in the internal plastic area, separate after exceeding the first plastic area, and the formation stress begins to change slowly at the boundary of elastic–plastic stress. If there are no TSS effects, as shown in Figure 17, the plastic radius is about 0.9 m. When the TSS extruding stress is

5 MPa, the elastic–plastic contact radius is obviously reduced from 0.9 m to 0.81 m. If the expansion performance and the stiffness of TSS materials are further improved, the strengthening effect of the wellbore wall will be greatly increased. Figure 19 shows the variation range of the plastic radius calculated after further improving the extruding stress of the TSS. Obviously, the external plastic radius decreases with the increase in expansion extruding stress.

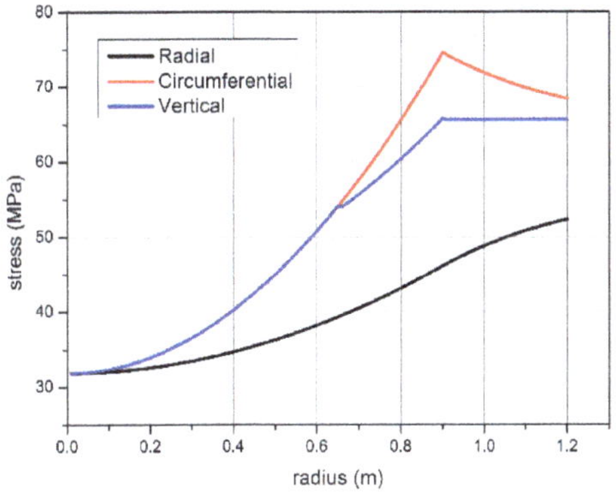

Figure 17. Plastic stress distribution (no extruding stress).

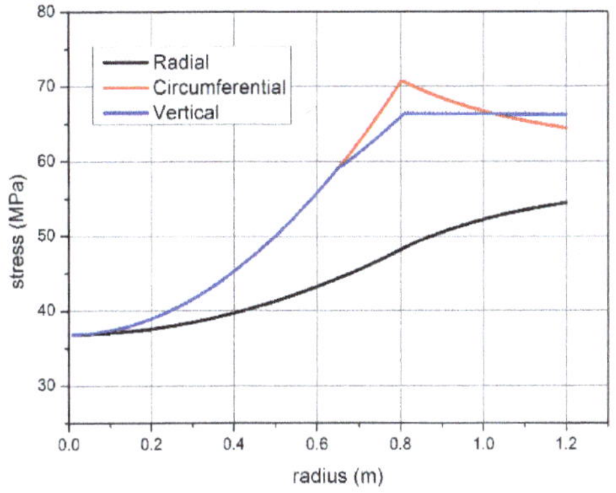

Figure 18. Plastic stress distribution (extruding stress 5 MPa).

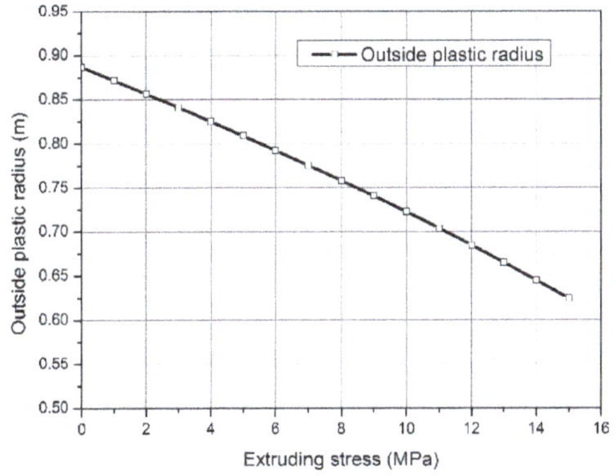

Figure 19. Variation in plastic radius with extruding stress.

According to the above formation stress distribution near the wellbore, if the expansion contact performance of the existing TSS is known, the effects of the TSS on formation stability can be predicted. Conversely, if the relevant parameters of the formation are known in advance, the mechanical properties that are required to be achieved after TSS expansion can be calculated according to the proposed method. The calculated mechanical parameters can help provide guidance for TSS material design to meet formation requirements. Therefore, this paper provides an optimized design approach combined with engineering practice for the performance design of TSSs in the future.

Upon expansion, the temperature-sensitive screen significantly increases the flow area while applying radial stress to the wellbore, thereby enhancing the stability of the near-wellbore zone. It is suitable for unconsolidated sandstones, weakly cemented formations, and other sand-prone reservoirs. Its expansion enables better adaptation to irregular wellbores, reducing the "crescent-shaped" gap between the screen and the wellbore, eliminating fluid channeling pathways, and ensuring more uniform production contribution. This offers significant advantages in wells with complex trajectories, such as horizontal and highly deviated wells. In harsh high-temperature and high-pressure environments like deepwater oil and gas or geothermal wells, the temperature-sensitive screen provides an innovative solution for addressing well integrity and long-term sand control challenges.

5. Conclusions

In this paper, we introduce thermosensitive screen technology to help stabilize a wellbore. The expansion extrusion force obtained by a simulation of the TSS expansion process can provide input parameters for borehole strengthening analysis. Combined with formation parameters, its influence on the elastic–plastic stress distribution near the wellbore can be evaluated, and the critical drawdown pressure and critical plastic radius under the influence of the TSS can be predicted. Some new conclusions can be drawn from the above analysis as follows:

1. An analysis method of significant expansion deformation is proposed, which can invert the model parameters according to the experimental data. The changes in stress and outer diameter during the expansion of the thermosensitive screen are analyzed and calculated with the method. The radial compressive stress after restoring the shape decreases with the increase in radial distance. After restoring the shape, the thermosensitive screen can provide extrusion force for the wellbore wall.
2. According to the elastic stress solution of formation and Mohr Coulomb failure criterion, the stress distribution around the wellbore considering the expansion of the thermosensitive screen is obviously improved. The stress difference in the formation near the wellbore is reduced due to the expansion extrusion force of the thermosensitive screen. The critical bottom flowing pressure increases with the increase in expansion extrusion force. If the extrusion force is 10 MPa, the critical drawdown pressure can be increased by 95% under the current formation conditions.
3. The plastic stress solution shows that if the expansion contact force of the TSS is large enough, the plastic yield radius of the formation will be effectively reduced, and the sand production radius can be reduced in the production well.

Author Contributions: Conceptualization, Y.Z. and X.Q.; methodology, Y.Z., Y.W. and X.Q.; software, Y.Z. and Y.W.; validation, Y.Z., Y.W. and X.Q.; formal analysis, X.Q. and Y.R.; investigation, Y.Z. and Y.R.; resources, Y.Z. and X.Q.; data curation, Y.Z. and Y.W.; writing—original draft preparation, Y.Z. and X.Q.; writing—review and editing, Y.R.; visualization, Y.Z., Y.W. and X.Q.; supervision, Y.Z.; project administration, X.Q.; funding acquisition, Y.Z. and X.Q. All authors have read and agreed to the published version of the manuscript.

Funding: The authors gratefully acknowledge the financial support from the National Science and Technology Major Project [grant number 2025ZD1401503], the National Natural Science Foundation of China [grant number 52004301], and the "Western young scholars" of Chinese Academy of Sciences [grant number 2021-XBQNXZ-033].

Data Availability Statement: The data that support the findings of this study are available from the corresponding author upon reasonable request.

Conflicts of Interest: Author Xing Qin was employed by the company Sinopec Research Institute of Petroleum Engineering. The remaining authors declare that the research was conducted in the absence of any commercial or financial relationships that could be construed as a potential conflict of interest.

Abbreviations

The following abbreviations are used in this manuscript:

TSS	Thermosensitive screen

References

1. Fuh, G.F.; Loose, P.K. Horizontal wellbore stability for open hole completions. In Proceedings of the SPE Annual Technical Conference and Exhibition, San Antonio, TX, USA, 8–11 October 1989. [CrossRef]
2. Tohidi, A.; Fahimifar, A.; Rasouli, V. Analytical solution to study depletion/injection rate on induced wellbore stresses in an anisotropic stress field. *Geotech. Geol. Eng.* **2018**, *36*, 1735–1744. [CrossRef]
3. Gholami, R.; Elochukwu, H.; Fakhari, N.; Sarmadivaleh, M. A review on borehole instability in active shale formations: Interactions, mechanisms and inhibitors. *Earth Sci. Rev.* **2018**, *177*, 2–13. [CrossRef]
4. Li, X.; Zhang, C.; Feng, Y.; Wei, Y.; Chen, X.; Weng, H.; Deng, J. An integrated geomechanics approach to evaluate and manage wellbore stability in a deep graben formation in Tarim Basin. *J. Pet. Sci. Eng.* **2022**, *208*, 109391. [CrossRef]
5. Qin, X.; Duan, Y.; Yue, H.; Ai, S.; Hou, Q.; Jing, B. Mechanical behavior of thermosensitive screen applied for wellbore collapse. In Proceedings of the SPE Annual Technical Conference and Exhibition, OnePetro, Virtual, 26–29 October 2020. [CrossRef]
6. Osunjaye, G.; Abdelfattah, T. Open hole sand control optimization using shape memory polymer conformable screen with inflow control application. In Proceedings of the SPE Middle East Oil & Gas Show and Conference, OnePetro, Manama, Bahrain, 6–9 March 2017; p. SPE-183947-MS. [CrossRef]
7. Wang, X.; Osunjaye, G. Advancement in openhole sand control applications using shape memory polymer. In Proceedings of the SPE Annual Technical Conference and Exhibition, OnePetro, Dubai, United Arab Emirates, 26–28 September 2016. [CrossRef]
8. Mcmillin, K. Using expandable sand screens in unconsolidated formations. *Offshore* **2000**, *60*, 52–53.
9. Metcalfe, P. Expandable sand screen technology increases production. *World Oil* **2000**, *221*, 94–95.
10. Carrejo, N.; Horner, D.N.; Johnson, M.H. Shape memory polymer as a sand management alternative to gravel packing. In Proceedings of the Canadian Unconventional Resources Conference, Calgary, AB, Canada, 15–17 November 2011. [CrossRef]
11. Carrejo, N.; Horner, D.N.; Johnson, M.H. The effects of dynamic loading on the sand management and permeability of shape memory polymer and gravel packs for sand management applications. In Proceedings of the SPE Brasil Offshore Conference, Macaé, Brazil, 7–10 June 2011. [CrossRef]
12. Xie, T. Tunable polymer multi-shape memory effect. *Nature* **2010**, *464*, 267–270. [CrossRef] [PubMed]
13. Mather, P.T.; Luo, X.; Rousseau, I.A. Shape memory polymer research. *Annu. Rev. Mater. Res.* **2009**, *39*, 445–471. [CrossRef]
14. Tobushi, H.; Hashimoto, T.; Hayashi, S.; Yamada, E. Thermomechanical Constitutive Modeling in Shape Memory Polymer of Polyurethane Series. *J. Intell. Mater. Syst. Struct.* **1997**, *8*, 711–718. [CrossRef]
15. Tobushi, H.; Okumura, K.; Hayashi, S. Thermomechanical constitutive model of shape memory polymer. *Mech. Mater.* **2001**, *33*, 545–554. [CrossRef]
16. Qi, H.J.; Nguyen, T.D.; Castro, F.; Yakacki, C.M.; Shandas, R. Finite deformation thermo-mechanical behavior of thermally induced shape memory polymers. *J. Mech. Phys. Solids* **2008**, *56*, 1730–1751. [CrossRef]
17. Ge, Q.; Luo, X.; Iversen, C.B.; Nejad, H.B.; Mather, P.T.; Dunn, M.L.; Qi, H.J. A finite deformation thermomechanical constitutive model for triple shape polymeric composites based on dual thermal transitions. *Int. J. Solids Struct.* **2014**, *51*, 2777–2790. [CrossRef]
18. Jia, H.; Gu, S.Y.; Chang, K. 3D printed self-expandable vascular stents from biodegradable shape memory polymer. *Adv. Polym. Technol.* **2018**, *37*, 3222–3228. [CrossRef]
19. Yakacki, C.M.; Shandas, R.; Lanning, C.; Rech, B.; Eckstein, A.; Gall, K. Unconstrained recovery characterization of shape-memory polymer networks for cardiovascular applications. *Biomaterials* **2007**, *28*, 2255–2263. [CrossRef] [PubMed]
20. Kim, N.H. *Introduction to Nonlinear Finite Element Analysis*; Springer Science & Business Media: Berlin/Heidelberg, Germany, 2014. [CrossRef]

21. Kim, B.; Lee, S.B.; Lee, J.; Cho, S.; Park, H.; Yeom, S.; Park, S.H. A comparison among Neo-Hookean model, Mooney-Rivlin model, and Ogden model for chloroprene rubber. *Int. J. Precis. Eng. Manuf.* **2012**, *13*, 759–764. [CrossRef]
22. Jaeger, J.C.; Cook, N.G.; Zimmerman, R. *Fundamentals of Rock Mechanics*; John Wiley & Sons: Hoboken, NJ, USA, 2009. [CrossRef]
23. Risnes, R.; Bratli, R.K.; Horsrud, P. Sand stresses around a wellbore. *Soc. Pet. Eng. J.* **1982**, *22*, 883–898. [CrossRef]

processes

MDPI

Article

Quantitative Evaluation Methods and Applications for Gravel Characteristics Distribution in Conglomerate Reservoirs

Zhenhu Lv [1,2], Jietao Xu [1], Tianbo Liang [1,*], Ping Li [3], Xiaolu Chen [2], Hao Cheng [3] and Yupeng Zhang [2]

1 State Key Laboratory of Petroleum Resources and Engineering, China University of Petroleum at Beijing, Beijing 102249, China; lvzhenhu2016@petrochina.com.cn (Z.L.); jt.xu@foxmail.com (J.X.)
2 The Oil Production Technology Research Institute, PetroChina Xinjiang Oilfield Company, Karamay 834000, China; chenxiaolu@petrochina.com.cn (X.C.); zhangyp2021@petrochina.com.cn (Y.Z.)
3 PetroChina Xinjiang Oilfield Development Company, Karamay 834000, China; li-ping66@petrochina.com.cn (P.L.); chenghao2@petrochina.com.cn (H.C.)
* Correspondence: liangtianboo@163.com

Abstract

Conglomerate reservoirs often exhibit chaotic internal structures and strong heterogeneity due to the influence of gravel, which seriously restricts the balanced initiation of multiple clusters and the balanced expansion of artificial fractures in the volume fracturing section of horizontal wells. Therefore, clarifying the distribution pattern of gravel in conglomerate reservoirs is of great significance for the design and parameter optimization of horizontal well segmentation and clustering. This work conducts research on the interpretation results of imaging logging, establishes a characterization model for the distribution characteristics of gravel around horizontal wells, develops gravel feature recognition and analysis software for conglomerate reservoirs using image processing technology, and effectively obtains the morphology of gravel in imaging logging. Based on this, a correlation model between conventional logging and imaging logging is constructed to predict the distribution of gravel in horizontal wells without imaging logging. Using the Kriging interpolation method, a "point line surface" gravel distribution prediction method is proposed. Through three methods of imaging logging, downhole eagle-eye camera, and on-site coring, the model accuracy is found to be greater than 80%, guiding segmented clustering to avoid high gravel areas. During the fracturing process, the wellhead pressure is lower than that of adjacent wells, enabling greater fluid savings per well. The production effect is better than that of adjacent wells in the same block, providing a reference for the study of gravel distribution characteristics in conglomerate oil reservoirs.

Keywords: conglomerate reservoir; imaging logging; prediction of gravel distribution; gravel line density; equivalent radius of gravel

1. Introduction

As the global economy continues to develop, society's demand for petroleum energy keeps rising, with the pressure on the supply–demand balance becoming increasingly apparent. Within global oil reserves, conventional reservoirs account for 30%, while unconventional reservoirs constitute 70%. It is generally acknowledged that most easily accessible reservoirs have begun exhibiting a decline in conventional oil production [1,2]. To meet the growing demand for petroleum, researchers will not only enhance oil recovery rates in existing fields but also develop new unconventional energy resources. Conglomerate reservoirs represent a typical type of coarse-grained sedimentary unconventional reservoir [3]. Following the initial discovery of multiple sandstone and conglomerate reservoirs in the Triassic Tupungato oil field and Jurassic–Cretaceous strata of Argentina's Cuyo Basin [4], similar reservoirs have since been identified in countries including the United States, Canada, Brazil, and the United Kingdom [5,6]. Domestically, effective utilization of conglomerate reservoirs has been achieved in regions such as the Mahu Depression in Xinjiang, the western depression of the Liaohe Oilfield, and the Erlian Basin of the Shengli Oilfield. Among these, the Mahu tight conglomerate reservoir—China's largest integrated conglomerate reservoir—has evolved over more than a decade of development into an annual production base yielding 3 million tons of unconventional oil. The development of gravel-packed reservoirs represents an iterative process of practice-based learning, progressing through stages including water injection development, horizontal well volume fracturing, intra-stage multi-cluster dense cutting, intra-stage multi-cluster temporary plug fracturing, three-dimensional well network fracturing, and pre-injection carbon dioxide fracturing for enhanced production. Both laboratory experiments and field practice indicate significant differences in mineral composition and rock mechanical properties between gravel and matrix within the reservoir [7]. This results in pronounced heterogeneity, where injected water during waterflooding tends to flow along the gravel–matrix–cementation interface, leading to premature waterflooding of oil wells. During horizontal well fracturing, artificial fractures exhibit two modes at gravel zones: penetration through gravel and bypassing gravel [8]. Post-fracturing downhole Eagle-Eye monitoring revealed low uniformity in fracture initiation across clusters within gravel sections, with some near-base perforation clusters receiving over 70% of the proppant load for that section [9]. Therefore, clarifying the gravel content and distribution patterns in gravel reservoirs is crucial for achieving uniform fracture initiation across multiple clusters within fractured sections and optimizing fracture parameters.

Predictions of gravel distribution characteristics currently rely primarily on sedimentary facies analysis, seismic attribute analysis, and well log curve feature inversion. These methods leverage differences in wave impedance and well log data between gravel and matrix to predict gravel content and size through differential characteristics. The fan delta foreland, fan delta plain, and pre-fan delta have been divided into nine distinct conglomerate facies based on sedimentary genesis, sedimentary structure, grain shape, arrangement patterns, support forms, cementation types, and other characteristics; based on sedimentary structures, grain shape, arrangement patterns, support forms, and cementation types, nine conglomerate facies have been further defined [10]. High-velocity conglomerates induce abnormally elevated overlay velocities below the conglomerate extinction point by investigating gravel's impact on underlying stratigraphic overlay velocities, enabling gravel-rich zone identification through these anomalies [11]. Five distinct seismic facies patterns have been listed for sandstone–conglomerate complexes based on the principle that sedimentary types dictate differing seismic facies characteristics [12]. Resistivity cutoff values have been proposed through core calibration based on resistivity differences in rock components within gravel reservoir electrical imaging [13]. Researchers have developed calculation methods for gravel, sandy, and muddy content, along with a grain size analysis method using pseudo-grain size probability cumulative curves, and others identified the characteristics of gravel and sandstone formations in the Kuche area of Tarim as low sonic velocity, high density, high resistivity, and low neutron content [14,15]. This was achieved by utilizing the principle that resistivity logging anomalies in gravel and sandstone strata correspond to high-resistivity anomalies and that the curve morphology is related to factors such as water influx and outflow cycles during deposition, distance from the source area, supply rate, flow energy, and sedimentary background. Logging curves of gravel reservoirs in the Yanjia-22 block have been characterized as exhibiting high gamma, medium-to-high natural potential, medium-to-low density, medium-to-low borehole diameter, and medium-to-low sonic characteristics [16]. Based on logging research, a quantitative calculation method was developed for gravel grain size using micro-resistivity imaging logging (MIRL) images [17]. By integrating digital image processing techniques, this approach established a method for calculating gravel grain size. Watershed algorithms were also introduced to investigate gravel identification within imaging logging images [18]. They enhanced the algorithm by adding filters, gradient processing, and region of interest (ROI) marking, thereby resolving the excessive segmentation issues inherent in watershed algorithms based on topological theory and simulated topography. Recently, a method was proposed for automatic identification and parameter calculation of gravel in electrical imaging based on a Saliency Object Detection Network [19–27]. This approach accurately calculates quantitative evaluation parameters such as grain size, roundness, and quantity using the identified gravel results. Seismic attribute analysis primarily relies on variations in wave impedance to infer gravel distribution. However, the complexity and heterogeneity of conglomerate reservoirs often cause wave impedance variations to be influenced by multiple factors, such as depositional environments and tectonic movements. This leads to potential errors in seismic data during the inversion process, thereby affecting the accuracy of gravel distribution predictions. Resistivity logging technology infers rock composition by measuring formation resistivity, but its accuracy is affected by multiple factors, including rock porosity, saturation, and fluid type. In conglomerate reservoirs, resistivity variations may not directly reflect gravel content and size, leading to imprecise quantitative assessments of gravel distribution. Furthermore, resistivity logging typically provides only local information, making comprehensive regional-scale predictions challenging.

In summary, while sediment facies analysis and seismic attribute analysis can yield gravel information, most approaches rely on inversion for prediction and cannot achieve a precise quantitative assessment. Logging-on-the-hole (LOH) feature analysis can capture continuous profile variations in gravel content along the wellbore axis, but no quantitative interpretation model linking LOH features to gravel content and size has been established [28–35]. Imaging logging currently represents the primary method for quantitative characterization of content and size [36–41]. However, it heavily relies on imaging logs, and quantitative calibration remains unattainable without such logs. Furthermore, imaging logging can only establish distribution characteristics on a single-well profile, failing to predict regional gravel content and size distribution [42–50]. Previous studies often failed to adequately account for the complexity of geological settings, such as variations in sedimentary facies and the influence of tectonic movements. The impact of these factors on gravel distribution may have been overlooked, leading to biased research outcomes. Many investigations employed a single methodology for analyzing gravel distribution, lacking a comprehensive approach. Integrating multiple data sources and analytical techniques could more effectively reveal the distribution characteristics of gravel.

To this end, building on the aforementioned research, this paper fully utilizes imaging logging interpretation results to establish a model that characterizes gravel distribution features around horizontal wells. It identifies two key parameters to represent gravel characteristics and employs image processing techniques to develop a method for characterizing the axial gravel distribution features based on imaging logging. To overcome the challenge of predicting distribution characteristics without imaging logs, a multivariate regression method is used to construct a correlation model between conventional logging and imaging logging. This enables the prediction of gravel distribution in horizontal wells without the need for imaging logs. The results are validated through three approaches, demonstrating high consistency. Building on this, a regional gravel distribution prediction method was developed using single-well prediction results, providing a basis for segmented and clustered optimization of conglomerate reservoirs and differentiated design of fracturing parameters.

2. Establishment of the Model for Characterizing the Distribution Characteristics of Gravel Around Wells

For sedimentary reservoirs formed by rapid accumulation of multi-phase alluvial fans, conglomerate reservoirs often exhibit complex internal structures and strong heterogeneity. During drilling in conglomerate reservoirs, horizontal wells frequently encounter multiple zones with varying gravel sizes and concentrations. To precisely characterize gravel distribution along the wellbore axis, consider an infinitesimal segment dx along the wellbore axis, as illustrated in Figure 1. When this infinitesimal element is sectioned axially, it forms a rectangle. The gravel radius and gravel density are defined on this rectangular infinitesimal element:

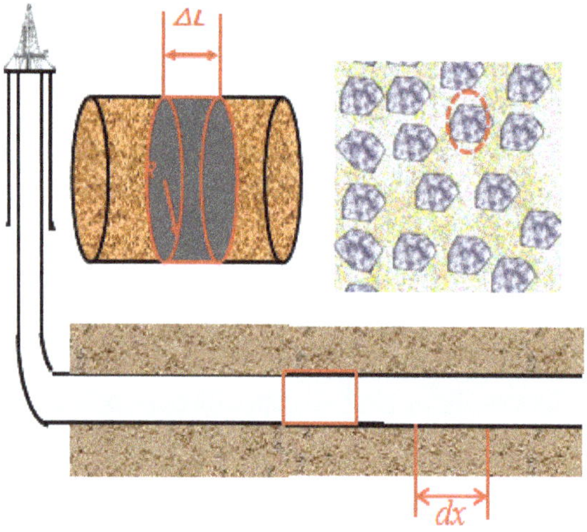

Figure 1. Rectangular Microelement Representation of Gravel Linear Density and Radius.

Due to variations in the size and shape of gravel particles within the micro-element rectangle, a uniform numerical value cannot be used for characterization. To achieve quantitative representation of gravel size, the area of gravel particles with different shapes is calibrated via imaging logging. The gravel radius within the micro-element rectangle is then characterized using an equivalent circle approach, with the specific formula shown below:

$$r = \sqrt{\frac{A}{\pi}} \tag{1}$$

In the formula, r denotes the radius of gravel on the infinitesimal rectangular body, mm, and A denotes the area of gravel calibrated on the infinitesimal rectangular body, mm^2.

To quantitatively characterize the gravel content along the axial direction of a horizontal wellbore, the gravel linear density is defined as the gravel content per unit area along the wellbore axis. The specific formula is as follows:

$$\mathrm{C} = 10^6 \times \frac{N}{2\pi R \Delta L} \tag{2}$$

In the formula, C represents the linear density of gravel along the borehole axis, units: pcs/m^2; N denotes the gravel content within the rectangular element, units: pcs; R is the borehole radius, units: mm; and L is the length of the element, units: mm.

3. Method for Assessing Gravel Distribution Characteristics

3.1. Method for Characterizing Axial Gravel Distribution in Boreholes Based on Imaging Logging

FMI imaging logging employs an electrode array to perform array scanning or rotary scanning around the wellbore to measure electrical conductivity. The acquired conductivity data undergoes depth correction, velocity correction, and balancing to generate a two-dimensional resistivity image of the wellbore wall. In FMI imaging logging interpretations, subtle color variations represent changes in lithology and physical properties, with colors ranging from light to dark indicating decreasing conductivity values. Due to the poor conductivity of gravel in conglomerate reservoirs, it appears as blocky bright areas in electrical images, distinguishable from the background color of the imaging log. Therefore, gravel distribution characteristics can be extracted from the imaging log through image processing. Effective gravel information extraction is achieved through grayscale processing, blind zone filling, and image segmentation of the imaging log, as illustrated in Figure 2. The specific workflow comprises the following four steps:

① Grayscale processing is applied to the two-dimensional resistivity images from imaging logging. This conversion preserves overall and local chromatic brightness characteristics while transforming the image into grayscale, where brighter areas indicate potential gravel zones.

② Blind zone filling is applied to eliminate white band blind zones caused by electrode plate occlusion in imaging logging, neutralizing the influence of high conductivity at occlusion points.

③ Image segmentation techniques are applied to effectively extract bright regions from the filled image. Adaptive thresholding is used to isolate gravel from the image.

④ The convex hull algorithm is employed to sequentially extract the envelope areas of bright regions in the segmented imaging picture. This process statistically determines the radial and linear density distribution characteristics of gravel along the wellbore axis.

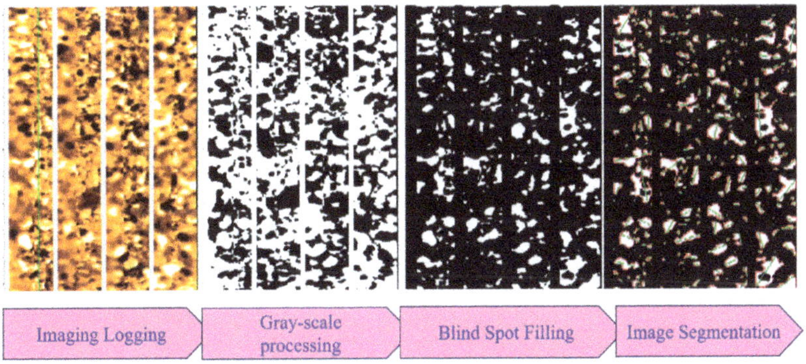

Figure 2. Imaging Log Image Processing Workflow.

Based on the aforementioned principles, the "Conglomerate Reservoir Gravel Characteristic Identification and Analysis Software V1.0" was developed. This software enables automatic processing and interpretation of imaging logging data, exporting the processed and interpreted results in tabular format. The software interface is shown in Figure 3, and the computational results are presented in Figure 4.

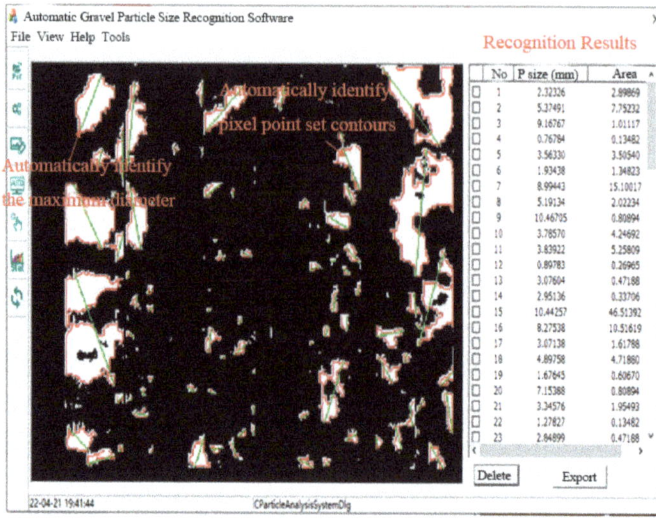

Figure 3. Interface of the Gravel Characteristic Identification Analysis Software for Conglomerate Reservoirs.

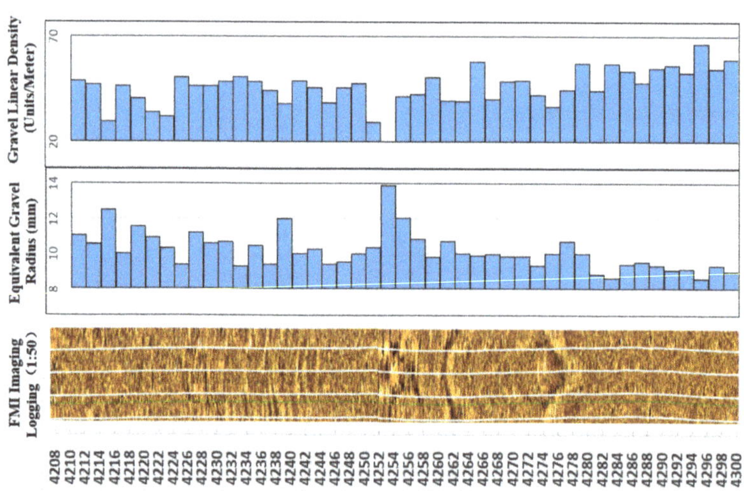

Figure 4. JLX Well FMI Gravel Identification Results (4208–4300 m).

3.2. Method for Characterizing Axial Gravel Distribution in Boreholes Based on Logging Data

Considering the field operation cycle and costs, FMI imaging logging is not performed on every well. Therefore, it is necessary to establish correlations between conventional logging data and gravel distribution characteristics based on imaging logging interpretation results, thereby characterizing the gravel distribution around horizontal wells under conditions where imaging logging is unavailable. The *p*-value from statistics is introduced to characterize the association between different types of logging data and gravel linear density and content. The *p*-value is the probability of observing a test statistic at least as extreme as the current result under the assumption that the null hypothesis is true. In hypothesis testing, the *p*-value is compared to the significance level α: if $p\text{-value} < \alpha$, the null hypothesis is rejected; otherwise, it is not rejected. Assume the null hypothesis is "There is no relationship between logging curve data and gravel equivalent radius or gravel linear density." Set the significance level $\alpha = 0.1$. If p-value < 0.1, reject the null hypothesis at the 0.1 significance level, indicating a statistically significant relationship exists between the logging curve and either the gravel equivalent radius or the gravel linear density. p-value analysis for the three indicators AC, CNL, and GR all yielded values less than 0.1. This indicates statistically significant relationships between these logging curves and gravel characteristics, as illustrated in Figures 5 and 6.

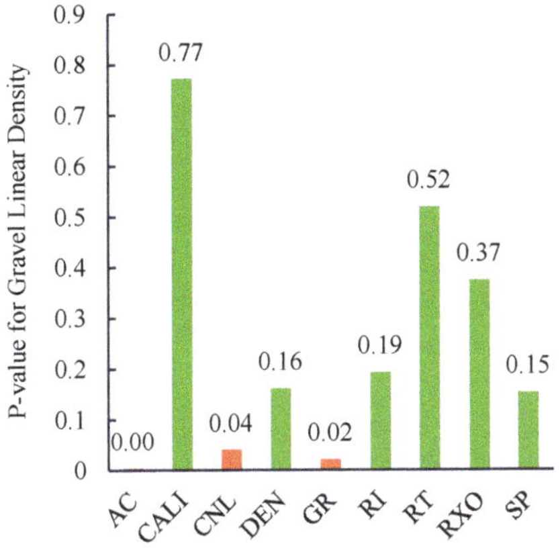

Figure 5. Logging Curve and p-value of Gravel Line Density (4208–4264 m).

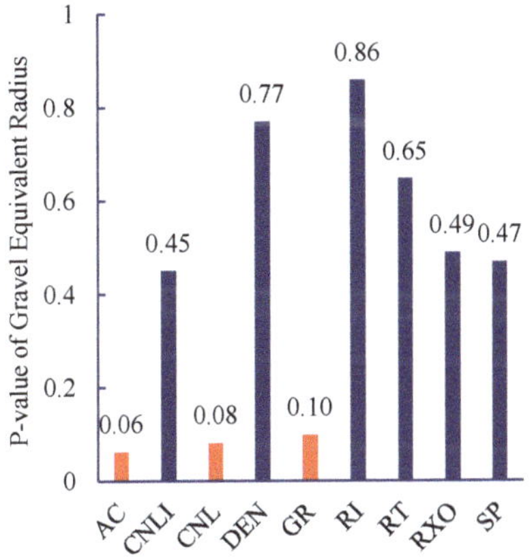

Figure 6. Log Curves and p-Values for Gravel Equivalent Radius (4208–4264 m).

Using multiple regression analysis, predictive models were established for conventional logging curves in the fan area of the JL1 wellbore, relating to gravel equivalent radius and gravel line density, as shown in Equations (3) and (4). Linear regression was performed between the predicted and observed results, verifying a fitting accuracy exceeding 80%, as illustrated in Figure 7.

$$\text{Equivalent Gravel Radius: } r = -0.884AC - 1.44CNL + 0.694GR + 72.03 \quad (3)$$

$$\text{Gravel Linear Density: } C = 0.152AC + 0.247CNL - 0.119GR + 5.64 \quad (4)$$

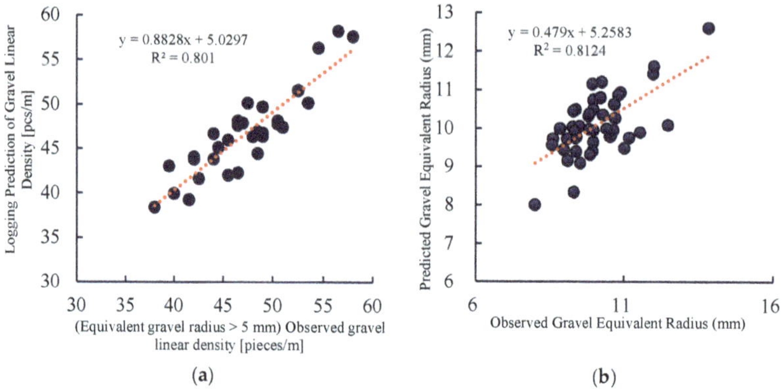

Figure 7. Fitting Curve:(**a**) Gravel Linear Density; (**b**) Gravel Equivalent Radius.

3.3. Data Standardization and Regression Diagnostics

(1) Data Standardization

The purpose of data standardization is to eliminate the influence of different units and dimensions on regression coefficients, enabling variables to be compared on a uniform scale. This study employs zero-mean unit variance standardization (Z-score standardization):

$$\text{X Standardization} = \frac{\text{X} - \mu}{\sigma} \tag{5}$$

Here, X represents the raw data, μ denotes the mean, and σ indicates the standard deviation.

(2) Regression Diagnosis

① R^2 (determination coefficient):

R^2 reflects the proportion of variance in the independent variable explained by the model, measuring the model's ability to fit the data. Its value ranges from 0 to 1; a higher value indicates that the model better explains the variability in the data. The closer R^2 is to 1, the more effectively the regression model predicts the dependent variable.

$$R^2 = 1 - \frac{\sum_{i=1}^{n}\left(y_i - \hat{y}_i\right)^2}{\sum_{i=1}^{n}(y_i - \overline{y})^2} \tag{6}$$

Among these, y_i is the actual value, $\hat{y}_i$ is the predicted value, and $\overline{y}$ is the mean of the target variable.

② RMSE (Root Mean Square Error):

RMSE is a standard measure of regression model error, representing the average deviation between predicted and actual values. A smaller RMSE indicates higher prediction accuracy of the model.

$$\text{RMSE} = \sqrt{\frac{1}{n}\sum_{i=1}^{n}\left(y_i - \hat{y}_i\right)^2} \tag{7}$$

Here, y_i represents the actual value, and $\hat{y}_i$ represents the predicted value.

③ Significance of the coefficient:

Significance tests for coefficients are used to assess whether independent variables have a significant effect on the dependent variable. A commonly used significance test is the t-test, whose test statistic is:

$$t = \frac{\hat{\beta}}{SE\left(\hat{\beta}\right)} \tag{8}$$

Here, $\hat{\beta}$ represents the estimated regression coefficient, and the standard error indicates the uncertainty in the coefficient estimate. If the absolute value of the t-statistic is large and the corresponding p-value is less than the pre-set significance level of 0.1, then the independent variable is considered to have a significant effect on the dependent variable.

3.4. Regional Gravel Characteristic Distribution Characterization

To further characterize regional gravel distribution patterns, based on the axial gravel distribution in horizontal wells, Kriging interpolation was employed. Kriging interpolation is a spatial interpolation method grounded in geostatistics. It leverages the spatial autocorrelation of data by establishing a variogram model to quantify and describe the spatial structure of data, thereby providing an optimal, unbiased estimate for unknown points. Its core advantage lies not only in providing predicted values but also in offering the prediction error or uncertainty for each predicted point. This makes it a more scientific and accurate method than simple interpolation for reflecting the true spatial variation of geographic phenomena.

Kriging interpolation characterizes regional gravel line density and equivalent gravel radius size, providing a basis for depicting fracture propagation patterns in horizontal wells within conglomerate reservoirs. forming a "point-line-surface" gravel distribution prediction method, as shown in Figure 8a,b. The vertical axis represents horizontal section length, while the horizontal axis denotes lateral displacement. By determining the color of coordinate points based on the specified horizontal section length and lateral displacement, gravel line density and equivalent gravel radius can be predicted.

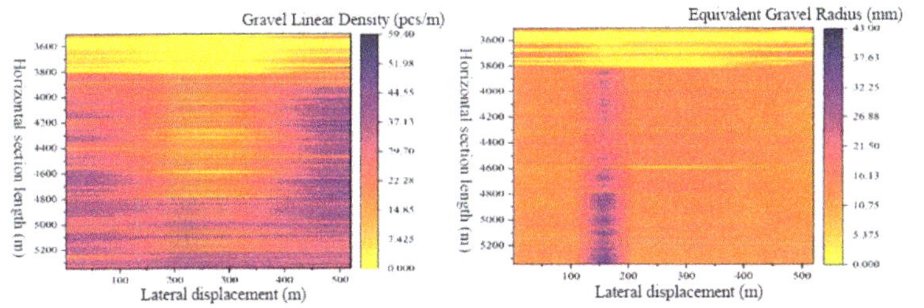

(a) (b)

Figure 8. Distribution Map:(**a**) Regional Gravel Line Density; (**b**) Regional Equivalent Gravel Radius.

4. Model Validation

To comprehensively evaluate the model's generalization capability and predictive accuracy, this study employed three independent validation methods: imaging logging results, Eagle Eye test outcomes, and core sampling data. These approaches validated the model from distinct perspectives, ensuring its reliability in practical applications.

4.1. Imaging Logging Verification

The model accuracy was validated using gravel identification results from FMI imaging logging of the remaining sections of Well J1, as shown in Table 1.

Table 1. Comparison of Gravel Equivalent Radius and Gravel Linear Density Identification Results.

Gravel Parameters	FMI Observation of Gravel Parameters	Frequency	Logging for Predicting Gravel Parameters	Frequency	Error	Standard Deviation
Equivalent radius of gravel (mm)	7–8	0	7–8	1	/	0.5099
	8–9	1	8–9	0	/	
	9–10	13	9–10	13	0%	
	10–11	16	10–11	18	13%	
	11–12	10	11–12	9	10%	
	12–13	3	12–13	3	0%	
	13–14	1	13–14	0	/	
	other	1	other	1	/	
Gravel Linear Density (pcs/m)	20–25	1	20–25	1	/	1.3595
	30–35	4	30–35	3	25%	
	35–45	10	35–45	9	10%	
	45–55	19	45–55	18	5%	
	55–60	9	55–60	10	11%	
	other	2	other	1	/	

Table 1 shows that for Well J1, the FMI observation indicates a gravel equivalent radius of 8–14 mm, while the logging model predicts a gravel equivalent radius of 9–12 mm, with an error range of 10–13% and an average error of 5.8%. The FMI-observed gravel linear density ranges from 20 to 60 grains per meter, while the logging model predicts a linear density of 20 to 60 grains per meter. The error ranges from 5% to 25%, with an average error of 12.8%. The standard errors for both predictions are relatively low, indicating that the coefficient estimates are more precise. The model-predicted parameter error is <15%, meeting engineering accuracy requirements.

4.2. Eagle-Eye Monitoring Results Comparison and Verification

The Downhole Eagle Eye is a high-precision downhole imaging technology. During model validation, it can serve as an independent verification data source, providing robust support for the accuracy and reliability of the model. Wells J2 and J1 are located within the same block. Through this model, the gravel equivalent radius and gravel linear density for Well J2 were obtained. Combined with Eagle Eye test results, single-stage gravel information (model prediction results) and annular erosion conditions (Eagle Eye test results) were derived, as shown in Table 2.

Table 2. Single-Stage Gravel Information and Pore Erosion Conditions.

Series	Average Gravel Linear Density (Pieces/m)	Average Gravel Radius (mm)	Gravel Linear Density Variance	Gravel Radius Variance	Average Erosion Area (mm)
2	8.0	16.6	27.3	0.8	1064.7
3	17.2	15.0	11.7	0.3	346.3
4	17.7	15.0	2.6	0.1	709.7
5	16.4	15.2	6.9	0.2	195.8

Table 2. *Cont.*

Series	Average Gravel Linear Density (Pieces/m)	Average Gravel Radius (mm)	Gravel Linear Density Variance	Gravel Radius Variance	Average Erosion Area (mm)
10	9.5	16.4	3.5	0.1	174.1
11	14.1	15.6	12.0	0.4	418.3
12	11.7	16.0	4.8	0.1	18.4
13	11.6	15.6	13.2	2.0	2308.8

The heterogeneity of conglomerate reservoirs is characterized by the product of single-stage gravel line density variance and gravel equivalent radius variance. The reservoir heterogeneity is fitted with the average eroded aperture area from the Eagle Eye test, with the fitting results shown in Figure 9.

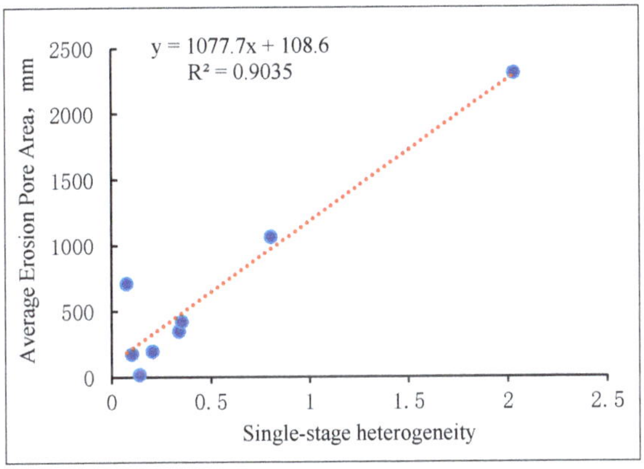

Figure 9. Heterogeneity of Single-Segment Gravel and Pore Erosion Area.

As shown in Figure 9, the correlation between pore-eye erosion area and single-segment heterogeneity exceeds 90%, demonstrating that this model can accurately reflect the reservoir characteristics of conglomerate reservoirs.

4.3. *Verification of Underground Core CT Scan Results*

Core data represent the most direct method for model validation, providing authentic gravel parameters. Based on the conventional logging method for characterizing gravelly sandstone reservoirs developed in this study and utilizing gravel information from the CT images of Well M1 cores, As shown in Figure 10. A multiple linear regression analysis was conducted to clarify the relationship between gravel parameters and logging data: The equivalent gravel radius showed significant correlation with AC, GR, and RT in the logging curves, as shown in Figure 11 The linear density of gravel shows a strong correlation with CNL, DEN, and RT in the logging curves, as illustrated in Figure 12.

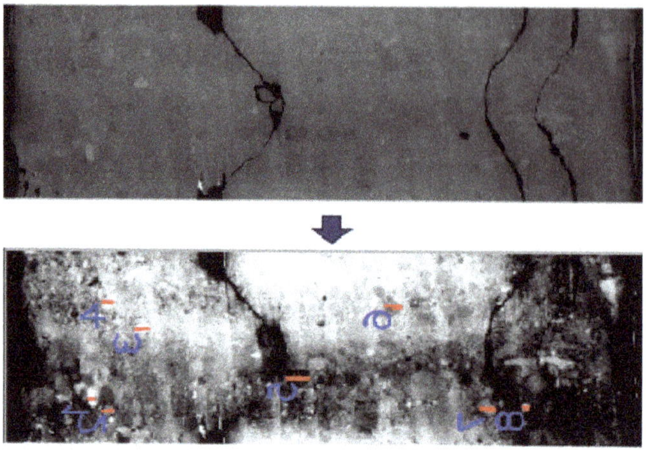

Figure 10. Gravel Parameter Extraction from CT Images of Well M1.

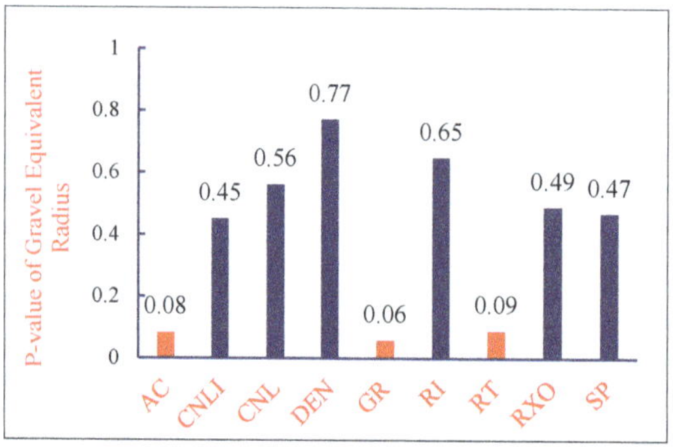

Figure 11. Logging Curves and p-value of Gravel Equivalent Radius.

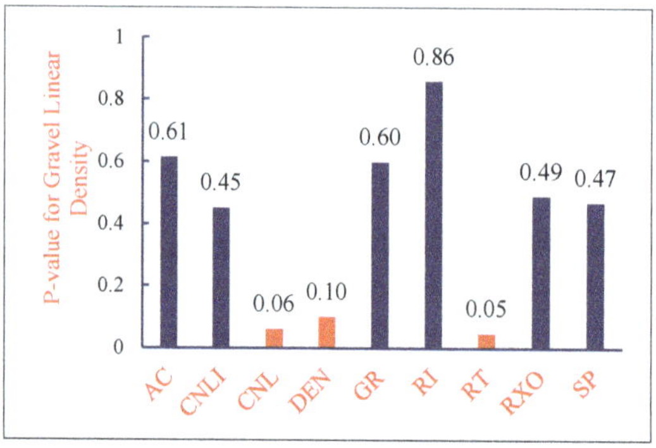

Figure 12. Logging Curves and Gravel Line Density p-value.

The M1 curve and the equivalent radius and line density of gravel are modeled as shown in Formulas (5) and (6).

$$r = 0.173 \times GR - 0.00497 \times RT - 0.102 \times AC + 1.33 \quad (9)$$

$$C = 67.78 \times DEN + 0.0742 \times RT + 0.1298 \times CNL - 161.414 \quad (10)$$

Verification was conducted using CT scan results from core samples taken from Well M1, as shown in Figure 13.

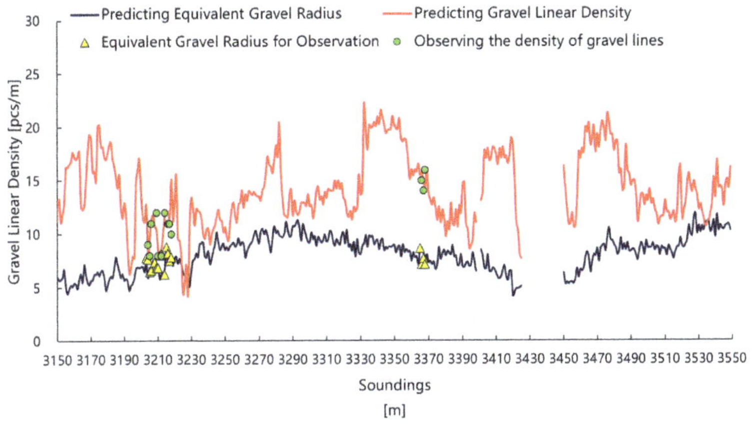

Figure 13. Comparison of FMI Imaging Logging and Predicted Gravel Parameters.

The linear density model achieved an R^2 value of 0.8135, while the equivalent conglomerate radius model attained an R^2 of 0.7044, indicating both models effectively fit the data. The linear density model yielded an RMSE of 0.26, and the equivalent conglomerate radius model produced an RMSE of 0.21, signifying increasingly precise model predictions.

As shown in Figure 13, the gravel parameters predicted by this method exhibit minimal deviation (<10%) from the results obtained from core CT scans at Well M1. Possible causes for discrepancies between predicted and actual values include insufficient resolution of imaging logging images and physical changes in cored rock layers, demonstrating the method's generalizability.

5. Field Application

During fracturing operations in conglomerate reservoirs, the presence of gravel fragments results in shorter perforation penetration depths compared to conventional sandstone formations, making fracturing more challenging. This often leads to difficulties in initiating fractures and To address this, the J2 well in a conglomerate reservoir employed the aforementioned method to precisely map gravel distribution along the wellbore axis. Fracturing design was segmented and clustered to avoid areas with high gravel density and large gravel radii, as illustrated in Figure 14.

Well J2 employed horizontal well volume fracturing with a pumping rate of 12–14 m^3/min. By avoiding areas with high gravel distribution, the operation difficulty was correspondingly lower. Sand injection was completed as designed, achieving a 10% fluid volume saving, as shown in Figure 15. The wellhead pressure was 3–5 MPa lower than that of adjacent wells, and the production performance exceeded that of neighboring wells in the block, as illustrated in Figure 16.

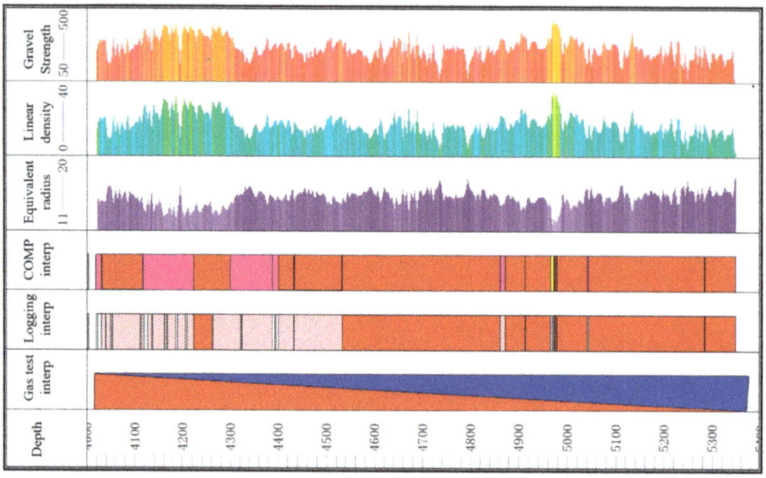

Figure 14. Gravel Content and Segmented Clustering Results for the Horizontal Section of JL2 Well.

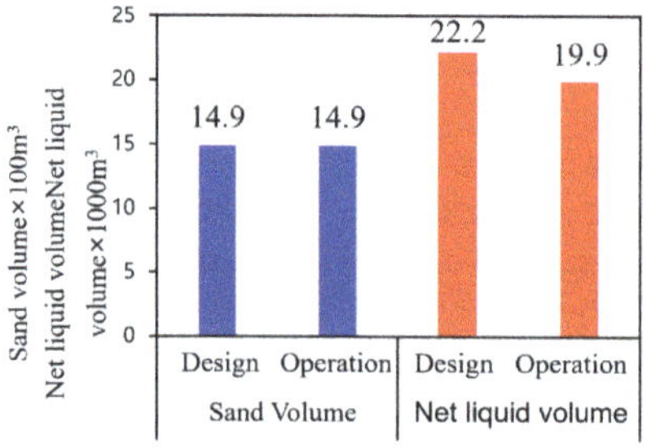

Figure 15. Statistical Data on Fluid and Sand Volumes for JL2 Well.

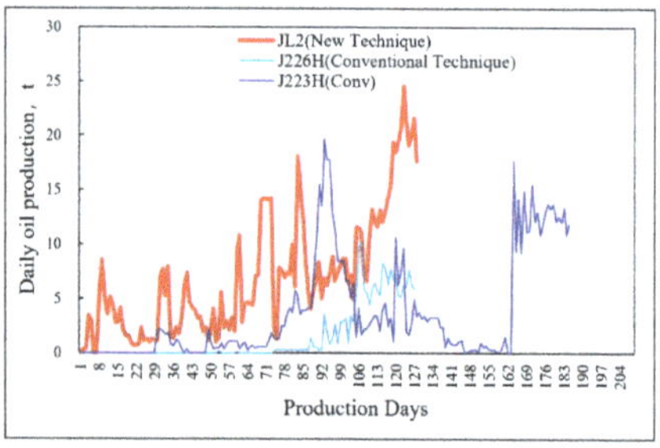

Figure 16. Production comparison between Wells.

6. Conclusions

This study uniquely combines multiple methodologies to establish a gravel distribution characteristic model, develops image-processing-based identification software for precise recognition, and effectively predicts gravel distribution without imaging logging—all with high model accuracy. Emphasizing practical engineering applications, it reduces construction pressure, conserves fluid volume, and enhances production efficiency. Specific conclusions are as follows:

(1) A model for characterizing gravel distribution around horizontal wells was established to quantitatively describe the axial distribution of gravel content and size within

wellbores. By systematically analyzing imaging log interpretation results and applying image processing techniques, gravel feature recognition software was developed, enabling effective extraction of gravel distribution characteristics from imaging logs and providing a robust basis for characterizing gravel in conglomerate reservoirs.

(2) The relationship between conventional logging curves and gravel content/size was clarified, leading to the establishment of a gravel distribution characterization method based on conventional logging data. Using Kriging interpolation, a "point-line-surface" approach for predicting gravel distribution was developed. Validation with imaging logging, downhole Eagle Eye data, and field coring confirmed model accuracy exceeding 80%.

(3) Detailed characterization of gravel distribution in horizontal wells guided staged and clustered fracturing operations, allowing bypass of high-gravel-content zones. This strategy effectively reduced fracturing complexity and lowered fracturing pressure by 3–5 MPa compared to neighboring wells. Single-well fluid consumption decreased by up to 10%, while production performance exceeded that of adjacent wells within the block. These findings provide a scientific basis for optimizing staged and clustered operations in gravel-rich reservoirs.

Based on existing research findings, future work can advance in multiple directions: the model accuracy requires improvement, so deep learning algorithms can be integrated to enhance gravel distribution prediction precision to over 90% [51–53]; beyond conglomerate reservoirs, the model can be extended to complex reservoir types, such as carbonates, to enhance applicability; real-time monitoring systems can be developed to optimize dynamic fracturing parameters; and green fracturing technologies can be explored to further reduce fluid consumption and environmental impact. These advancements will collectively drive the intelligent and sustainable development of oil and gas extraction.

Author Contributions: Investigation, Z.L., J.X., T.L., P.L., X.C., H.C. and Y.Z.; Data curation, J.X.; Writing—original draft, Z.L. and J.X.; Writing—review & editing, T.L.; Supervision, T.L. All authors have read and agreed to the published version of the manuscript.

Funding: This research was funded by the General Program Grant from the National Natural Science Foundation of China (No. 52274051) and Xinjiang Uygur Autonomous Region "Tianshan Elite" Cultivation Program (No. 2024TSYCJC0010).

Data Availability Statement: The original contributions presented in this study are included in the article. Further inquiries can be directed to the corresponding author.

Conflicts of Interest: Authors Zhenhu Lv, Ping Li, Xiaolu Chen, Hao Cheng and Yupeng Zhang were employed by the PetroChina Xinjiang Oilfield Development Company. The remaining authors declare that the research was conducted in the absence of any commercial or financial relationships that could be construed as a potential conflict of interest.

References

1. Lu, T.; Li, Z.; Zhou, Y.; Zhang, C. Enhanced oil recovery of low-permeability cores by SiO_2 nanofluid. *Energy Fuels* **2017**, *31*, 5612–5621. [CrossRef]
2. Dai, C.; Wang, X.; Li, Y.; Lv, W.; Zou, C.; Gao, M.; Zhao, M. Spontaneous imbibition investigation of self-dispersing silica nanofluids for enhanced oil recovery in low-permeability cores. *Energy Fuels* **2017**, *31*, 2663–2668. [CrossRef]
3. Wentworth, C.K. A scale of grade and class terms for clastic sediments. *J. Geol.* **1922**, *30*, 377–392. [CrossRef]
4. Urien, C.M.; Zambrano, J.J. Petroleum Systems in the Neuquén Basin, Argentina. In *The Petroleum System—From Source to Trap*; Magoon, L.B., Dow, W.G., Eds.; American Association of Petroleum Geologists: Tulsa, OK, USA, 1994; Volume 60, ISBN 978-1-62981-092-8.
5. Xiao, M.; Wu, S.; Yuan, X.; Xie, Z. Conglomerate Reservoir Pore Evolution Characteristics and Favorable Area Prediction: A Case Study of the Lower Triassic Baikouquan Formation in the Northwest Margin of the Junggar Basin, China. *J. Earth Sci.* **2021**, *32*, 998–1010. [CrossRef]

6. Mello, M.R.; Koutsoukos, E.A.M.; Mohriak, W.U.; Bacoccoli, G. Selected Petroleum Systems in Brazil. In *The Petroleum System—From Source to Trap*; Magoon, L.B., Dow, W.G., Eds.; American Association of Petroleum Geologists: Tulsa, OK, USA, 1994; Volume 60, ISBN 978-1-62981-092-8.
7. Zou, Y.; Shi, S.; Zhang, S.; Yu, T.; Tian, G.; Ma, X.; Zhang, Z. Experimental Modeling of Sanding Fracturing and Conductivity of Propped Fractures in Conglomerate: A Case Study of Tight Conglomerate of Mahu Sag in Junggar Basin, NW China. *Pet. Explor. Dev.* **2021**, *48*, 1383–1392. [CrossRef]
8. Yi, Y.; Wang, L.; Li, J.; Chen, S.; Tian, H.; Tian, G. Optimization of Re-Fracturing Method and Fracture Parameters for Horizontal Well in Mahu Conglomerate Oil Reservoir. *Front. Energy Res.* **2022**, *10*, 856524. [CrossRef]
9. Zhang, Z.; Du, J.; Mavko, G.M. Reservoir Characterization Using Perforation Shots: Anisotropy, Attenuation and Uncertainty Analysis. *Geophys. J. Int.* **2019**, *216*, 470–485. [CrossRef]
10. Wu, J.; Yang, S.; Gan, B.; Cao, Y.; Zhou, W.; Kou, G.; Wang, Z.; Li, Q.; Dong, W.; Zhao, B. Pore Structure and Movable Fluid Characteristics of Typical Sedimentary Lithofacies in a Tight Conglomerate Reservoir, Mahu Depression, Northwest China. *ACS Omega* **2021**, *6*, 23243–23261. [CrossRef] [PubMed]
11. Han, B.; Wang, B.; Luo, F.; Zhang, Y.; Han, B.; Xia, Y.; Ding, W. Analysis and Application of Seismic Velocity Characteristics in Areas with near Surface Conglomerate Development: A Case Study in YKB Area in Kuqa Piedmont Zone. *China Pet. Explor.* **2023**, *28*, 117–125. [CrossRef]
12. Fu, Y.; Cheng, R.; Gao, Y.; Zhou, Y.; Xu, Z. Early Cretaceous Sedimentary Records of the Early-Stage Continental Rifting in the Songliao Basin, NE China. *J. Asian Earth Sci.* **2024**, *259*, 105913. [CrossRef]
13. Yuan, R.; Yang, B.; Huang, L.; Hu, W. Utilizing Borehole Electrical Images to Evaluate Sorting of Conglomeratic Formation: Example from Lower Triassic Baikouquan Formation in Well M152, Mahu Depression, Junggar Basin, China. *Arab. J. Geosci* **2020**, *13*, 408. [CrossRef]
14. Lai, J.; Wang, G.; Chai, Y.; Xin, Y.; Wu, Q.; Zhang, X.; Sun, Y. Deep Burial Diagenesis and Reservoir Quality Evolution of High-Temperature, High-Pressure Sandstones: Examples from Lower Cretaceous Bashijiqike Formation in Keshen Area, Kuqa Depression, Tarim Basin of China. *AAPG Bull.* **2017**, *101*, 829–862. [CrossRef]
15. Zhou, Z.; Hilterman, F.J. A Comparison between Methods That Discriminate Fluid Content in Unconsolidated Sandstone Reservoirs. *Geophysics* **2010**, *75*, B47–B58. [CrossRef]
16. Yuan, R.; Zhang, L.; Xu, Q.; Feng, Y.; An, Z.; Wu, J.; Zhao, K.; Liu, S.; Huang, P. Utilizing Borehole Electrical Image and Conventional Logs to Characterize Petrology of Mixed Volcanic and Sedimentary Rocks in Jiamuhe Formation at JL2 Wellfield, Zhongguai Uplift, Junggar Basin, NW China. *Arab. J. Geosci.* **2020**, *13*, 1209. [CrossRef]
17. Kumara, G.H.A.J.J.; Hayano, K.; Ogiwara, K. Image Analysis Techniques on Evaluation of Particle Size Distribution of Gravel. *GEOMATE J.* **2012**, *3*, 290–297. [CrossRef]
18. Yu, S.Y.; Li, S.H.; Li, J.; Song, D.W.; Shi, J.H. An Algorithm for Particle Size Analysis of Gravel Image by Using Gray-Value Vibration Frequency. *Appl. Mech. Mater.* **2013**, *318*, 253–256. [CrossRef]
19. Wang, L.; Lu, J.; Luo, Y.; Ren, B.; Li, A.; Zhao, N. An Automated Quantitative Methodology for Computing Gravel Parameters in Imaging Logging Leveraging Deep Learning: A Case Analysis of the Baikouquan Formation within the Mahu Sag. *Processes* **2024**, *12*, 1337. [CrossRef]
20. Xie, F.; Xiao, C.; Liu, R.; Zhang, L. Multi-threshold de-noising of electrical imaging logging data based on the wavelet packet transform. *J. Geophys. Eng.* **2017**, *14*, 900–908. [CrossRef]
21. Guimaraes, C.; Piazza, R. Optimizing Image Logging Acquisition with Deep Learning. In Proceedings of the SPE Annual Technical Conference and Exhibition, New Orleans, LA, USA, 23–25 September 2024. [CrossRef]
22. de Jesus, C.M.; Compan, A.L.M.; Surmas, R. Permeability Estimation Using Ultrasonic Borehole Image Logs in Dual-Porosity Carbonate Reservoirs. *Petrophysics* **2016**, *57*, 620–637. Available online: https://onepetro.org/petrophysics/article/57/06/620/171448/Permeability-Estimation-Using-Ultrasonic-Borehole (accessed on 9 November 2025).
23. Aghli, G.; Moussavi-Harami, R.; Mohammadian, R. Reservoir Heterogeneity and Fracture Parameter Determination Using Electrical Image Logs and Petrophysical Data (a Case Study, Carbonate Asmari Formation, Zagros Basin, SW Iran). *Pet. Sci.* **2020**, *17*, 51–69. [CrossRef]
24. Xie, F.; Zhang, C.; Liu, R.; Xiao, C. Production Prediction for Fracture-Vug Carbonate Reservoirs Using Electric Imaging Logging Data. *Pet. Explor. Dev.* **2018**, *45*, 369–376. [CrossRef]
25. Zohreh, M.; Junin, R.; Jeffreys, P. Evaluate the Borehole Condition to Reduce Drilling Risk and Avoid Potential Well Bore Damages by Using Image Logs. *J. Pet. Sci. Eng.* **2014**, *122*, 318–330. [CrossRef]
26. Wood, D.A. Expanding Role of Borehole Image Logs in Reservoir Fracture and Heterogeneity Characterization: A Review. *Adv. Geo-Energy Res.* **2024**, *12*, 194–204. [CrossRef]
27. Dias, L.O.; Bom, C.R.; Faria, E.L.; Valentín, M.B.; Correia, M.D.; de Albuquerque, M.P.; de Albuquerque, M.P.; Coelho, J.M. Automatic Detection of Fractures and Breakouts Patterns in Acoustic Borehole Image Logs Using Fast-Region Convolutional Neural Networks. *J. Pet. Sci. Eng.* **2020**, *191*, 107099. [CrossRef]

28. Lai, J.; Wang, G.; Wang, S.; Cao, J.; Li, M.; Pang, X.; Han, C.; Fan, X.; Yang, L.; He, Z.; et al. A Review on the Applications of Image Logs in Structural Analysis and Sedimentary Characterization. *Mar. Pet. Geol.* **2018**, *95*, 139–166. [CrossRef]
29. Li, N.; Liu, P.; Wu, H.; Li, Y.; Zhang, W.; Wang, K.; Feng, Z.; Wang, H. Development and Prospect of Acoustic Reflection Imaging Logging Processing and Interpretation Method. *Pet. Explor. Dev.* **2024**, *51*, 839–851. [CrossRef]
30. Du, L.; Lu, X.; Li, H. Automatic Fracture Detection from the Images of Electrical Image Logs Using Mask R-CNN. *Fuel* **2023**, *351*, 128992. [CrossRef]
31. Fernández-Ibáñez, F.; DeGraff, J.M.; Ibrayev, F. Integrating Borehole Image Logs with Core: A Method to Enhance Subsurface Fracture Characterization. *AAPG Bull.* **2018**, *102*, 1067–1090. [CrossRef]
32. Xiao, L.; Li, J.; Mao, Z.; Yu, H. A Method to Evaluate Pore Structures of Fractured Tight Sandstone Reservoirs Using Borehole Electrical Image Logging. *AAPG Bull.* **2020**, *104*, 205–226. [CrossRef]
33. Huang, J.; Peng, S.; Wang, X.; Xiao, K. Applications of Imaging Logging Data in the Research of Fracture and Ground Stress. *Acta Pet. Sin.* **2006**, *27*, 65–69. [CrossRef]
34. Roussel, N.P.; Sharma, M.M. Optimizing Fracture Spacing and Sequencing in Horizontal-Well Fracturing. *SPE Prod. Oper.* **2011**, *26*, 173–184. [CrossRef]
35. Waters, G.; Dean, B.; Downie, R.; Kerrihard, K.; Austbo, L.; McPherson, B. Simultaneous Hydraulic Fracturing of Adjacent Horizontal Wells in the Woodford Shale. In Proceedings of the SPE Hydraulic Fracturing Technology Conference, The Woodlands, TX, USA, 19–21 January 2009. [CrossRef]
36. Xu, Z.; Zhao, H.; Liu, X.; Liu, X.; Luo, P.; Liang, L. Numerical Analysis of the Hydraulic Fracture Propagation Behavior Encountering Gravel in Conglomerate Reservoirs. *Sci. Rep.* **2025**, *15*, 1869. [CrossRef]
37. Tian, F.; Liu, Z.; Zhou, J.; Shao, J. Rock Cracking Simulation in Tension and Compression by Peridynamics Using a Novel Contact-Friction Model with a Twin Mesh and Potential Functions. *J. Rock Mech. Geotech. Eng.* **2025**, *17*, 3395–3419. [CrossRef]
38. Zheng, Y.; Baudet, B.A. Evolution of Nano-Pores in Illite-Dominant Clay during Consolidation. *Acta Geotech.* **2024**, *19*, 71–83. [CrossRef]
39. Al-Kobaisi, M.; Ozkan, E.; Kazemi, H. A Hybrid Numerical/Analytical Model of a Finite-Conductivity Vertical Fracture Intercepted by a Horizontal Well. *SPE Reserv. Eval. Eng.* **2006**, *9*, 345–355. [CrossRef]
40. Raghavan, R.S.; Chen, C.-C.; Agarwal, B. An Analysis of Horizontal Wells Intercepted by Multiple Fractures. *SPE J.* **1997**, *2*, 235–245. [CrossRef]
41. Daneshy, A. Analysis of Horizontal Well Fracture Interactions, and Completion Steps for Reducing the Resulting Production Interference. In Proceedings of the SPE Annual Technical Conference and Exhibition, Dallas, TX, USA, 24–26 September 2018. [CrossRef]
42. Chen, J. The Influence of Tensile Strength Difference with Variable Gravel Sizes on the Hydraulic Fracture Propagation in the Conglomerate Reservoir. *Geofluids* **2022**, *2022*, 5290189. [CrossRef]
43. Ozkan, E.; Brown, M.; Raghavan, R.; Kazemi, H. Comparison of Fractured Horizontal-Well Performance in Conventional and Unconventional Reservoirs. In Proceedings of the SPE Western Regional Meeting, San Jose, CA, USA, 24–26 March 2009. [CrossRef]
44. Ozkan, E.; Brown, M.; Raghavan, R.; Kazemi, H. Comparison of Fractured-Horizontal-Well Performance in Tight Sand and Shale Reservoirs. *SPE Reserv. Eval. Eng.* **2011**, *14*, 248–259. [CrossRef]
45. Daneshy, A.; King, G. Horizontal Well Frac-Driven Interactions: Types, Consequences, and Damage Mitigation. *J. Pet. Technol.* **2019**, *71*, 45–47. [CrossRef]
46. Kumar, D.; Ghassemi, A. A Three-Dimensional Analysis of Simultaneous and Sequential Fracturing of Horizontal Wells. *J. Pet. Sci. Eng.* **2016**, *146*, 1006–1025. [CrossRef]
47. Lei, Q.; Xu, Y.; Cai, B.; Guan, B.; Wang, X.; Bi, G.; Li, H.; Li, S.; Ding, B.; Fu, H.; et al. Progress and Prospects of Horizontal Well Fracturing Technology for Shale Oil and Gas Reservoirs. *Pet. Explor. Dev.* **2022**, *49*, 191–199. [CrossRef]
48. Li, Y.; Hubuqin; Wu, J.; Zhang, J.; Yang, H.; Zeng, B.; Xiao, Y.; Liu, J. Optimization Method of Oriented Perforation Parameters Improving Uneven Fractures Initiation for Horizontal Well Fracturing. *Fuel* **2023**, *349*, 128754. [CrossRef]
49. Liu, N.; Zhang, Z.; Zou, Y.; Ma, X.; Zhang, Y. Propagation Law of Hydraulic Fractures during Multi-Staged Horizontal Well Fracturing in a Tight Reservoir. *Pet. Explor. Dev.* **2018**, *45*, 1129–1138. [CrossRef]
50. Liu, X.; Zhang, A.; Tang, Y.; Wang, X.; Xiong, J. Investigation on the Influences of Gravel Characteristics on the Hydraulic Fracture Propagation in the Conglomerate Reservoirs. *Nat. Gas Ind. B* **2022**, *9*, 232–239. [CrossRef]
51. Liu, Z.; Yu, J.; Ren, C.; Elbaz, K.; Zhu, D.; Cai, Y. Fatigue Behaviour Characteristics and Life Prediction of Rock under Low-Cycle Loading. *Int. J. Min. Sci. Technol.* **2025**, *35*, 737–752. [CrossRef]

52. Fan, X.; Wang, G.; Dai, Q.; Li, Y.; Zhang, F.; He, Z.; Li, Q. Using Image Logs to Identify Fluid Types in Tight Carbonate Reservoirs via Apparent Formation Water Resistivity Spectrum. *J. Pet. Sci. Eng.* **2019**, *178*, 937–947. [CrossRef]
53. Lai, J.; Pang, X.; Xiao, Q.; Shi, Y.; Zhang, H.; Zhao, T.; Chen, J.; Wang, G.; Qin, Z. Prediction of Reservoir Quality in Carbonates via Porosity Spectrum from Image Logs. *J. Pet. Sci. Eng.* **2019**, *173*, 197–208. [CrossRef]

processes MDPI

Article

Pre-Crosslinked Gel Particles Enhanced by Amphiphilic Nanocarbon Dots in Harsh Reservoirs: Synthesis and Deep Stimulation Mechanism

Guorui Xu [1,2,3], Xiaoxiao Li [1,2,3,*], Jinzhou Yang [1,2,3], Chunyu Tong [1,2,3], Xiaolong Wang [1,2,3] and Tengfei Wang [4,*]

1 Tianjin Key Laboratory of Offshore Difficult to Recovery Reserves Exploitation, Tianjin 300452, China; xugr@cosl.com.cn (G.X.); yangjzh14@cosl.com.cn (J.Y.); tongchy2@cosl.com.cn (C.T.); wangxl151@cosl.com.cn (X.W.)
2 Production Optimization, China Oilfield Service Limited, Tianjin 300452, China
3 State Key Laboratory of Offshore Oil Exploitation, Tianjin 300452, China
4 College of Chemistry & Environmental Engineering, Yangtze University, Jingzhou 434000, China
* Correspondence: lixx49@cosl.com.cn (X.L.); wangtengfeiforever@126.com (T.W.)

Abstract

To address the issues of easy degradation, dehydration, and insufficient deep plugging strength of traditional pre-crosslinked gel particles (PPGs) in high-temperature and high-salinity reservoirs, this study innovatively introduced amphiphilic carbon dots (CDs) with both hydrophilic and hydrophobic structures as multifunctional modifiers. The carbon dot-reinforced PPGs (CD-PPGs) were successfully prepared through in situ polymerization. Through systematic characterization, microscopic visualization experiments, and macroscopic oil displacement evaluation, the performance enhancement mechanism and profile control behavior were deeply explored. The results show that the amphiphilic carbon dots significantly enhanced the material's temperature resistance (up to 110 °C), salt resistance (up to 15×10^4 mg/L salinity), and mechanical properties by constructing a "hydrogen bond-hydrophobic association" dual crosslinking system within the PPG network. More importantly, it was found that CD-PPGs exhibit a unique "self-aggregation" ability in deep reservoirs, which enables the in situ formation of high-strength plugging micelles at the target location while ensuring excellent injectability. At a permeability range of 539.0–2988.6 mD, the sealing rate of 0.5 PV CD-PPGs was greater than 95%. With permeabilities of 490.1 mD and 3020.5 mD under heterogeneous reservoir simulation conditions, the total recovery degree after the CD-PPGs was displaced was 52.6%, which was 20.5% higher than that of single water flooding. This study not only developed a high-performance profile control nanomaterial but also elucidated its strengthening mechanism, providing new insights and a theoretical basis for advancing deep profile control technology.

Keywords: pre-crosslinked gel particles; amphiphilic carbon dots; self-aggregation; high-temperature and high-salt reservoirs; enhanced oil recovery

1. Introduction

With the continuous growth of the global energy demand and the gradual depletion of conventional oil and gas resources, the efficient development of harsh reservoirs with high temperatures and high salinity has become an important strategic direction for ensuring national energy security. Such reservoirs typically undergo long-term water injection development and enter a high-water-cut period, during which the inherent strong heterogeneity of the reservoir becomes increasingly pronounced [1–3]. The permeability differences formed in the reservoir due to sedimentation, diagenesis, and later tectonic movements cause the injected fluid to circulate ineffectively along high-permeability channels for a long time, seriously restricting the expansion of water flooding sweep volume and leaving a large amount of remaining oil trapped in low-permeability areas, which seriously affects the effect of reservoir fracturing and the final recovery rate. Therefore, achieving flow direction control and improving displacement efficiency in the deep part of the reservoir, i.e., deep profile control and displacement, are crucial to overcoming a development bottleneck in high-temperature and high-salinity reservoirs [4–6].

Among various profile control and displacement agents, pre-crosslinked gel particles (PPGs) have received extensive attention and application due to their advantages, including high gelation reliability, strong shear resistance, and reduced susceptibility to the compatibility of formation minerals and fluids [7–9]. However, in the face of the harsh conditions of many oil fields in China, such as high temperatures (>110 °C) and high salinity (>10 × 10^4 mg/L), traditional PPGs have two major bottleneck problems: (1) Insufficient stability of the body: High temperatures will intensify the thermal motion and chemical degradation of polymer molecules, and high-valent metal ions will severely compress the double electric layer on the surface of PPGs, causing them to dehydrate and shrink and lose their profile control and displacement function. (2) The contradiction between deep plugging strength and injectability: Small-sized particles designed for good injectability often cannot effectively block larger pores and holes when they reach deep formations due to their small size; that is, "injectability" and "plugging strength" are difficult to balance [10–12].

The rise of nanotechnology provides new ideas for solving the above problems. Carbon dots (CDs), as a new type of fluorescent carbon nanomaterial, have shown great potential in energy, biology, and catalysis due to their small size, large specific surface area, rich surface functional groups, and easy modification [13–15]. Inspired by this, this work designs an amphiphilic carbon dot and introduces it as a multifunctional nanomodifier into the synthesis of PPG. We expect that the amphiphilic carbon dots can not only serve as physical crosslinking points to enhance the stability of the PPG network but also, by virtue of their unique hydrophilic–lipophilic balanced structure, endow PPGs with excellent "intelligent behavior"—that is, maintaining a dispersed state during injection to ensure injectability, and upon reaching the target area and remaining there, they can self-agglomerate under the trigger of the external environment (such as temperature), thereby achieving "self-amplification" of the plugging strength at the required location [16–18].

Based on this, this work is dedicated to designing and synthesizing carbon dots with a clear amphiphilic structure and utilizing them as the basis for preparing a new generation of intelligent PPG (CD-PPG). Through comprehensive structural characterization and performance testing of CD-PPG, the intrinsic mechanism of its performance enhancement is clarified. Physical simulation experiments of sand-filled tubes are conducted to quantitatively evaluate the deep profile control and displacement effect, as well as the enhanced oil recovery capacity of the CD-PPG system under simulated reservoir conditions. The results of this work are expected to provide innovative material solutions and a solid theoretical basis for overcoming the technical bottlenecks of enhanced oil recovery in harsh reservoirs (high-temperature and high-salt).

2. Materials and Methods

2.1. Main Reagents and Instruments

2.1.1. Reagents

Reagents include acrylamide (AM, analytical purity), 2-acrylamido-2-methylpropanesulfonic acid (AMPS, analytical purity), N,N′-methylene bisacrylamide (MBA, chemical purity), ammonium persulfate (APS, analytical purity), citric acid (analytical purity), dodecylamine (analytical purity), sodium chloride, calcium chloride (both analytical purity, used for preparing simulated formation water), simulated crude oil (with certain viscosity), anhydrous ethanol, and deionized water.

2.1.2. Instrumentation and Equipment

(1) The following synthesis and processing equipment was used: microwave reactor (for carbon dot synthesis), heat-collecting constant-temperature magnetic stirrer, precision electronic balance, vacuum drying oven, high-speed grinder, and standard inspection sieve.
(2) In addition, the characterization equipment included a Fourier transform infrared spectrometer (FT-IR, VERTEX70), scanning electron microscope (SEM, ZEISS EV0MA15), thermogravimetric analyzer (TGA, STA449 F3), laser particle size analyzer, and advanced rotational rheometer.
(3) Oil displacement experimental equipment: Constant temperature box, sand-filling tube model (stainless steel, length 60 cm, inner diameter 2.5 cm), double plunger constant-speed pump, pressure data acquisition system, liquid collection and measurement device [19,20].

2.2. Preparation of Amphiphilic Carbon Dots (A-CDs)

The improved microwave-assisted solvothermal method was adopted. The procedure is as follows: accurately weigh 2.10 g of citric acid and 2.41 g of dodecylamine in a beaker, add 20 mL of deionized water, and stir with a magnetic stirrer until complete dissolution and clarity. Transfer the mixed solution to the inner lining of the microwave reactor, seal it, and place it in the microwave reactor. Maintain at 180 °C for 30 min. After the reaction, let it cool naturally to room temperature, obtaining a brownish A-CDs crude solution. Subsequently, transfer this crude solution to a pre-treated dialysis bag (with a molecular weight cut-off of 1000 Da) and perform dialysis in deionized water for 48 h to remove unreacted small molecules. Finally, freeze-dry the pure A-CDs solution, obtaining a brownish spongy solid. Grind it and store it in a desiccator for future use [21,22].

2.3. Synthesis and Optimization of CD-PPG

This study employed an integrated synthesis process of aqueous solution polymerization, in situ compounding, and freeze grinding. The specific steps are as follows:

In a 250 mL three-necked flask, 100 mL of deionized water was added, and the monomers were weighed according to a molar ratio of AM to AMPS of 7:3, with the total monomer concentration controlled at 28 wt%. The monomers were completely dissolved under an ice-water bath and continuous stirring. A total of 0.4 wt% of A-CD powder was added to the above solution, and then it was ultrasonicated in an ultrasonic disperser at a power of 400 W for 40 min to form a uniform and stable precursor dispersion solution. An appropriate amount of crosslinking agent MBA (0.05 wt% of the total mass) was added, and the mixture was stirred for another 30 min [23–25]. Subsequently, high-purity nitrogen gas was continuously introduced into the system for 45 min to completely remove dissolved oxygen. The three-necked flask was transferred to a 45 °C constant temperature water bath. After the system temperature was balanced, the initiator APS solution (0.08 wt%

of the total mass) was added all at once. The reaction was carried out under nitrogen protection for 6 h, then the temperature was raised to 60 °C for 2 h to gel, resulting in a colorless, transparent, highly resilient blocky composite gel. The gel block was immersed in liquid nitrogen for deep freeze-crystallization treatment. Immediately after that, it was crushed using a high-speed grinder and screened through a standard inspection sieve to obtain particles of 100–120 mesh. The wet particles were repeatedly washed with a 70% ethanol solution three times to remove residual monomers. Finally, the washed particles were placed in a 50 °C vacuum drying oven to dry to constant weight, resulting in the final CD-PPG product. As a blank control, the ordinary PPGs (denoted as C-PPGs) were synthesized identically without adding A-CDs [26–30].

2.4. Characterization and Testing Methods

(1) SEM observation: After the sample was fully swollen and quenched with liquid nitrogen, it was subjected to a gold-spraying treatment and then observed for its internal microstructure under a 15 kV acceleration voltage.
(2) Thermal stability analysis (TGA): In the presence of a nitrogen gas flow (50 mL/min), the sample was heated at a rate of 10 °C/min from room temperature to 800 °C, and the change in sample mass with temperature was recorded.
(3) Expansion and temperature/alkali resistance performance test: We accurately weighed 0.100 g (denoted as Wd) of the dry CD-PPG and C-PPG, and immersed them, respectively, in different mineralization degrees (fresh water, 5×10^4 mg/L, 10×10^4 mg/L, 15×10^4 mg/L $NaCl/CaCl_2$ mixed salt water (with Ca^{2+} concentration fixed at 500 mg/L)). We took them out regularly, dried the surface free of water with filter paper, and weighed them again (denoted as Ws). The expansion ratio (Q) calculation formula is $Q = (Ws - Wd)/Wd$. At the same time, we placed the fully swollen samples in a constant temperature oven at 95 °C for a long-term aging experiment. We regularly observed and recorded their morphological and weight changes [31].
(4) Texture analyzer analysis: The test was conducted using the CT3 texture analyzer from Brookfield USA. The main body gel blocks were cut into cuboid pieces with dimensions of 100 mm × 80 mm × 20 mm. To ensure uniform force application, the surface of the samples had to be flat, smooth, and free of visible bubbles or defects. The compression and TPA tests were carried out under indoor conditions (room temperature 15 °C, relative humidity 38%).
(5) Particle steady-state viscoelasticity: During the migration process in the formation, the actual shear stress direction that PPGs were subjected to by the porous medium of the formation remained unchanged. The elastic behavior exhibited under such conditions is referred to as the steady-state viscoelasticity of PPGs. The steady-state viscoelasticity of particles can reflect the actual viscoelasticity of the viscoelastic particle system during the migration process in the formation. Steady-state rate scanning experiments were conducted using a rotational rheometer to determine the elastic behavior of viscoelastic particles during the steady-state shear process. The main indicators used for characterization were the normal stress difference (N1) and the Wiesenberg number. The concentration of viscoelastic particles was 0.2%, and the test temperature was 85 °C [32,33].

The magnitude of the normal stress difference is the main manifestation of the elastic behavior of viscous elastic fluids during flow. The normal stress difference was divided into the first normal stress difference and the second normal stress difference. Due to the very small second normal stress difference in the oil-displacing viscoelastic particle solution, the current evaluation of the elasticity of the viscoelastic particle steady flow under the oil-displacing condition mainly relies on the first normal stress difference. The description of the Wiesenberg effect of the viscoelastic particle solution can be expressed by the Wiesenberg number, which reflects the relative size of the solution's elasticity. When We is large, the flow characteristics are mainly determined by the first normal stress difference, that is, elasticity plays a dominant role; when We is small, the flow characteristics are mainly determined by the viscous force. Thus, through We, the roles of elasticity and viscosity in the viscous elastic fluid during the flow process can be known [34,35].

2.5. Plugging Performance Test

Plugging experiments were conducted using a single sand-filling pipe model to evaluate the injection performance and deep-seated sealing performance of PPGs. The length of the sand-filling pipe used in the experiment was 0.6 m, and its diameter was 2.5 cm. The basic parameters are shown in Table 1. The simulated formation water used in the experiment had a salinity of 150,000 mg/L and a temperature of 110 °C. The sand-filling pipe was saturated with water, and the permeability was measured. Then, 0.5 PV of the dispersion liquid of C-PPG and CD-PPG with a mass fraction of 0.2 wt% was injected. Every 0.05 PV, the pressure change at the injection end of the sand-filling pipe and the production volume at the outlet end were recorded, and the pressure gradient, resistance factor, and sealing rate were calculated. The experiment was stopped after the pressure stabilized following subsequent water injection [36,37].

Table 1. Basic parameters of single sand-filling pipe model.

Gel Particles	Foam Agent Mass Fraction/wt%	Injection Rate/(mL/min)	Porosity/%	Permeability/mD
C-PPG	0.2	1	30.8	1073.1
		1	33.2	3014.6
CD-PPG	0.2	1	30.7	1009.3
		1	33.6	2988.6

2.6. Water Control Flooding Experiment for Oil Recovery

The CD-PPG water control flooding effect was evaluated using a dual-tube sand-filling model (to simulate heterogeneous reservoir conditions). Sand-filled tubes with different permeabilities were connected in parallel. The injection rate was set at 1 mL/min. After water flooding for 0.9 PV, 0.5 PV of the CD-PPG dispersion liquid (with a mass fraction of 0.2%) was injected. Then, 0.8 PV of secondary water flooding was carried out. The water production rate, oil production rate, and pressure changes at the outlet end were recorded to evaluate the oil displacement effect of CD-PPGs under the simulated heterogeneous reservoir conditions. The basic parameters of the dual-sand-filling tube model are shown in Table 2 [26,38].

Table 2. T Basic parameters of double sand-filling pipe models.

Reservoir Model	Porosity/%	Permeability/mD
High-permeability tube	35.6	3020.5
Low-permeability tube	28.9	490.1

3. Experimental Results and Discussion

3.1. Characterization of Amphiphilic Carbon Dots (A-CDs)

The transmission electron microscope (TEM) results show that the synthesized A-CDs are nearly spherical in shape, with good monodispersity, and the average particle size is approximately 2.3 nm [15,39]. A transmission electron microscope image of A-CDs is shown in Figure 1.

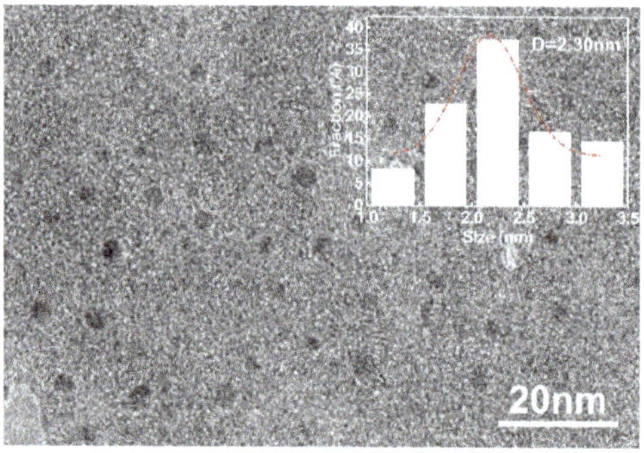

Figure 1. Transmission electron microscope image of A-CDs.

The FT-IR spectrum confirmed that the surface of the A-CDs simultaneously contained hydrophilic functional groups (such as -OH/-COOH) and hydrophobic long-chain alkanes, successfully achieving the expected amphiphilic structure [34]. The FT-IR spectrum of the A-CDs is shown in Figure 2.

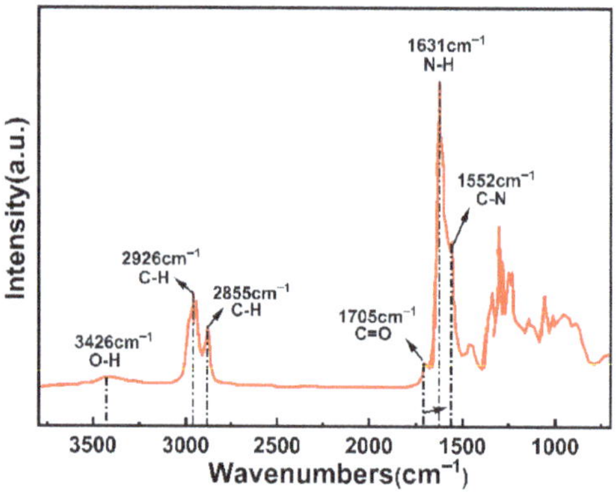

Figure 2. The FT-IR spectrum of the A-CDs.

3.2. Structural Characterization and Formation Mechanism of CD-PPGs

The SEM and microscopic morphology analysis results of CD-PPGs are shown in Figure 3. The SEM images provide compelling morphological evidence. C-PPGs show a relatively loose and non-uniform three-dimensional network structure. CD-PPGs exhibit a significantly denser and more uniform honeycomb-like or porous structure [11,40].

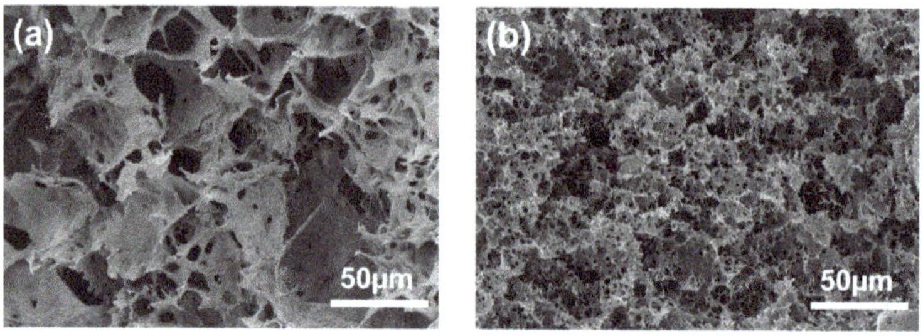

Figure 3. The results of SEM and microscopic morphology analysis.

Mechanism explanation: A-CDs play multiple crucial roles in the polymer system: ① Nano-enhancing core. The solid core of A-CDs can serve as an ideal nano-filler, directly enhancing the polymer matrix. ② Multifunctional physical crosslinking center. Hydrogen bond crosslinking: The numerous functional groups, such as carboxyl and amino groups, on the surface of A-CDs form a dense hydrogen bond network with the amide and sulfonic acid groups on the P(AM-AMPS) molecular chains. Hydrophobic association crosslinking: The hydrophobic termini of A-CDs and the hydrophobic microregions induced by them can generate strong hydrophobic association interactions. The synergistic effect of these two mechanisms jointly constructs a robust and stable "dual crosslinking network". This structure is the primary reason for the excellent temperature resistance, salt resistance, and mechanical properties of CD-PPGs [41].

3.3. *Systematic Evaluation of Thermal Resistance, Salt Resistance, and Mechanical Properties*

3.3.1. Thermal Stability (TGA)

The TGA curve clearly shows that the initial thermal decomposition temperature of the C-PPGs is 274.1 °C. The initial decomposition temperature of the CD-PPGs is as high as 315.2 °C, and the residual carbon content at 600 °C is also significantly higher than that of the C-PPGs. This incontrovertibly proves that the introduction of A-CDs greatly improved the thermal stability of PPGs [14]. The TGA test results of C-PPGs and CD-PPGs is shown in Figure 4.

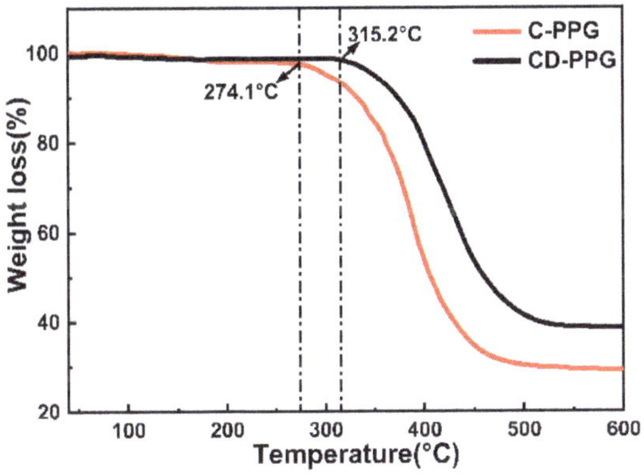

Figure 4. The TGA test graphs of C-PPGs and CD-PPGs.

3.3.2. Salt-Tolerance and Temperature Tolerance Properties

(1) Salt-tolerance at room temperature

In freshwater, the equilibrium expansion multiples of C-PPGs and CD-PPGs are similar (approximately 10 times). However, in 10 × 10^4 mg/L of highly mineralized salt water, the expansion multiple of C-PPGs drops sharply to approximately 4 times, showing severe salt sensitivity. In contrast, CD-PPGs can still maintain an expansion capacity of about 8 times in the same salt water, demonstrating excellent salt resistance. The evaluation results of the normal-temperature salt resistance of C-PPGs and CD-PPGs are shown in Figure 5 [19].

(2) Long-term stability at high temperatures

The samples were placed in an extreme environment of 110 °C and a mineralization degree of 15 × 10^4 mg/L for a 14-day aging test. The C-PPGs showed significant shrinkage, fragmentation, and even dissolution. However, after 3 days of aging, the CD-PPGs began to aggregate into particle clusters, forming viscoelastic gels. The formed gels have excellent viscoelastic properties and can enhance the sealing ability of the particle system in porous media. This "self-aggregation" ability of CD-PPGs can, under the premise of ensuring excellent injectability, form high-strength sealing clusters in the target location in situ, effectively resolving the contradiction between injectability and sealing ability of the particle-based sealing agent [42].

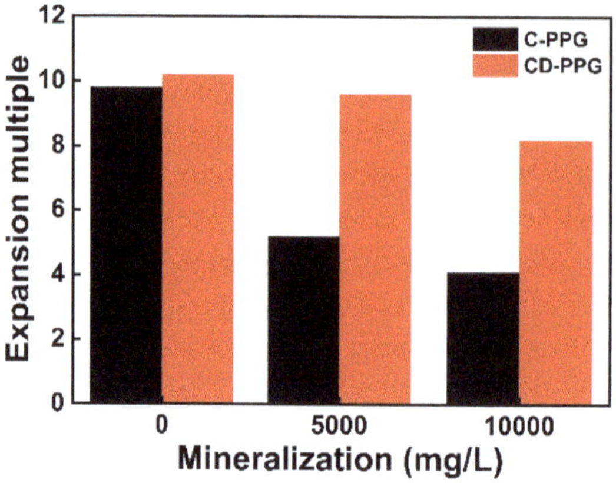

Figure 5. The evaluation results of the normal-temperature salt resistance of C-PPGs and CD-PPGs.

The evaluation results of the long-term stability under high temperatures of C-PPGs and CD-PPGs are shown in Figure 6.

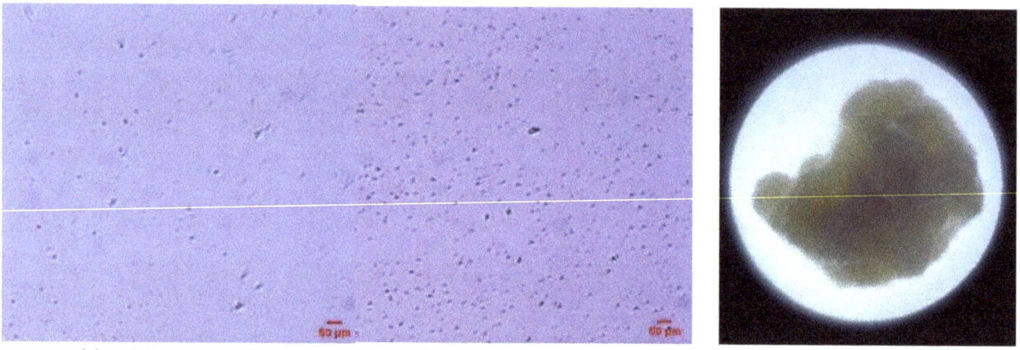

(**a**) C-PPGs aged for 7 days (**b**) CD-PPGs aged for 7 days (**c**) CD-PPGs self-agglomerate to form micelles

Figure 6. The evaluation results of long-term stability under high temperatures of C-PPGs and CD-PPGs.

3.3.3. Mechanical Properties of the Matrix Adhesive

The mechanical properties of the bulk gel blocks of two types of particles were measured using a texture analyzer. The results are shown in Table 3. Since the addition of the A-CD cores can effectively dissipate mechanical energy, the fracture stress, elastic modulus, and toughness of the gel structure of the CD-PPG bulk gel significantly increased [33].

Table 3. The influence of A-CDs on the compression performance of the matrix adhesive.

Type of the Main Body Adhesive Block	Fracture Strain/%	Breaking Stress/MPa	Elasticity Modulus/MPa	Tenacity/(MJ·m^{-3})
CD-PPG	79.37	11.91	14.23	10.76
C-PPG	64.42	2.66	11.78	5.46

3.3.4. CD-PPG Steady-State Viscoelasticity

Based on the comparison experiment results of the first normal stress difference and the Wenzinger number of two types of particles, it can be seen that both the N1 representing the elasticity of CD-PPGs and the Wenzinger number We, reflecting the relative size of the solution's elasticity, have significantly increased compared to C-PPGs. This indicates that CD-PPGs have higher elasticity during the formation movement process and have a stronger sealing ability compared to conventional viscoelastic particles [18,30]. The influence of A-CDs on the compression performance of the matrix adhesive is shown in Table 4.

Table 4. The comparison results of the N_1 and the We for the two types of particles.

Particle Type	N1/kPa	We
CD-PPG	132	0.51
C-PPG	35	0.22

3.4. Comprehensive Evaluation of Blocking and Oil Expulsion Effects

3.4.1. Plugging Performance

The relationship curves of pressure gradient, resistance factor, and injection volume for the two types of particles under different permeabilities are shown in Figure 7. As can be seen from Figure 7a, under the simulated reservoir conditions with permeabilities of 1073.1 mD and 3014.6 mD, after injecting 0.5 PV of C-PPGs with a mass fraction of 0.2 wt%, the breakthrough pressure gradient and resistance factor were 1.89 MPa/m and 0.19 MPa/m and 58.93 and 16.96. In contrast, after injecting 0.5 PV of CD-PPGs with a mass fraction of 0.2 wt%, the breakthrough pressure gradient and resistance factor were higher, reaching 2.86 MPa/m and 0.80 MPa/m and 85.43 and 68.32. After injecting CD-PPGs, it was found that the resistance factor increased significantly after water flooding was initiated. CD-PPGs migrated towards the deeper part of the reservoir through compression deformation and achieved secondary sealing, and their anti-erosion stability was very good [28,42].

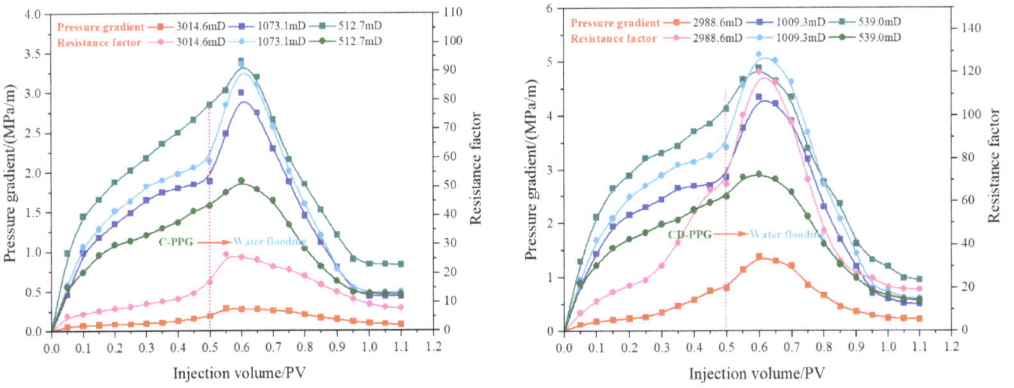

(**a**) C-PPGs (**b**) CD-PPGs

Figure 7. Pressure gradient and resistance factor curves with injection volume under varying permeabilities.

Under the simulation reservoir conditions of permeabilities of 3014.6 mD and 2988.6 mD, after injecting CD-PPGs with a mass fraction of 0.2 wt% and 0.5 PV, the sealing rate reached over 95%. The sealing performance met the production adjustment and increase requirements for high water cut blocks in offshore oil fields [34,40].

Pre-crosslinked gel particles form a stable three-dimensional network during the synthesis stage under the combined effect of surface crosslinking agents and internal network structures. This design enables them to mainly undergo elastic deformation rather than structural rupture when subjected to high shear pumping in the wellbore and near-wellbore regions, thereby avoiding the risk of near-wellbore blockage caused by excessive crosslinking or shear degradation during injection. When the particles migrate to the target deep reservoir area, they moderately hydrate/soften on the surface under the influence of high temperatures, electrolytes, and rock surface affinity. Relying on their inherent viscoelasticity and deformation ability, they adaptively deform and undergo limited volume expansion at pore throats. This characteristic of "rigid transportation and flexible plugging" enables them to form effective blockages in deep, high-permeability channels through bridging, accumulation, and physical adsorption mechanisms, thereby dynamically adjusting the flow direction of subsequent fluids.

3.4.2. Water Control and Oil Recovery Effect

The curves showing the relationship between injection pressure, production degree, water content, and injection volume in the high- and low-permeability tubes during the oil displacement process are, respectively, shown in Figures 8 and 9. The oil displacement experiment indicates that, in the single water flooding process, the crude oil mainly flows out through the high-permeability tubes, and the production degree of the high-permeability tubes is 8 times higher than that of the low-permeability tubes. After the single water flooding process, the production degrees of the high- and low-permeability tubes were 51.3% and 6.2%, respectively, and the total production degree was 32.1%. The production degree of the low-permeability tube was lower. After injecting CD-PPGs, the production degrees of the high- and low-permeability tubes were 68.9% and 30.7%, respectively, which increased by 17.6% and 24.5%, respectively. After CD-PPG displacement, the total production degree was 52.6%, which was 20.5% higher than that of the single water flooding. After injecting CD-PPGs, the particle-blocking agent could effectively seal the high-permeability channels through bridging and aggregation in the pores. In the subsequent water flooding, the remaining oil in the low-permeability area could be effectively utilized, and the sweep efficiency of the low-permeability area was increased by changing the liquid flow direction [43,44].

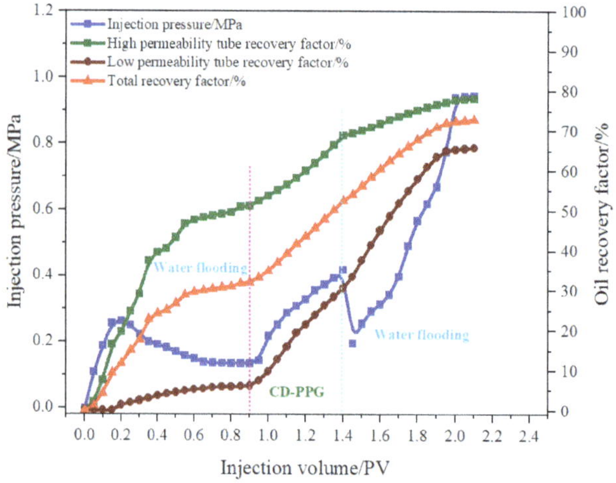

Figure 8. The oil recovery factor and injection pressure curves with injection volume at varying flooding stages.

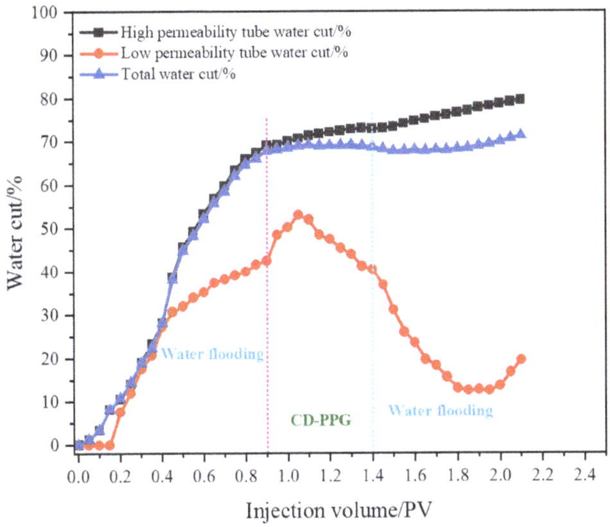

Figure 9. The water cut with injection volume at varying flooding stages.

From Figures 8 and 9, it can be seen that during the first water flooding stage, the water content of the high-permeability wells significantly increased, reaching up to 69.15%, while the water content of the low-permeability wells rose to 41.54%. During this stage, the high-permeability wells had a high production rate, while water migration was severe, whereas the low-permeability wells had a low utilization rate. During the CD-PPG displacement stage, the increase in water content in the high-permeability wells slowed down significantly, while the water content in the low-permeability wells first rose and then dropped. The CD-PPGs entered the flowing channel and preferentially blocked the high-permeability pores and throats [39]. Therefore, during the early stage of gel particle displacement, the water content of the low-permeability channels continued to rise. After the gel particles entered the hypotonic tube, the water content dropped sharply, and the degree of crude oil utilization significantly increased. After the high-permeability pores and throats were blocked, the gel particle displacement stage could more effectively activate the low-permeability well's crude oil, resulting in an increase in crude oil content in the produced fluid and a decrease in water content. During the secondary water flooding stage, the gel particles migrated towards the deeper part of the reservoir through compression deformation. After achieving secondary sealing, the water saturation of the high-permeability zones showed a relatively small increase, while the water saturation of the low-permeability zones first decreased and then rose. Compared with the hypertonic pipe, the low-hypertonic pipe had a smaller amount of gel particles. After long-term flushing, some of the gel particles were extracted, resulting in a reduced effect of profile adjustment and water control. Therefore, after an injection volume of 2.0 PV, the water content of the low-hypertonic pipe increased. CD-PPGs could effectively control water production and enhance the oil recovery rate under heterogeneous reservoir conditions [23,37].

This study designs the moderate crosslinking degree and elastic modulus of particles, making their particle size distribution slightly larger than the mainstream pore throat size in a deep target reservoir. This design has the following rationale: (a) Particles are not rigid spheres. Good deformability allows them to pass through throats slightly smaller than their particle size by 'squeezing deformation' and to partially restore their shape after passing through. (b) Deep plugging mainly occurs at abrupt changes in pore-throat structure or by the bridging and accumulation of multiple particles in larger pore spaces. This kind of plugging is permeability-dependent; that is, the high-permeability channels that fluids preferentially enter due to their fast flow rate and large throats are more likely to capture and accumulate particles, thereby forming stronger flow resistance and achieving fluid flow diversion. In contrast, low-permeability areas, with their small throats, are less likely to have particles entering or passing through quickly and thus are less affected. Therefore, the seemingly contradictory concepts of "larger-sized particles" and "deep migration plugging" are unified through the deformability of particles and the heterogeneity of formation permeability.

4. Conclusions

In this work, amphiphilic carbon dots (A-CDs) were introduced into the synthesis system of pre-crosslinked gel particles (PPGs), successfully developing a CD-PPG system with self-aggregation characteristics. The following main conclusions were achieved:

(1) A novel PPG based on a dual crosslinking network of "hydrogen bonding—hydrophobic association" was successfully developed, which significantly improved the temperature and salt resistance as well as the mechanical stability of the particle plugging agent. At 110 °C and a salinity of 15×10^4 mg/L, CD-PPGs began to aggregate after aging for 3 days, forming viscoelastic gel clusters, effectively enhancing the plugging capacity of the particle system in porous media.

(2) Under simulated heterogeneous reservoir conditions with permeabilities ranging from 539.0 to 2988.6 mD, the plugging efficiency of 0.5 PV CD-PPGs was above 95%. The gel particles could achieve secondary plugging by migrating deep into the reservoir under pressure deformation, and they had good anti-erosion stability.

(3) Under simulated heterogeneous reservoir conditions with permeabilities of 490.1 mD and 3020.5 mD, the total recovery after displacement by CD-PPGs was 52.6%, which was 20.5% higher than that of primary water flooding. During secondary water flooding, the expansion of gel particles in the pores could effectively plug high-permeability channels, increase the sweep efficiency, and, through fluid flow diversion, enhance the oil recovery efficiency in low-permeability zones.

Author Contributions: Conceptualization, G.X. and T.W.; methodology, X.L.; validation, J.Y. and X.W.; formal analysis, C.T.; investigation, C.T. and T.W.; resources, X.W.; data curation, G.X.; writing—original draft preparation, T.W.; writing—review and editing, J.Y. and X.L. All authors have read and agreed to the published version of the manuscript.

Funding: This research was funded by Tianjin Key Laboratory of Offshore Difficult to Recovery Reserves Exploitation Open Research Project (202516411500).

Data Availability Statement: Data are contained within the article.

Acknowledgments: The authors would like to extend their heartfelt thanks to Yangtze University for the invaluable educational foundation and academic support that have significantly contributed to the development of this work, and express their sincere gratitude to China Oilfield Service Limited for their material support and for providing access to laboratory facilities that were essential for the successful completion of this study. Additionally, the authors also express their sincere appreciation to the anonymous reviewers for their valuable and constructive comments.

Conflicts of Interest: Authors Guorui Xu, Xiaoxiao Li, Jinzhou Yang, Chunyu Tong, and Xiaolong Wang were employed by the China Oilfield Service Limited. The remaining authors declare that the research was conducted in the absence of any commercial or financial relationships that could be construed as a potential conflict of interest.

References

1. Song, T.; Bai, B.J.; Huang, R.; Zhang, S.; Liu, P.D.; Eriyagama, Y.; Tian, X.Y.; Ahdaya, M.; Schuman, T. Development and evaluation of lysine-crosslinked re-crosslinkable particle gel for water control in high-temperature reservoirs. *J. Mol. Liq.* **2024**, *407*, 125133. [CrossRef]
2. Li, M.; Gong, W.H.; Wang, F.; Li, L.; Nie, Y.H.; Chen, Q.B. A novel carbon dot-based degradable preformed particle gel for sensitive detection of iron ions. *Colloids Surf. A* **2025**, *711*, 136414. [CrossRef]
3. Lai, N.J.; Chen, S.F.; Tang, L.; Huang, Y.J.; Xu, H.W. Migration characteristics and profile control capabilities of preformed particle gel in porous media. *Petroleum* **2022**, *8*, 483–498. [CrossRef]
4. Zhu, D.Y.; Zhang, H.J.; Qin, J.H.; Zhao, Q.; Wang, G.Q.; Su, Z.H.; Shi, C.Y. Experimental evaluation of temporary plugging performance and mechanism analysis of degradable preformed particle gel based on NMR technology. *Geoenergy Sci. Eng.* **2025**, *244*, 213480. [CrossRef]
5. Rezaei, A.; Khodabakhshi, A.; Esmaeili, A.; Razavifar, M. Effects of initial wettability and different surfactant-silica nanoparticles flooding scenarios on oil-recovery from carbonate rocks. *Petroleum* **2022**, *8*, 499–508. [CrossRef]
6. Du, L.; Xiao, Y.Y.; Jiang, Z.C.; Zeng, H.B.; Li, H.Z. High-temperature stable dispersed particle gel for enhanced profile control in carbon capture, utilization, and storage (CCUS) applications. *Engineering* **2025**, *48*, 128–140. [CrossRef]
7. Li, W.Z.; Luo, P.Y.; Wang, X.D. Preparation and performance evaluation of graphene oxide-based self-healing gel for lost circulation control. *Polymers* **2025**, *17*, 1999. [CrossRef]
8. Jin, F.Y.; Jiang, T.T.; Varfolomeev, M.A.; Yuan, C.D.; Zhao, H.Y.; He, L.; Jiao, B.L.; Du, D.J.; Xie, Q. Novel preformed gel particles with controllable density and its implications for EOR in fractured-vuggy carbonated reservoirs. *J. Pet. Sci. Eng.* **2021**, *205*, 108903. [CrossRef]
9. Li, M.; Liu, J.; Xia, Y. Risk prediction of gas hydrate formation in the wellbore and subsea gathering system of deep-water turbidite reservoirs: Case analysis from the South China Sea. *Reserv. Sci.* **2025**, *1*, 52–72. [CrossRef]
10. Wu, W.P.; Hou, J.R.; Qu, M.; Yang, Y.L.; Zhang, W.; Wu, W.M.; Wen, Y.C.; Liang, T.; Xiao, L.X. A novel polymer gel with high-temperature and high-salinity resistance for conformance control in carbonate reservoirs. *Pet. Sci.* **2022**, *19*, 3159–3170. [CrossRef]
11. Wang, L.L.; Wang, T.F.; Wang, J.X.; Tian, H.T.; Chen, Y.; Song, W. Enhanced oil recovery mechanism and technical boundary of gel foam profile control system for heterogeneous reservoirs in Changqing. *Gels* **2022**, *8*, 371. [CrossRef] [PubMed]
12. Pu, J.Y.; Johnston, K.P.; Wu, P.; Ahmad, M.; Luo, M.L.; Zhang, N.; He, J.T. Facile synthesis of chromium chloride/poly(methyl methacrylate) core/shell nanocapsules by inverse miniemulsion evaporation method and application as delayed crosslinker in secondary oil recovery. *Pet. Sci.* **2023**, *20*, 396–406. [CrossRef]
13. Kurnia, I.; Fatchurrozi, M.; Anwary, R.; Zhang, G. Lessons learned from coreflood experiments with surfactant-polymer and alkali-surfactant-polymer for enhanced oil recovery. *Petroleum* **2023**, *9*, 487–498. [CrossRef]
14. Liu, Q.; Zhou, K.M.; Liu, Y.J.; Zhang, Y.F.; Chen, W.; Tang, S. Exploring the potential applications of amphiphilic carbon dots based nanocomposite hydrogel in liquid chromatographic separations. *Anal. Chim. Acta* **2024**, *1299*, 342445. [CrossRef]
15. Gao, P.L.; Xie, Z.G.; Zheng, M. Small nanoparticles bring big prospect: The synthesis, modification, photoluminescence and sensing applications of carbon dots. *Chin. Chem. Lett.* **2022**, *33*, 1659–1672. [CrossRef]
16. Cai, M.X.; Lai, L.; Zhao, M.W. Synthesis, properties, and applications of amphiphilic carbon dots: A review. *Carbon* **2025**, *233*, 119921. [CrossRef]
17. Xiao, C.Q.; Wu, J.; Lai, L. Demicellization thermodynamics of amphiphilic carbon dots via isothermal titration calorimetry. *Thermochim. Acta* **2025**, *745*, 179935. [CrossRef]
18. Cao, L.; Lv, M.; Li, C.; Sun, Q.; Wu, M.; Xu, C.; Dou, J. Effects of crosslinking agents and reservoir conditions on the propagation of fractures in coal reservoirs during hydraulic fracturing. *Reserv. Sci.* **2025**, *1*, 36–51. [CrossRef]
19. Zhao, M.W.; Yang, X.Y.; Ma, Z.F.; Li, Y.; Meng, X.J.; Dai, C.L.; Wang, Y.N. A novel nanosurfactant for EOR: Amphiphilic carbon quantum dots (ACDs). *Langmuir* **2025**, *41*, 24004–24015. [CrossRef] [PubMed]
20. Sazzad Parveg, A.S.M.; Zhou, Y.Q.; Leblanc, R.M.; Ratner, A. Effects of gel-like carbon dots (G-CDs) and surfactant (Span80) on the droplet combustion dynamics of liquid fuels. *Fuel* **2025**, *381*, 133385. [CrossRef]
21. Zhai, Y.Y.; Fang, Z.; Feng, J.; Sun, C.J.; Deng, W.J.; Wen, Y.B. Development and performance evaluation of re-crosslinkable preformed particle gel under high temperature and high salt conditions. *Polymer* **2024**, *296*, 126824. [CrossRef]
22. Yang, C.H.; Xiao, D.; Wang, J.; Liang, T. Evaluation and experiment of high-strength temperature- and salt-resistant gel system. *Gels* **2025**, *11*, 669. [CrossRef]
23. Amraeiniya, A.; Shojaei, S.; Mohseni, A.; Mahani, B.; Saatchi, S.; Bahari, A.; Sisakht, S. Experimental study on nanoparticles-assisted low-salinity water for enhanced oil recovery in asphaltenic oil reservoirs. *Petroleum* **2023**, *9*, 395–402. [CrossRef]
24. Liang, L.; Li, J.L.; Li, D.X.; Xu, J.; Zheng, B.; Zhao, J.K. Urea delays high-temperature crosslinking of polyacrylamide for in situ preparation of an organic/inorganic composite gel. *Gels* **2025**, *11*, 256. [CrossRef]

25. Zhou, G.W.; Zhang, X.; Yan, W.J.; Qiu, Z.S. Synthesis, characteristics, and field applications of high-temperature and salt-resistant polymer gel tackifier. *Gels* **2025**, *11*, 378. [CrossRef]
26. Jia, H.; Chen, H.; Zhao, J.Z. Development of a highly elastic composite gel through novel intercalated crosslinking method for wellbore temporary plugging in high-temperature reservoirs. *SPE J.* **2020**, *25*, 2853–2866. [CrossRef]
27. Xin, H.; Fang, B.; Yu, L.Y.; Lu, Y.J.; Xu, K.; Li, K.J. Rheological performance of high-temperature-resistant, salt-resistant fracturing fluid gel based on organic-zirconium-crosslinked HPAM. *Gels* **2023**, *9*, 151. [CrossRef] [PubMed]
28. Bai, Y.Y.; Wen, Y.; Pu, W.F.; Fang, W.; Jin, X. Preparation and self-growth characteristics of in situ self-growing dispersed gel system for offshore large-well-distance reservoirs. *ACS Omega* **2025**, *10*, 40450–40459. [CrossRef] [PubMed]
29. Zhang, J.; Liang, D.; Wang, L.Q.; Tang, E.G.; Li, X.J.; Zhao, W.S. Application of horizontal well discontinuous chemical flooding technology in Bohai oilfield. *Front. Energy Res.* **2023**, *10*, 1036835.
30. Xin, X.K.; Liu, Q.; Liu, S.J.; Yu, G.M.; Wan, Q.S. Optimization of gel flooding during the high water cut stage in a conglomerate reservoir of the Xinjiang A Oilfield. *Polymers* **2023**, *15*, 1809. [CrossRef]
31. Ge, L.Z.; Chen, X.M.; Wang, G.; Zhang, G.H.; Li, J.Y.; Liu, Y.; Xiao, L.X.; Wen, Y.C.; Yuan, W.F.; Qu, M.; et al. Analysis of the distribution pattern of remaining oil and development potential after weak gel flooding in the offshore LD Oilfield. *Gels* **2024**, *10*, 236. [CrossRef]
32. Wang, F.; Kobina, F. The Influence of geological factors and transmission fluids on the exploitation of reservoir geothermal resources: Factor discussion and mechanism analysis. *Reserv. Sci.* **2025**, *1*, 3–18. [CrossRef]
33. Xie, K.; Su, C.; Liu, C.L.; Cao, W.J.; He, X.; Ding, H.N.; Mei, J.; Yan, K.; Cheng, Q.; Lu, X.G. Synthesis and performance evaluation of an organic/inorganic composite gel plugging system for offshore oilfields. *ACS Omega* **2022**, *7*, 12870–12878. [CrossRef]
34. Gallego, J.F.; Medina, O.E.; Tengono, J.A.; Mazo, C.; Maya, A.F.; Caro, C.; Cortés, F.B.; Franco, C.A. Demulsification of Water-in-oil emulsion with carbon quantum dot (CQD)-enhanced demulsifier. *Processes* **2025**, *13*, 575. [CrossRef]
35. Qu, J.; Wang, P.; You, Q.; Zhao, G.; Sun, Y.; Liu, Y. Soft movable polymer gel for controlling water coning of horizontal well in offshore heavy oil cold production. *Gels* **2022**, *8*, 352. [CrossRef] [PubMed]
36. Gao, J.; Wang, X.; Zhang, D.; Wang, J. The formation mechanism of residual oil and methods of enhanced oil recovery in a fractured low-permeability metamorphic rock reservoir in Bohai Bay. *Processes* **2024**, *12*, 2281. [CrossRef]
37. Li, Y.; Liu, C.; Zhang, Y.; Xue, B.; Lv, J.; Miao, C.; Li, Y.; Liu, Z. Experimental investigation on the application of polymer agents in offshore sandstone reservoirs: Optimization design for enhanced oil recovery. *Polymers* **2025**, *17*, 673. [CrossRef]
38. Lai, N.; Wang, D.; Wang, J.; Tang, L. Synthesis, optimization, and characterization of fluorescent particle preformed gel. *Energies* **2022**, *15*, 8712. [CrossRef]
39. Rahman, A.; Torabi, F.; Shirif, E. Surfactant and nanoparticle synergy: Towards improved foam stability. *Petroleum* **2023**, *9*, 255–264. [CrossRef]
40. Zhang, L.; Gao, Z.; Liu, Y.; Li, Y.; Zhou, K.; Wang, P.; Zhang, K.; Wang, C.; Zhang, M. Study on rapid screening method for different chemical flooding methods in heavy-oil reservoirs. *Processes* **2025**, *13*, 2992. [CrossRef]
41. Wang, J.; Wang, R.; Liu, P.; Xu, H. Gel-Forming and plugging performance evaluation of emulsion polymer crosslinking system in fractured carbonate rock. *Processes* **2022**, *10*, 1550. [CrossRef]
42. Tang, E.; Zhang, J.; Jin, Y.; Li, L.; Xia, A.; Zhu, B.; Sun, X. Optimization of discontinuous polymer flooding processes for offshore oilfields using a novel PSO–ICA algorithm. *Energies* **2024**, *17*, 1971. [CrossRef]
43. Shen, M.; Qi, L.; Zhang, L.; Ma, P.; Liu, W.; Ngai, T.; Jiang, H. Dual-responsive hybrid microgels enabling phase inversion in pickering emulsions. *Polymers* **2025**, *17*, 2762. [CrossRef] [PubMed]
44. Wang, R.; Wu, H.; Liu, Z.; Wang, H.; Zhang, Y. Development of high-performance composite cementitious materials for offshore engineering applications. *Materials* **2025**, *18*, 3324. [CrossRef]

processes

MDPI

Article

Biochar-Enhanced Inorganic Gel for Water Plugging in High-Temperature and High-Salinity Fracture-Vuggy Reservoirs

Shiwei He [1] and Tengfei Wang [2,*]

1 No.2 Oil Production Plant, Northwest Oilfield Company, SINOPEC, Luntai 841604, China; hesw@sinopec.com
2 College of Chemistry & Environmental Engineering, Yangtze University, Jingzhou 434000, China
* Correspondence: wangtengfeiforever@126.com

Abstract

With the expansion of global oil and gas resource exploration and development into deep and ultra deep layers, the efficient development of deep carbonate rock fracture cave reservoirs has become the key to ensuring energy security. However, this type of reservoir commonly faces high temperatures, high salinity, and extremely strong heterogeneity, leading to increasingly severe water content spikes caused by dominant water flow channels. Although the existing traditional inorganic plugging agent has good temperature resistance, it has the defects of great brittleness and easy cracking, while the organic polymer gel is prone to degradation failure under high temperature and high salt environments. In order to solve the above problems, a new biochar-enhanced inorganic composite gel system was constructed by using biochar prepared from agricultural and forestry waste pyrolysis as a functional enhancement component. Through rheological testing, high-temperature and high-pressure mechanical experiments, long-term thermal stability evaluation, and dynamic sealing experiments of fractured rock cores, the reinforcement and toughening laws and rheological control mechanisms of biochar on inorganic matrices were systematically studied. Research has found that a biochar content of 0.5 wt% can significantly improve the micro pore structure of the matrix. By utilizing its micro aggregate filling effect and interfacial chemical bonding, the compressive strength of the solidified body can be increased to over 2 MPa, and there is no significant decline in strength after aging at 130 °C for 30 days. More importantly, the unique "adsorption slow-release" mechanism of biochar effectively stabilizes the hydration reaction kinetics at high temperatures, extending the solidification time of the system to 15 h and solving the problem of flash condensation in deep well pumping. This system exhibits excellent shear thinning characteristics and crack sealing ability, and presents a unique "yield reconstruction" toughness sealing feature. This study elucidates the multidimensional strengthening mechanism of biochar in inorganic cementitious materials, providing technical reference for stable oil and water control in deep fractured reservoirs.

Keywords: deep oil reservoir; fracture cave type carbonate rock; water blocking and profile control; biochar; inorganic gel; microscopic mechanism

1. Introduction

With the increasing depletion of shallow oil and gas resources, the focus of global oil and gas exploration and development is accelerating towards deep (4500–6000 m) and ultra deep (>6000 m) layers [1]. The Tarim Basin and Sichuan Basin in northwest and southwest China are rich in oil and gas resources of deep carbonate rocks. However, the geological conditions of such reservoirs are extremely complex, and their prominent characteristics are the development of fracture and karst cave systems, with strong heterogeneity [2]. During the process of water injection development, the injected water is prone to form dominant flow channels along highly permeable cracks or caves, leading to a rapid increase in water content in oil wells in a short period of time, with some well areas even having water content exceeding 90% [3]. This not only causes a huge waste of injected water resources, but also seriously restricts the improvement of crude oil recovery rate and significantly increases the environmental pressure on produced water treatment [4]. Therefore, efficient sealing of cracks and caves, cutting off the dominant channels of water flow, and forcing liquid flow to turn towards oil containing fractures and caves are the core key technologies for achieving stable production and increased oil production in such reservoirs [5].

Although water plugging technology has made significant progress in the past few decades, the special "high temperature, high salt, large-scale fracture cavity" environment of deep fracture cavity reservoirs has put forward extremely demanding requirements for plugging materials [6]. At present, the main water blocking materials used in the industrial sector are divided into two systems: organic polymers and inorganic particles [7], but they have obvious limitations under extreme working conditions in deep wells. Organic plugging agents mainly include polyacrylamide gel, phenolic resin gel, etc. [8]. These materials have good viscoelasticity and injectivity, and perform well in the adjustment of heterogeneity [9]. However, the thermal stability of organic polymer materials is their fatal weakness. Relevant research shows that when the bottom hole temperature is more than 120 °C and the salinity is more than 20×10^4 mg/L, the polymer molecular chain is prone to serious hydrolysis, thermal degradation or oxidative degradation, leading to dehydration and shrinkage of gel structure, significant attenuation of plugging strength in a short time, and it is difficult to achieve long-term plugging [10,11]. In addition, commonly used organic crosslinking agents (such as chromium and phenolic) have potential environmental toxicity, and their application is increasingly restricted with the increasingly strict environmental regulations.

In contrast, inorganic plugging agents such as oil well cement, ultra-fine cement, fly ash, water glass, and phosphate cement have natural advantages such as wide raw material sources [12], low cost, high temperature and corrosion resistance, and play an irreplaceable role in high-temperature deep wells. However, traditional inorganic cementitious materials are essentially brittle materials with low flexural strength and poor toughness [13]. Under the intense alternating stress or continuous flushing of high-speed fluids in deep oil reservoirs, rigid plugs are prone to brittle fracture or detachment within fractures [14], leading to the reopening of water breakthrough channels. In addition, the hydration reaction rate of inorganic materials at high temperatures is extremely fast [15], and the thickening time is difficult to accurately control, making it prone to "flash condensation" accidents during pumping, resulting in catastrophic consequences such as pipe blockage [16]. Therefore, how to overcome the "brittle defects" of inorganic materials and achieve "precise control" of curing time at high temperatures is an urgent bottleneck in the research of deep well water blocking materials [17].

In recent years, with the development of materials science, the use of functional fillers to modify inorganic matrices has become a research hotspot [18]. Biochar, as a carbon-rich porous material prepared by pyrolysis of agricultural and forestry waste (such as straw, fruit shells, etc.) under anaerobic conditions, has been widely used in environmental governance and soil improvement due to its large specific surface area, rich hierarchical pore structure, and active oxygen-containing functional groups (hydroxyl, carboxyl, etc.)

 https://doi.org/10.3390/pr14061014

on the surface [19]. In the field of building materials, cutting-edge research has explored the introduction of biochar into cement-based composite materials. It has been found that an appropriate amount of biochar can not only fill the pores as micro aggregates, but also play an "internal curing" role by adsorbing moisture, significantly reducing the self-shrinkage cracks of concrete and improving the microstructure of the interface transition zone. However, there have been no reports on introducing biochar into the field of oilfield chemistry, especially utilizing its unique pore structure and surface chemical properties to construct high-temperature resistant, high-strength, and high toughness deep water plugging systems [20].

Based on the above background, this study aims to solve the difficult problem of balancing the "strength toughness temperature resistance" of water blocking materials in deep carbonate reservoirs [21]. An innovative design approach for composite materials has been proposed: utilizing multiphase composite inorganic admixture (MCA) as a high-temperature-resistant cementitious matrix and introducing biochar as a functional enhancing component [22,23]. This study hypothesizes that biochar can exhibit triple synergistic effects: firstly, utilizing its porous structure to adsorb slow-release activators and regulate high-temperature hydration kinetics; secondly, as a micro aggregate to fill matrix defects and improve density; and the third is to form chemical cross-linking with hydration products through surface functional groups, endowing the material with toughness [24]. This article will verify the adaptability of this new composite system in extreme environments of deep wells through systematic indoor experiments, including rheological characterization, high temperature and high pressure, and long-term aging testing, as well as dynamic sealing evaluation of simulated fracture core. Combined with microscopic analysis methods, its strengthening mechanism will be revealed in order to provide technical support for the efficient development of deep oil and gas resources [25].

2. Materials and Methods

We simulated the actual working conditions of a deep oil reservoir in northwest China, and conducted experimental research at a temperature of 130 °C and a mineralization degree of 20×10^4 mg/L.

2.1. Experimental Materials and Instruments

The experimental core cementitious material selected is an industrial grade solid-phase composite admixture (MCA), mainly composed of SiO_2 (36.29%), CaO (34.30%), and Al_2O_3 (13.07%), supplemented by a small amount of Fe_2O_3 and Na_2O (see Table 1 for details). This component has typical pozzolanic activity and can generate C-S-H gel with cementation strength through hydration reaction in an alkaline environment [26,27]. To improve early strength, a high-performance activator with a mesh size of 600–800 is formulated (Shandong Jiuqi Building Materials Co., Ltd., Weifang, Shandong, China). The functional enhancement material selected is micro scale biochar prepared by pyrolysis of agricultural and forestry waste (Jinan Shangjiu Material Technology Co., Ltd., Jinan, Shandong, China). To ensure the uniform suspension of solid particles in the liquid phase, an efficient suspension stabilizer (Wuhan Wuxin New Building Materials Co., Ltd., Wuhan, Hubei, China) was selected. The experimental water is prepared based on the ion composition of the target reservoir formation water, with a total mineralization of 200,000 mg/L. The main experimental instruments include: high-temperature and high-pressure viscosity meter (OFITE 2025 (OFI Testing Equipment, Houston, TX, USA; 130 °C, 60 MPa, 150 rpm), 130 °C, 60 MPa, 150 rpm), used to simulate the thickening process under bottomhole temperature and pressure conditions; a six-speed rotational viscometer (ZNN-D6 (Qingdao Haitongda Instrument Co., Ltd., Qingdao, Shandong, China), speeds:

 https://doi.org/10.3390/pr14061014

3, 6, 100, 200, 300, 600 rpm; 25 °C), used for rheological testing; a universal material testing machine (Instron 3369 (Instron Corp., Norwood, MA, USA; loading rate 1 mm/min, 25 kN load cell), loading rate 1 mm/min, 25 kN load cell), used for mechanical strength testing; a core displacement device, used for dynamic sealing evaluation; and an environmental scanning electron microscope, used for microstructure characterization.

Table 1. Chemical composition of raw materials (wt%).

Material	SiO_2	CaO	Al_2O_3	MgO	Fe_2O_3	Na_2O	K_2O	TiO_2	MnO
MCA	36.29	34.30	15.28	8.79	0.63	1.09	0.63	2.19	0.52
Suspension concentrate	69.68	0.11	1.85	-	0.16	27.67	0.30	0.03	-

Note: The composition of the suspension concentrate was determined from the dried solid residue of the liquid slurry. The high SiO_2 and Na_2O contents are attributed to sodium silicate-based additives used in the formulation.

The chemical composition of all raw materials (MCA powder and the dried residue of the suspension concentrate) was determined by X-ray fluorescence (XRF) spectroscopy using a PANalytical Axios MAX X-ray fluorescence spectrometer (Malvern Panalytical, Almelo, The Netherlands). Samples were prepared as fused beads prior to analysis to ensure homogeneity and accurate quantification.

2.2. Preparation of Composite Gel System

Considering the differences in physical and chemical properties of each component, in order to ensure the homogeneity of the system, this study optimized the preparation process and adopted the "graded dispersion high-speed shearing" method: firstly, water and MCA were weighed at a water cement ratio of 0.25 (water to MCA mass ratio 4:1) at room temperature, and vigorously sheared and dispersed using a high-speed mixer. After homogenization, an 8 wt% suspension agent was added, and high-speed shearing was continued until a stable base liquid was formed. Then, a predetermined proportion of biochar (0.5–1%) was added and stirred again. During this process, the strong water absorption of biochar is utilized to pre adsorb some free water and chemical activators dissolved in water, forming a microcapsule structure; finally, high performance activator and residual biochar are added, stirred at a low speed to no dry powder, and then sheared at high speed for 3 min to prepare the final biochar-enhanced inorganic composite gel plugging agent [28,29].

2.3. Performance Evaluation Methods

(1) Mechanical performance testing

Inject the prepared slurry into a standard mold (25 mm × 25 mm × 25 mm), place it in a high-temperature and high-pressure reactor (Parr 4598 (Parr Instrument Company, Moline, IL, USA), 130 °C, 60 MPa, stirring speed 200 rpm), and cure it in a sealed environment at 130 °C and 60 MPa for 24 h. After removal, demold and use a universal testing machine to measure its uniaxial compressive strength [30]. Through orthogonal experimental design, investigate the interactive effect of MCA dosage (8–15%) and biochar dosage (0.5–4%) on strength.

(2) Rheological characterization

Use a six-speed rotational viscometer to measure the apparent viscosity of the slurry at 25 °C and the readings at different rotational speeds. Test the rheological parameters of the initial state and after 4 h of standing, and use the Herschel–Bulkley model to fit the rheological curve to evaluate its pumping performance and suspension stability [31].

(3) Determination of setting time

https://doi.org/10.3390/pr14061014

Use a high-temperature and high-pressure viscosity meter to simulate the bottom-hole conditions at 130 °C and 60 MPa, record the changes in slurry viscosity over time, determine the thickening start time (initial setting) and pumping limit time, and evaluate construction safety [32].

(4) Thermal stability evaluation

Place the cured sample in an aging tank containing simulated formation water and conduct a static aging experiment for up to 30 days in a 130 °C constant temperature chamber. Regularly test the compressive strength and evaluate its long-term temperature resistance performance.

(5) Sealing and anti-erosion performance

An artificial fracture core (diameter 25 mm, fracture width 1–4 mm) is used to simulate the leakage channel of the formation. Inject the plugging agent into the crack and cure it at 130 °C, conduct a water drive experiment, record the breakthrough pressure and subsequent pressure fluctuations, calculate the breakthrough pressure gradient, and evaluate its plugging strength and toughness. The schematic diagram of the crushed rock core and the flowchart of the blocking experiment are shown in Figures 1 and 2 respectively.

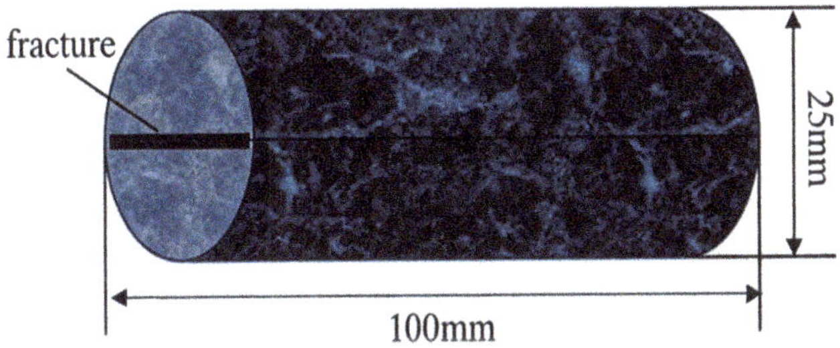

Figure 1. Schematic diagram of fractured rock core.

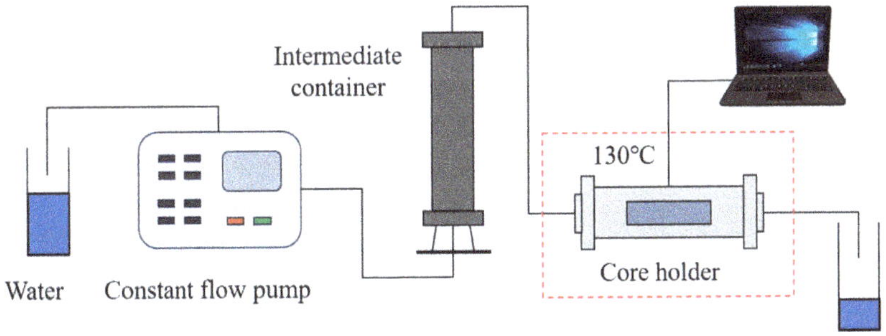

Figure 2. Flow chart of blocking experiment.

3. Results and Discussion

3.1. Mechanical Properties and Influencing Factors

The compressive strength of solidified materials is the primary indicator for evaluating whether water blocking materials can maintain integrity under high pressure differentials in deep formations. As shown in Figure 3, the dosage of MCA and biochar has a completely different impact on the strength of the system. As the main gelling component, the increase in MCA content (from 8% to 15%) directly increases the density of hydration products (C-S-H gel) in unit volume, thus linearly improving the compressive strength. However, the impact of biochar on strength exhibits significant nonlinear characteristics. When the

https://doi.org/10.3390/pr14061014

content of biochar increases from 0 to 0.5%, the compressive strength reaches a peak of over 2 MPa; when the dosage continues to increase to 1% or even 4%, the strength actually decreases significantly. This phenomenon reveals a dual mechanism of biochar in inorganic matrices: on the one hand, at low dosages (0.5%), micron-sized biochar particles exhibit a "micro aggregate filling effect". Due to its rich hierarchical pore structure, it can effectively fill the micrometer-sized pores formed by the accumulation of MCA hydration product particles, significantly reducing the overall porosity of the solidified body. At the same time, it absorbs and sustains the release of the alkali activator to avoid internal stress caused by an early hydration rate which is too fast, and its stable carbon skeleton can inhibit the collapse of C-S-H gel during the high-temperature environment of 130–140 °C [13]; oxygen containing functional groups such as hydroxyl and carboxyl groups on the surface of biochar can not only coordinate with Ca^{2+} and Al^{3+} in the hydration products to form Ca-O-C and Al-O-C chemical bonds and strengthen the three-dimensional cross-linking network, but also act as the hydration reaction nucleation site to induce the directional growth of C-S-H gel, and can also absorb free Na^{+} and Ca^{2+} ions in the high salt environment, reducing the ion exchange corrosion of hydration products; in addition, the porous structure of biochar can form a "mortise and tenon" mechanical interlocking with the pores of the rock core.

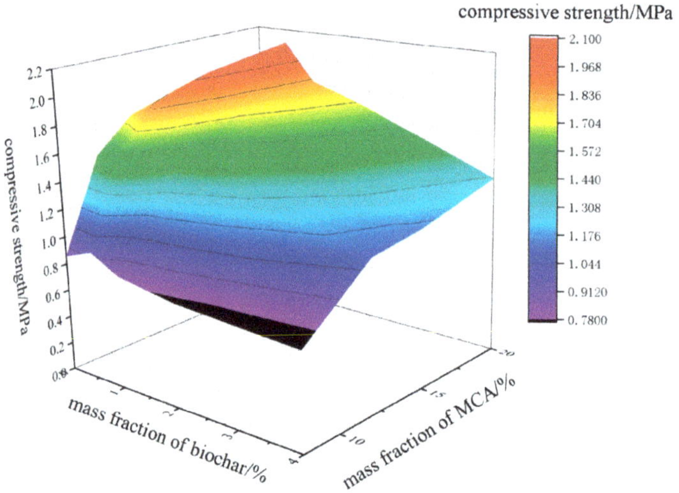

Figure 3. Compressive strength of biochar and MCA water plugging agent with different mass fractions.

Its surface functional groups can also condense with the hydroxyl groups on the surface of the rock core to form Si-O-C bonds, enhancing the interfacial adhesion between the water blocking agent and the rock core, and ultimately synergistically improving the strength of the water blocking agent (Figure 4). On the other hand, at high dosages (>0.5%), due to the large specific surface area and hydrophobicity of biochar, excess particles are difficult to uniformly disperse in the aqueous phase and are prone to agglomeration. These loose aggregates cannot be effectively encapsulated by hydration products, forming a large number of structural defect points inside the solidified body. Under external forces, these defect points first become stress concentration zones, inducing the initiation and propagation of microcracks, resulting in a sharp decrease in the overall structural strength. Therefore, the optimal dosage for biochar is determined to be 0.5% [33].

https://doi.org/10.3390/pr14061014

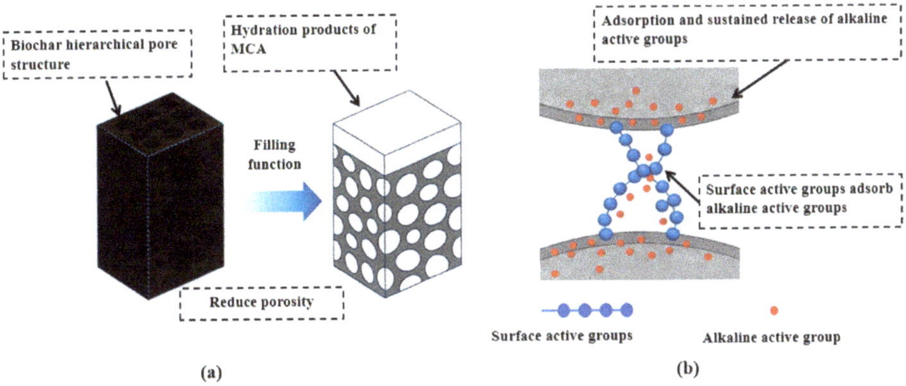

Figure 4. Adsorption and slow-release of alkaline activator by biochar (**a**,**b**).

3.2. Curing Performance and Control Mechanism

In a high-temperature environment of 130 °C, the hydration reaction of traditional inorganic materials is extremely intense, often completing condensation within a few hours, which poses a great safety hazard ("flash condensation" risk) for long-distance pumping in deep wells. The experiment found (Figure 5) that under the same high temperature and high-pressure conditions, the solidification time of the system without biochar was about 9 h, while the addition of 0.5% biochar significantly prolonged the solidification time to 15 h.

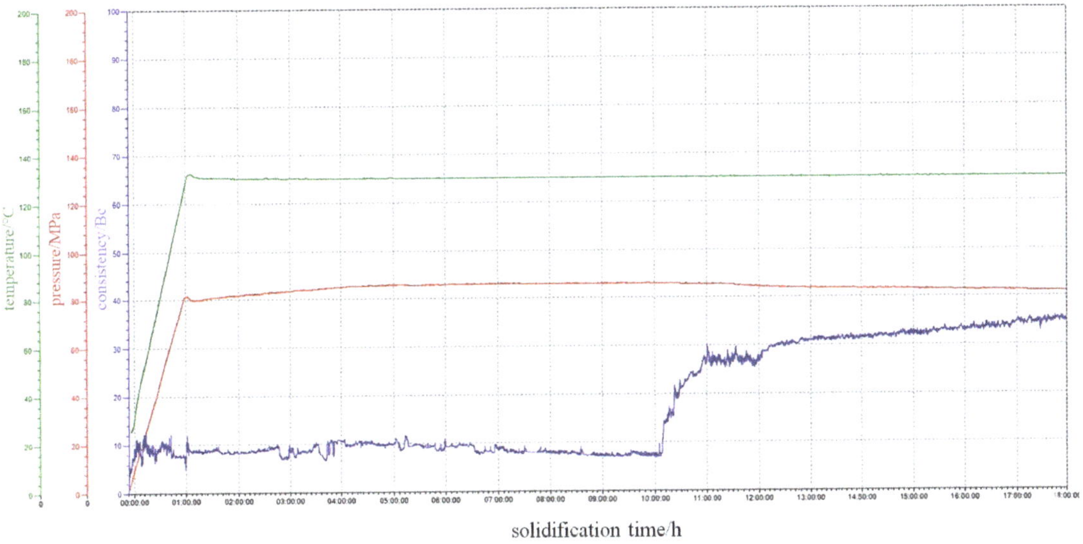

Figure 5. Thickening curve of water blocking agent.

This significant retarding effect can be attributed to the unique "adsorption slow-release" mechanism of biochar [19–21]. The well-developed mesoporous structure of biochar enables it to rapidly adsorb free water and catalytic alkaline ions from the pore solution during the initial mixing stage. This physical adsorption effect reduces the effective water cement ratio and ion concentration of the system in the early stages of the reaction, effectively suppressing the intensity of early hydration reactions and smoothing out the concentrated release of hydration heat. As the reaction progresses, the ion concentration

https://doi.org/10.3390/pr14061014

inside the system decreases, and the water and ions adsorbed by biochar are slowly released under the driving force of concentration gradient, supplementing the reaction system and maintaining the continuous progress of the later hydration reaction. This mechanism not only successfully extended the construction safety window by 6 h, ensuring the safety of deep well operations, but also helped reduce temperature shrinkage cracks caused by rapid early response, improving the integrity of the solidified body.

3.3. Rheological Properties and Injection Performance

The rheological parameters are the key to determining whether the plugging agent can be smoothly pumped to the deep part of the formation and enter the flowing fractures. The rheological test results (Table 2, Figure 6) indicate that the composite system has excellent shear dilution (pseudoplasticity) characteristics. By fitting with the Herschel–Bulkley model, the flow index n of the system is always less than 1 (initial state 0.91, 0.87 after 4 h of standing). This means that when the fluid flows at high speed (high shear rate) inside the wellbore, its apparent viscosity will significantly decrease, thereby significantly reducing pumping friction and lowering construction pressure; when the fluid enters the formation fractures or stops pumping at a low shear rate, its viscosity can quickly recover, exhibiting high yield stress, which is conducive to carrying solid particles to prevent settlement and effectively staying in the target layer. In addition, the initial viscosity of the system was less than 25 mPa·s, and it maintained good flowability after 4 h of standing without significant thickening or solidification. This rheological property endows the plugging agent with excellent deep injection capability and construction flexibility, which can adapt to the complex construction process requirements of deep wells.

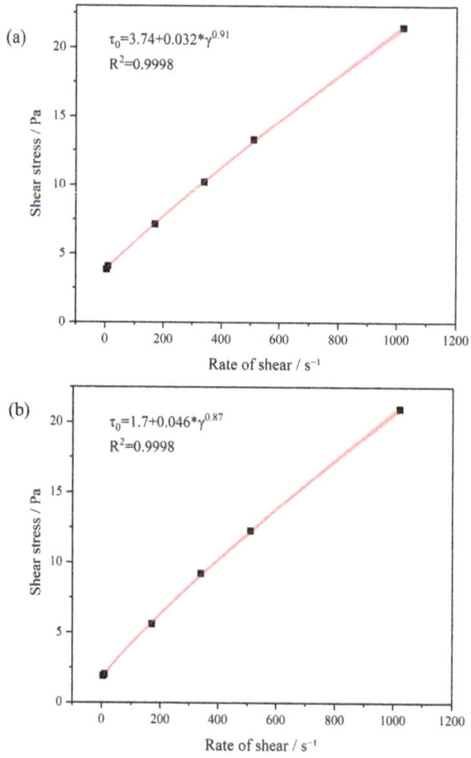

Figure 6. Rheological curve of water blocking agent: (**a**) initial state (0 h), (**b**) standing for 4 h.

https://doi.org/10.3390/pr14061014

Table 2. Measurement results of rheological parameters of a six-speed rotational viscometer.

Time, h	Speed, r/min	Numerical Value	Shear Rate, s^{-1}	Static Shear Force, Pa	Shear Stress, Pa
0	3	7.5	5.1	3.83	3.83
	6	8	10.2		4.09
	100	14	170.3		7.15
	200	20	340.6		10.22
	300	26	510.9		13.29
	600	42	1021.8		21.46
4	3	3.8	5.1	1.94	1.94
	6	4	10.2		2.04
	100	11	170.3		5.62
	200	18	340.6		9.2
	300	24	510.9		12.26
	600	41	1021.8		20.95

3.4. Long-Term Thermal Stability

Temperature resistance is the core indicator for evaluating whether deep well water blocking materials can achieve long-term sealing. In the 130 °C high-temperature aging experiment (Figure 7), the compressive strength of the composite cured body showed a slight decrease in the initial stage (first 5 days) (from 2.0 MPa to 1.9 MPa), which may be related to the crystal transformation or recrystallization of metastable hydration products under high temperature and pressure. However, in the following 25 days, the trend of strength decline slowed down significantly and stabilized, ultimately remaining at around 1.75 MPa, with a strength retention rate exceeding 75%. Different from the "tofu dregs" phenomenon of traditional polymer gel caused by molecular chain breaking and hydrolysis at high temperature, the inorganic composite system constructed in this study shows excellent thermal stability. This is due to the stable carbon skeleton structure of biochar. In a high-temperature environment, biochar does not degrade itself, and its three-dimensional network skeleton can also limit the shrinkage deformation of C-S-H gel and inhibit the generation of microcracks. This long-term stability indicates that the system can meet the engineering requirements of long-term stable oil and water control in deep oil reservoirs [34].

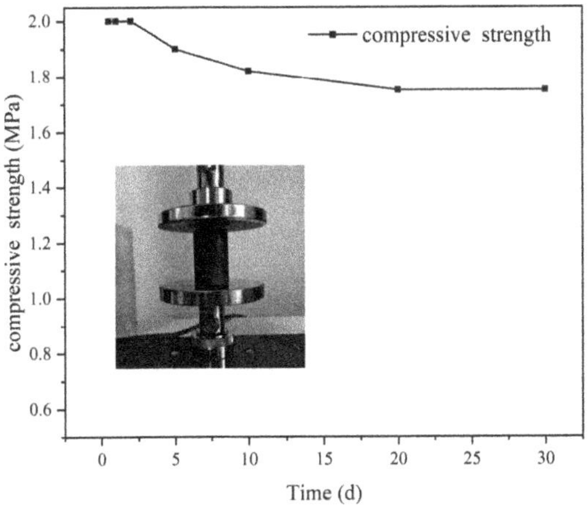

Figure 7. High temperature aging test results of water blocking agent.

https://doi.org/10.3390/pr14061014

3.5. Dynamic Sealing Characteristics and Toughening Mechanism of Cracks

Regarding the common problem of fracture leakage in fractured reservoirs, the results of core displacement experiments (Figure 8) show that the plugging agent exhibits excellent plugging ability for fractures with a width of 1–3 mm, breaking through pressure gradients greater than 10 MPa/m. As the subsequent water injection continues, the displacement pressure rises to its peak, and then the water blocking agent is broken through, and the pressure begins to decrease and gradually remains at a relatively stable value. However, compared with the breakthrough pressure, the decrease is not significant and there is still a certain degree of sealing strength, indicating that the water blocking agent can still maintain sealing for a long time under fluid erosion and has good anti erosion performance [35].

The SEM microstructure analysis is shown in Figure 9. Under the electron microscope, it is clearly visible that the biochar particles are uniformly dispersed in the inorganic hydration products, forming a reinforced structure similar to the "micro steel bars" in concrete. The interface between biochar and the matrix is tightly bonded, with no obvious interface delamination, indicating the formation of strong chemical bonds (such as Si-O-C bonds). This structure effectively hinders the rapid propagation of cracks and dissipates fracture energy. In addition, the addition of biochar significantly reduces the porosity of the water blocking agent. This is because biochar is filled inside the hydration products of MCA, reducing the porosity of the water blocking agent and making its structure more compact, allowing it to adapt to cracks of various sizes and form effective sealing [19,36].

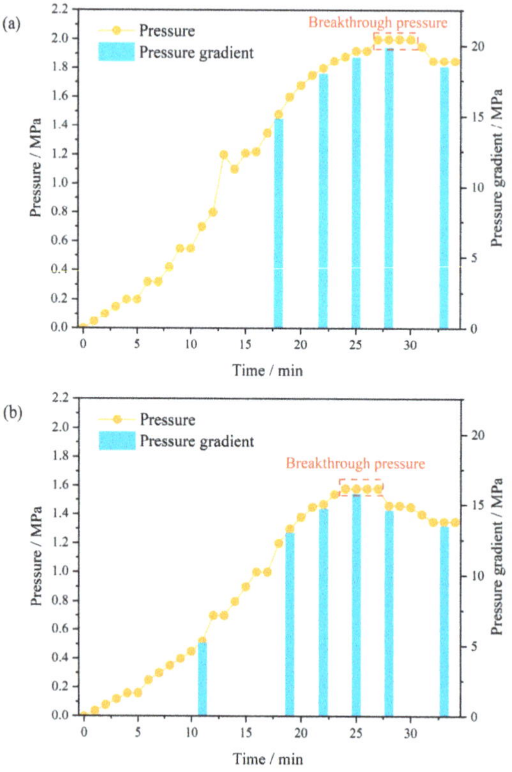

Figure 8. *Cont.*

https://doi.org/10.3390/pr14061014

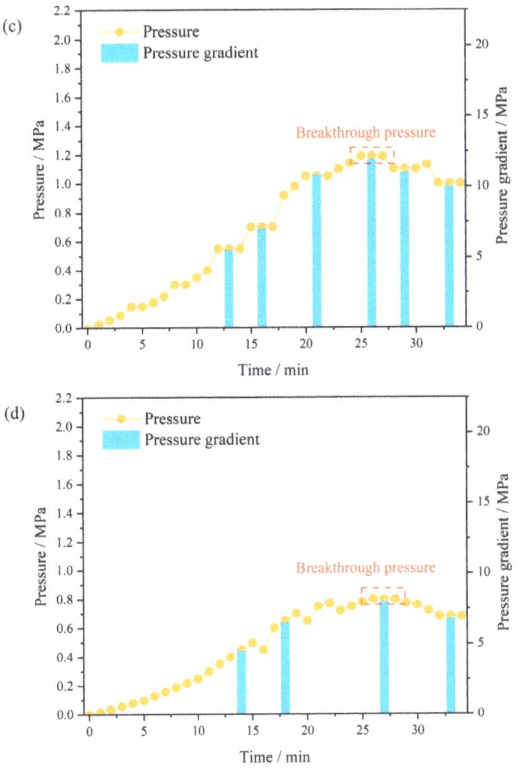

Figure 8. Time pressure relationship curves of erosion at different fracture scales: (**a**) 1 mm, (**b**) 2 mm, (**c**) 3 mm, (**d**) 4 mm.

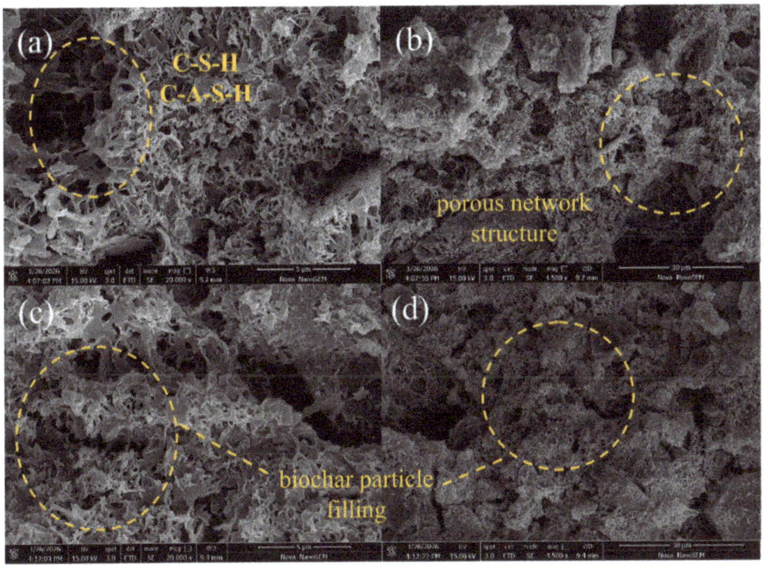

Figure 9. SEM results of water blocking agent: (**a**) observation point 1 20,000× magnification, (**b**) observation point 1 4500× magnification, (**c**) observation point 2 20,000× magnification, (**d**) observation point 2 4500× magnification.

https://doi.org/10.3390/pr14061014

4. Conclusions

(1) The optimal formula based on 15% MCA, 0.5% biochar, and high-performance activator has been determined. Under this formula, the system successfully overcame the brittleness of traditional inorganic materials, achieving a compressive strength of 2 MPa and a significant improvement in mechanical properties.

(2) The multi-dimensional mechanism of biochar in an inorganic gel system was clarified. At the micro level, biochar enhances the density and toughness of the matrix through the micro aggregate filling effect and interfacial chemical bonding; macroscopically, by utilizing its unique "adsorption slow-release" mechanism, the kinetics of hydration reactions at high temperatures have been effectively regulated, eliminating the risk of flash condensation in deep well operations.

(3) This system has rheological characteristics of low viscosity, high strength, and shear dilution, making it easy to pump and inject deep into the system; after aging in a high temperature and high salt environment at 130 °C for 30 days, it still maintains high strength and has excellent weather resistance; it exhibits high sealing ability for cracks of different scales.

Author Contributions: Conceptualization, S.H. and T.W.; methodology, S.H. and T.W.; validation, S.H. and T.W.; formal analysis, S.H. and T.W.; investigation, S.H. and T.W.; resources, S.H. and T.W.; data curation, S.H. and T.W.; writing—original draft preparation, T.W.; writing—review and editing, S.H. and T.W. All authors have read and agreed to the published version of the manuscript.

Funding: This research received no external funding.

Data Availability Statement: The original contributions presented in this study are included in the article. Further inquiries can be directed to the corresponding author.

Acknowledgments: The authors would like to extend their heartfelt thanks to Yangtze University for the invaluable educational foundation and academic support that have significantly contributed to the development of this work, and express their sincere gratitude to No.2 Oil Production Plant, Northwest Oilfield Company, SINOPEC for material support and for providing access to laboratory facilities that were essential for the successful completion of this study. Additionally, the authors also express their sincere appreciation to the anonymous reviewers for their valuable and constructive comments.

Conflicts of Interest: Author Shiwei He was employed by the No.2 Oil Production Plant, Northwest Oilfield Company, SINOPEC. The remaining authors declare that the research was conducted in the absence of any commercial or financial relationships that could be construed as a potential conflict of interest.

References

1. Zeng, L.; Song, Y.; Han, J.; Han, J.; Yao, Y.; Huang, C.; Zhang, Y.; Tan, X.; Li, H. Control of structure and fluid on ultra-deep fault-controlled carbonate fracture-vug reservoirs in the Tarim Basin, NW China. *Petrol. Explor. Dev.* **2025**, *52*, 143–156. [CrossRef]
2. Guo, W.; Fu, S.; Li, A.; Xie, H.; Cui, S.; Nangendo, J. Experimental research on the mechanisms of improving water flooding in fractured-vuggy reservoirs. *J. Petrol. Sci. Eng.* **2022**, *213*, 110383. [CrossRef]
3. Yuan, C.; Pu, W.; Jin, F.; Zhang, Y.; Jia, H.; Zhao, T. Performance of oil-based cement slurry as a selective water-plugging agent in high-temperature and high-salinity cave-fractured carbonate reservoirs. *Ind. Eng. Chem. Res.* **2014**, *53*, 6137–6149. [CrossRef]
4. Cheng, H.; Yuan, F.; Zhang, S.; Li, L.; Luo, X.; Chen, B. Investigation on water invasion mode and remaining oil utilization rules of fractured-vuggy reservoirs: A case study of the intersection region of S99 unit in Tahe Oilfield. *Processes* **2023**, *11*, 1833. [CrossRef]
5. Song, Z.; Wu, W.; Qu, M.; Xi, J.; Yang, M.; Jia, X.; Zhao, Y.; Liu, L.; Cui, H. Research on the distribution characteristics of Urea-Formaldehyde resin gel influenced by structural development in fractured-vuggy reservoirs. *Gels* **2025**, *11*, 868. [CrossRef]
6. Xu, Z.; Sun, M.; Tao, L.; Bai, J.; Shi, W.; Zhang, N.; Peng, Y. Mechanisms of mobility control and enhanced oil recovery of weak gels in heterogeneous reservoirs. *Gels* **2025**, *11*, 854. [CrossRef]
7. Wang, D.; Li, Y.; Hu, Y.; Li, B.; Deng, X.; Liu, Z. Integrated dynamic evaluation of depletion-drive performance in naturally fractured-vuggy carbonate reservoirs using DPSO–FCM clustering. *Fuel* **2016**, *181*, 996–1010. [CrossRef]

https://doi.org/10.3390/pr14061014

8. Lenji, A.; Haghshenasfard, M.; Sefti, V.; Salehi, M.; Heidari, A. Experimental study of swelling and rheological behavior of preformed particle gel used in water shutoff treatment. *J. Petrol. Sci. Eng.* **2018**, *169*, 739–747. [CrossRef]
9. Yao, Y.; Wei, M.; Bai, B. Descriptive statistical analysis of experimental data for wettability alteration with surfactants in carbonate reservoirs. *Fuel* **2022**, *310*, 122110. [CrossRef]
10. Hou, Y.; Ma, S.; Hao, J.; Lin, C.; Zhao, J.; Sui, X. Construction and ion transport-related applications of the hydrogel-based membrane with 3D nanochannels. *Polymers* **2022**, *14*, 4037. [CrossRef]
11. Wang, T.; Wang, L.; Song, W.; Fan, H.; Cheremisin, A.; Yuan, C. Temperature- and salt-tolerant 2-acrylamide-2-methylpropanesulfonic acid/AM@SiO_2 microgel particles: Synthesis and enhanced oil recovery performance evaluation. *Ind. Eng. Chem. Res.* **2023**, *62*, 14879–14890. [CrossRef]
12. Liu, H.; Fei, G.; Tian, K.; Zhu, Z.; Ji, D.; Luo, H. Synergy of low injection resistance and high plugging in a high-strength nanobentonite/amphoteric polymer gel. *Gels* **2025**, *11*, 847. [CrossRef]
13. Zhu, B.; Tang, H.; Yin, S.; Chen, G.; Zhao, F.; Li, L. Experimental and numerical investigations of particle plugging in fracture-vuggy reservoir: A case study. *J. Petrol. Sci. Eng.* **2022**, *208*, 109610.
14. Yang, G.; Gao, F.; Guo, X.; Gao, Q.; Xu, Y.; Yang, H.; Xu, Z.; Leng, L. Improving temperature and salinity tolerance of polymer gels by modifying with lignin-containing nanocellulose. *Fuel* **2025**, *384*, 134019. [CrossRef]
15. Liu, Z.; Xu, J.; Peng, W.; Yu, X.; Chen, J. The development and deployment of degradable temporary plugging material for ultra-deepwater wells. *Processes* **2023**, *11*, 1685. [CrossRef]
16. Telin, A.; Sergeeva, N.; Asadullin, R.; Gusarova, E.; Yakubov, R.; Dokichev, V.; Politov, A.; Sunagatova, E.; Gibadullina, N.; Teptereva, G.; et al. Polyacrylamide and polyacrylamide/polysaccharide hydrogels for well water shutoff in high-temperature reservoirs. *Gels* **2025**, *11*, 862. [CrossRef] [PubMed]
17. Tian, L.; Bu, Y.; Liu, H.; Guo, S. Study on the penetration of strengthening material for deep-water weakly consolidated shallow formation. *J. Petrol. Sci. Eng.* **2022**, *210*, 109862. [CrossRef]
18. Tang, X.; He, B.; Li, J.; Qin, G. One-step preparation of Cu/BC catalyzed the upgrading of heavy oil assisted by microwave. *J. Petrol. Sci. Eng.* **2022**, *208*, 109683. [CrossRef]
19. Ji, C.; Wang, S.; Zhang, M.; Sun, H.; Meng, Q.; Zhao, K.; Li, Y.; Su, S.; Han, B.; Liu, W. Long-term aging of oxychar and biochar: Structural, adsorptive, and microbial transformations. *Fuel* **2026**, *415*, 138501. [CrossRef]
20. Faga, M.G.; Duraccio, D.; Di Maro, M.; Pedraza, R.; Bartoli, M.; d'Ayala, G.G.; Torsello, D.; Ghigo, G.; Malucelli, G. Ethylene-Vinyl Acetate (EVA) containing waste hemp-derived biochar fibers: Mechanical, electrical, thermal and tribological behavior. *Polymers* **2022**, *14*, 4171. [CrossRef]
21. Verma, V.; Akash, T.; Maddirala, R.; Pal, N. A Sustainable approach to enhanced oil recovery via cerium-modified sugarcane bagasse biochar-stabilized nanoemulsion. *Ind. Eng. Chem. Res.* **2026**, *65*, 1210–1227. [CrossRef]
22. Shagiakhmetov, A.; Yushchenko, S. Substantiation of in situ water shut-off technology in carbonate oil reservoirs. *Energies* **2022**, *15*, 5059. [CrossRef]
23. Wang, T.; Fan, H.; Yang, W.; Meng, Z. Stabilization mechanism of fly ash three-phase foam and its sealing capacity on fractured reservoirs. *Fuel* **2020**, *264*, 116832. [CrossRef]
24. Farzad, P.; Ghasem, Z.; Ali, K.; Ahmad, R.; Ali, M.; Sheikh, Z. Experimental evaluation of designed and synthesized Alkaline-Surfactant-polymer (ASP) for chemical flooding in carbonate reservoirs. *Fuel* **2022**, *321*, 124090.
25. Yang, H.; Zhang, J.; Wang, Z.; Li, S.; Wei, Q.; He, Y.; Li, L.; Zhao, J.; Xu, C.; Zhang, Z. In situ-prepared nanocomposite for water management in high-temperature reservoirs. *Gels* **2025**, *11*, 405. [CrossRef] [PubMed]
26. Zhu, D.; Bai, B.; Hou, J. Polymer gel systems for water management in high-temperature petroleum reservoirs: A chemical review. *Energy Fuels* **2017**, *31*, 13063–13087. [CrossRef]
27. Gao, Y.; Cheng, H.; Li, G.; Li, H.; Tan, H.; Zhang, J.; Zhu, D. Gelation and reservoir conformance control performance of in situ crosslinked polymer gels prepared with different crosslinkers. *Pet. Sci.* **2026**, *in press*. [CrossRef]
28. Zhu, D.; Zhang, J.; Li, G.; Gao, Y.; Tan, H.; Li, H.; Cheng, H.; Lu, C. Development of oil sludge preformed particle gels (OS-PPG) for petroleum reservoir conformance control. *Energy Fuels* **2026**, *40*, 2027–2037. [CrossRef]
29. Zhang, J.; Tan, H.; Zhang, T.; Gao, Y.; Cheng, H.; Li, H.; Zhu, D. New injection method using composite gel system and gel breaker to improve sweep efficiency and mitigate formation damage. *Pet. Sci.* **2026**, *23*, 383–392. [CrossRef]
30. Liang, C.; Qi, N.; Zhao, L.; Li, X.; Li, Z. Development of a Water-Sensitive Self-Thickening Emulsion Temporary Plugging Diverting Agent for High-Temperature and High-Salinity Reservoirs. *Polymers* **2025**, *17*, 1543. [CrossRef]
31. Wang, J.; Feng, Y.; Cao, A.; Zhang, J.; Chen, D. Co-injection of foam and particles: An approach for bottom water control in fractured-vuggy reservoirs. *Processes* **2024**, *12*, 447. [CrossRef]
32. He, S.; Xue, C.; Du, C.; Mao, Y.; Li, S.; Zhong, J.; Guo, L.; Wang, S. Understanding plugging agent emplacement depth with polymer shear thinning: Insights from experiments and numerical modeling. *Processes* **2024**, *12*, 893. [CrossRef]

 https://doi.org/10.3390/pr14061014

33. Khoshkar, A.; Fatemi, M.; Ghazanfari, H. Static and dynamic evaluation of the effect of nanomaterials on the performance of a novel synthesized PPG for water shut-off and improved oil recovery in fractured reservoirs. *J. Petrol. Sci. Eng.* **2020**, *189*, 107019. [CrossRef]
34. Yu, T.; Zheng, W.; Guan, X.; Li, A.; Chen, D.; Chu, W.; Xia, X. Development and mechanism of the graded polymer profile-control agent for heterogeneous heavy oil reservoirs under water flooding. *Gels* **2025**, *11*, 856. [CrossRef]
35. Ma, B.; Zheng, K.; Zhang, C.; Pu, L.; Feng, B.; Shi, Q.; Fu, Q.; Lin, Q.; Wang, Y.; Xu, P. Review of cement-based plugging systems for severe lost circulation in deep and ultra-deep formations. *Processes* **2026**, *14*, 76. [CrossRef]
36. Zhu, D.; Yang, Y.; Tan, H.; Guo, S.; Lu, J.; Zhang, T.; Zhang, J.; Gao, Y.; Li, H.; Cheng, H. CO_2 conformance control and carbon storage performance of particle gels in fractured reservoirs using nuclear magnetic resonance methods. *SPE J* **2026**, *31*, 588–601. [CrossRef]

https://doi.org/10.3390/pr14061014

processes

MDPI

Article

A Method for Predicting Bottomhole Pressure Based on Data Augmentation and Hyperparameter Optimisation

Xiankang Xin [1,2,3,4], Xuecheng Jiang [1], Saijun Liu [1,2,3,4,*], Gaoming Yu [1,2,3,4,*] and Xujian Jiang [5]

1 School of Petroleum Engineering, Yangtze University, Wuhan 430100, China; xiankang.xin@hotmail.com (X.X.); jiang_stu2023@163.com (X.J.)
2 State Key Laboratory of Low Carbon Catalysis and Carbon Dioxide Utilization, Yangtze University, Wuhan 430100, China
3 Hubei Key Laboratory of Oil and Gas Drilling and Production Engineering, Yangtze University, Wuhan 430100, China
4 National Engineering Research Center for Oil & Gas Drilling and Completion Technology, School of Petroleum Engineering, Yangtze University, Wuhan 430100, China
5 Research Institute of Petroleum Exploration and Development, Tarim Oilfield Company, PetroChina, Korla 841000, China; jiangxujian123@126.com
* Correspondence: liusj@yangtzeu.edu.cn (S.L.); ygm@yangtzeu.edu.cn (G.Y.)

Abstract

With the continuous development of the petroleum industry, bottomhole pressure prediction technology, which exerts a significant impact on oil production and recovery, has become a key research direction in the current oil and gas field. To enhance the accuracy and robustness of bottomhole pressure prediction under transient and variable operating conditions, a method based on data augmentation strategies and hyperparameter optimization was proposed in this paper. Addressing challenges such as limited data volume and significant disturbances in actual oilfield production, a data augmentation strategy incorporating noise perturbation and sliding windows was introduced to expand training samples and improve model generalization. In terms of model architecture, a deep network integrating CNN, BiGRU, and Multi-Head Attention mechanisms was proposed in this paper, which is referred to as the CNN-BiGRU-Multi-Head Attention model. By introducing Bayesian optimization for automatic hyperparameter search, the performance of the temporal model was further enhanced, achieving efficient extraction and dynamic focusing of wellbore pressure temporal features. Prediction results demonstrated that the proposed method outperforms existing mainstream forecasting models in metrics such as Mean Absolute Error (MAE) and Coefficient of Determination (R^2), with R^2 reaching 0.9831, which confirms its strong generalization capability and engineering applicability. Practical guidance for intelligent oilfield production management and bottomhole pressure forecasting, along with a novel prediction method, is provided by this study, which holds significant importance for extending well life and stabilizing hydrocarbon production.

Keywords: bottomhole pressure; predictive model; deep learning; data augmentation; fusion time-series model

1. Introduction

Bottomhole Pressure (BHP) serves as a crucial indicator of fluid dynamics within oil and gas reservoirs, playing an irreplaceable role in reservoir development [1,2]. Accurate BHP monitoring and prediction are essential for effectively guiding production control as well as for revealing internal reservoir pressure distributions and fluid flow patterns, both of which are crucial for ensuring operational continuity and stability [3,4]. Particularly during the mid-to-late development stages of oil and gas fields, where non-steady-state flow and complex, variable operating conditions prevail, the spatiotemporal variations in BHP become increasingly intricate [5]. Traditional measurement and prediction methods struggle to meet the demands of practical production management [6,7].

Currently, two methods are employed to obtain bottomhole pressure data in field operations. One involves direct measurement using downhole pressure gauges. While this method captures pressure data, it suffers from issues such as high equipment costs, complex maintenance, and short instrument lifespans. Additionally, the data obtained is lagging, suitable only for monitoring and unable to support subsequent production decisions [8]. The other method employs theoretical calculations combining existing data with empirical formulas. This approach calculates bottomhole pressure using previously constructed multiphase flow models of the wellbore [9,10]. While less costly than direct measurement, it suffers from numerous model assumptions, limited applicability, difficulties in accurately obtaining key parameters during actual operation, and insufficient prediction accuracy [11–13]. It struggles to meet the demands of modern intelligent oilfields for high-precision, real-time predictions.

In recent years, with the advancement of machine learning and deep learning technologies [14–18], data-driven bottomhole pressure prediction methods have emerged as a research hotspot [9,19–23]. These approaches can automatically extract complex temporal features, enhancing model predictive capability and generalization performance [24–27]. However, existing methods often overlook practical challenges in oil and gas fields, such as limited data volume, strong disturbances, and highly variable operating conditions, leading to insufficient model robustness in non-steady-state and complex environments, which ultimately limits their widespread engineering application.

In response to the aforementioned challenges, Ternyik et al. [28] employed an artificial neural network (ANN) model to predict pipeline pressure drop under various operating conditions, representing an early study utilizing artificial intelligence methods for pressure prediction. After that, Osman et al. [29] developed a vertical multiphase flow ANN model using a multilayer feedforward network approach, which was validated using field data encompassing a wide range of variables. Sami et al. [30] proposed employing three machine learning techniques to predict multiphase flow bottomhole pressure. Results indicate that artificial neural networks outperform other machine learning methods in terms of error control. These studies effectively validate the feasibility of machine learning in bottomhole pressure prediction tasks.

With the advancement of technology, bidirectional models are gradually being applied in the petroleum sector. Mohammadpoor et al. [10] developed an ANN model based on the backpropagation learning algorithm to investigate correlations under different flow conditions. Al-Shammari et al. [31] combined the backpropagation learning algorithm with least-squares function theory to propose an adaptive neural-fuzzy inference system model suitable for two-phase flow systems. Li et al. [20] enhanced the backpropagation neural network model to accommodate segmented computational workflows, achieving improved performance under steady-state flow conditions. These studies further validate the advantages of bidirectional models.

The introduction of fusion models represents a major advancement in foundational model research. Sun et al. [32] proposed a GA-XGBoost model for predicting bottomhole pressure in karstic reservoirs, achieving a prediction accuracy of 0.84 R^2 on the test set. Zhang et al. [6] proposed a fusion model combining CNN and GRU with Bayesian hyperparameter optimization for bottomhole pressure prediction, demonstrating the superiority of CNN-based fusion models in this task. While these studies further validate the advan-

https://doi.org/10.3390/pr14081194

tages of fusion models in bottomhole pressure prediction, challenges remain, including suboptimal prediction performance and limited data availability.

To capture the inherent relationship between long-term dependencies and key features, Li et al. [33] proposed applying the CNN-BiLSTM-Multi-Head Attention model for coalbed methane production forecasting, achieving remarkable results that demonstrated the effectiveness of bidirectional models and multi-head attention mechanisms in enhancing the stability and accuracy of time-series predictions. Pang et al. [34] employed Bayesian optimization to tune model hyperparameters, with empirical tests confirming the necessity of Bayesian hyperparameter optimization.

However, the aforementioned methods still exhibit certain limitations. Most notably, the aforementioned research often inadequately addresses the issues of limited data and strong noise interference, leading to insufficient robustness in complex operating environments. Furthermore, even in studies that employ attention mechanisms or hyperparameter optimization, there is a prevalent lack of multiscale joint modeling designed to simultaneously capture short-term dynamics and long-term trends.

A deep learning-based bottomhole pressure prediction method, which integrates data augmentation strategies with hyperparameter optimization, was proposed in this paper.

By introducing noise perturbation and sliding window techniques to expand the training data, coupled with a multi-scale temporal input mechanism, the model was able to capture both short-term dynamic variations and long-term trends in bottomhole pressure. A composite architecture combining Convolutional Neural Networks (CNN), Bidirectional Gated Recurrent Units (BiGRU), and Multi-Head Attention achieved efficient extraction and dynamic focusing of temporal features. Simultaneously, a Bayesian optimization algorithm was employed to automatically tune the model hyperparameters, further enhancing prediction accuracy and stability. The proposed method was subsequently validated using actual production data from Oilfield B and compared against multiple conventional machine learning models. The results demonstrated that the proposed approach significantly outperformed the baseline models in both prediction accuracy and stability, accurately capturing the temporal evolution characteristics of bottomhole pressure. This study provided a novel technical pathway for accurate bottomhole pressure prediction and offered important references for production optimization and safety management in intelligent oil and gas fields.

The core innovations of this study were summarized as follows: (1) A targeted data augmentation strategy was proposed to alleviate the problem of insufficient oilfield data samples. (2) A multi-scale time series input design was constructed to capture the dynamic temporal characteristics of bottomhole pressure. These two improvements effectively enhanced the prediction performance of the model in this study.

2. Materials and Methods

2.1. Framework for Deep Learning Models

The deep learning framework developed in this study was structured around two main components: model development and data preprocessing (Figure 1). The data preprocessing stage consists of several sequential steps, including data collection and analysis, data cleaning, feature selection and dataset partitioning. Model development was the core focus of this research. Performance evaluation metrics including MAE, RMSE, SMAPE, and R2 are employed to assess common deep learning models alongside the proposed CNN-BiGRU-Multi-Head Attention model. Finally, visualization techniques were employed to display and compare the prediction results of each model.

https://doi.org/10.3390/pr14081194

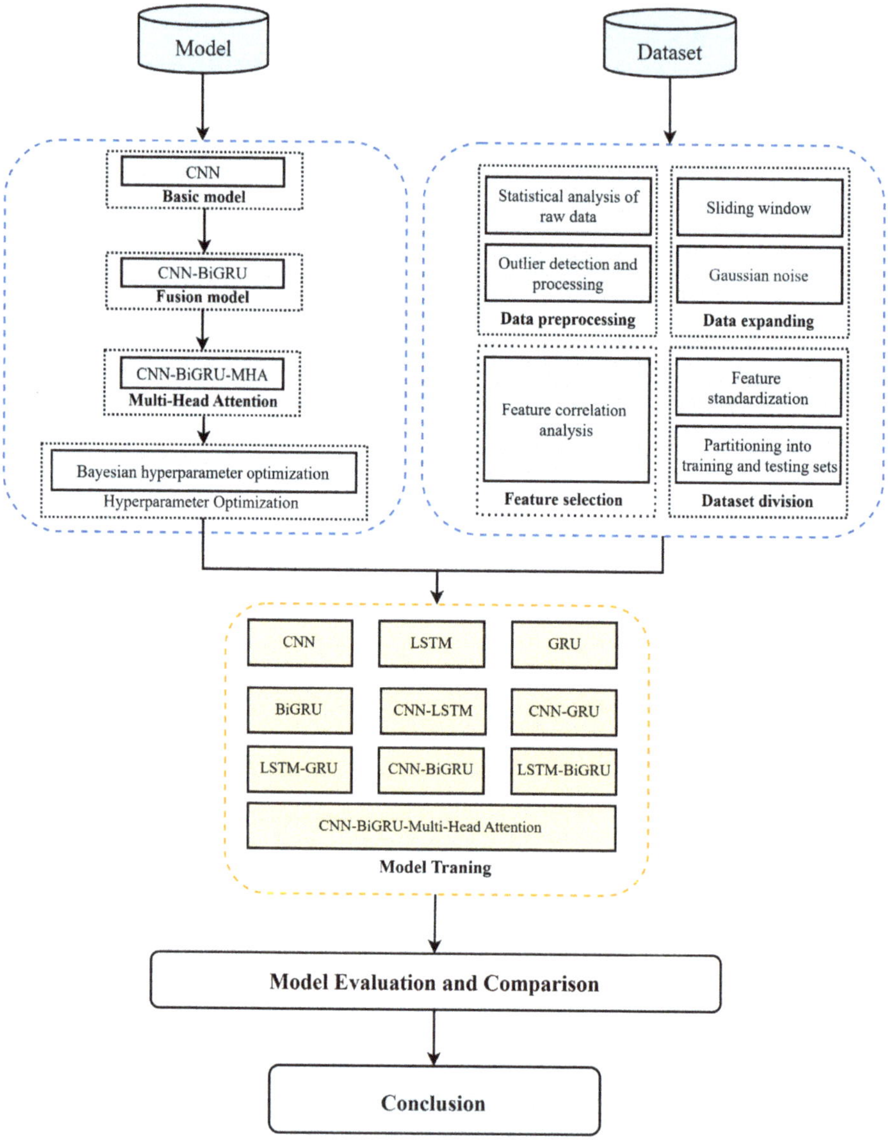

Figure 1. The Overall Workflow of Machine Learning Research.

2.2. Model Construction and Design

2.2.1. CNN Model

In this study, the convolutional neural network (CNN) extracts local temporal features from the input sequence through two convolutional layers. By scanning the data using convolutional kernels (filters), it effectively captures local temporal dependencies, providing crucial foundational data for subsequent sequence modeling.

$$f_t = \mathrm{ReLU}(W_{conv} * X_t + b), \tag{1}$$

https://doi.org/10.3390/pr14081194

In the equation, f_t represents the feature map output at the time step, which is the result processed by the convolutional layer; W_{conv} denotes the convolutional weights, serving as trainable parameters for feature extraction; $*$ signifies the convolutional operator, indicating the convolution computation; X_t denotes the input data at the time step; b represents the bias term, a learnable scalar or vector used to shift the convolution result; ReLU denotes the modified linear unit activation function.

2.2.2. BIGRRU Model

The output feature map f_t from the convolutional layer is fed into a bidirectional GRU (BiGRU). The BiGRU layer captures long-term dependencies in the time series and combines them with the local features extracted by the CNN.

$$h_t^{(f)} = GRU\left(x_t, h_{t-1}^{(f)}\right), \tag{2}$$

$$h_t^{(b)} = GRU\left(x_t, h_{t-1}^{(b)}\right), \tag{3}$$

$$h_t = \left[h_t^{(f)}, h_t^{(b)}\right], \tag{4}$$

In the equation, $h_t^{(f)}$ is the output of the forward GRU unit at time step t (i.e., the forward hidden state); x_t is the input data at time step t; $h_{t-1}^{(f)}$ is the output of the forward GRU unit at time step $t-1$; $h_t^{(b)}$ is the output of the backward GRU unit at time step t (i.e., the backward hidden state); $h_{t-1}^{(b)}$ is the output of the backward GRU unit at time step $t-1$; GRU is the operation of the gated recurrent unit; h_t is the output of the bidirectional GRU, which is the concatenation of the forward and backward GRU outputs.

2.2.3. Multi-Head Attention

Self-attention is a key technique in deep learning that dynamically assigns attention weights to different parts of an input sequence, enabling the model to capture correlations between data at various positions. The multi-head attention mechanism not only focuses on a finite set of features within the input sequence but also comprehensively captures information across different positions within the sequence. The framework of the multi-head attention mechanism is shown in Figure 2.

Multi-Head Attention applies Softmax and linear layer transformations to BiGRU outputs, enabling weighted attention across different time steps within the model. This mechanism empowers the model to capture long-term dependencies and critical temporal information.

$$\alpha_t^{(i)} = \frac{\exp\left(W^{(i)^T} h_t\right)}{\sum_{j=1}^{T} \exp\left(W^{(i)^T} h_j\right)}, \tag{5}$$

$$c^{(i)} = \sum_t \alpha_t^{(i)} h_t, \tag{6}$$

In the equation, $\alpha_t^{(i)}$ represents the attention weight of the i-th attention head at time step t; $W^{(i)}$ denotes the weight matrix of the i-th attention head; h_t is the hidden state output by the bidirectional GRU at time step t, representing the feature representation at the current time step; exp is the exponential function, used to amplify larger values and enhance the effectiveness of the Softmax operation; $c^{(i)}$ is the context vector for the i-th attention head, containing key information relevant to prediction; $\alpha_t^{(i)}$ is the attention weight for the i-th attention head at time step t, reflecting the model's focus on information.

https://doi.org/10.3390/pr14081194

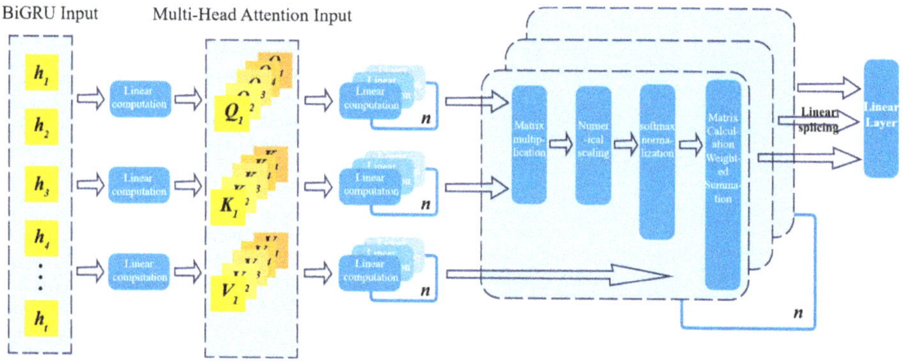

Figure 2. Overall framework of Multi-Head Attention.

2.2.4. Bayesian Hyperparameter Optimization

Machine learning models typically set fixed hyperparameters before training. Default parameters can easily lead to models being overly complex or overly simplistic. Improper regularization parameters may cause overfitting or underfitting, potentially increasing sensitivity to noise and outliers. Hyperparameter optimization enhances the performance of the CNN-BiGRU-Multi-Head Attention model by systematically searching the hyperparameter space to identify an optimal set of hyperparameters.

$$\lambda = \left\{ \left[k_{\text{CNN}}, d_{\text{BiGRU}}, H_{\text{att}}, \eta, B \right]^{\text{T}} \right\}, \tag{7}$$

In the equation, λ is the hyperparameter vector to be optimized in this study; k_{CNN} denotes the size of the convolutional kernel in the CNN section; d_{BiGRU} is the dimension of the BiGRU hidden layer; H_{att} represents the number of heads in the multi-head attention component; η is the learning rate; and B is the batch size.

2.2.5. Predicted Value Calculation

$$\hat{y} = W_{\text{fc}} \text{Pool}\left(\left\{ c^{(1)}, c^{(2)}, \ldots, c^{(H_{\text{att}})} \right\} \right) + b_{\text{fc}}, \tag{8}$$

In the equation, $\hat{y}$ is the final predicted value; W_{fc} and b_{fc} represent the weights and biases of the fully connected layer, respectively; H_{att} represents the number of heads in the multi-head attention component; and Pool is the pooling operation.

2.3. Data Processing

2.3.1. Data Sources and Feature Descriptions

The data used in this study were derived from single-well production records of Oilfield B, comprising a total of 2027 daily sampling entries.

Fifteen original features were utilized in this paper. These consisted of daily process parameters (Daily liquid production, Daily oil production, Daily water production, Daily gas production, Water cut, Oil pressure, Tubing pressure, Wellhead back pressure, Wellhead temperature, Post-choke temperature) and cumulative production metrics (Total gas production, Total oil production, Total water production, Gas-oil ratio). The "Bottomhole pressure" is the target for prediction.

The predictive performance of statistical analysis and machine learning models is highly dependent on data quality. Data analysis can reveal patterns and relationships within the data, helping to understand its structure and thereby optimize model selection and tuning. Through preprocessing, noise removal, and missing value imputation, model

https://doi.org/10.3390/pr14081194

stability and accuracy can be enhanced. Simultaneously, data analysis aids in identifying outliers, preventing them from negatively impacting model results. Sound data analysis not only improves model effectiveness but also provides robust support for research.

2.3.2. Data Description

This study conducted meticulous analysis and preprocessing of the data, and preliminary statistical results are shown in Table 1. Skewness is prevalent across all variables, with some variables—such as Daily water production and Wellhead back pressure—exhibiting anomalies or extreme values. To enhance modeling accuracy, the raw data require analysis and appropriate processing.

Table 1. Data Description of Relevant Production Characteristics.

Feature	Mean	Standard Deviation	Minimum	First Quartile	Median	Third Quartile	Maximum	Skewness	Abundance
Daily liquid production	77.70	18.43	0.90	68.90	78.36	84.26	161.23	0.37	2.05
Daily oil production	76.43	17.60	0.90	68.90	77.62	83.84	155.04	0.28	2.27
Daily water production	1.32	1.84	0.00	0.00	0.00	3.02	23.04	1.69	9.79
Daily gas production	45.42	10.34	0.56	40.32	47.29	50.27	123.86	0.94	8.08
Water cut	1.49	2.02	0.00	0.00	0.00	3.80	23.34	1.42	6.67
oil pressure	58.44	17.82	0.00	46.59	50.12	82.91	90.90	0.65	−1.10
Tubing pressure	16.39	13.68	0.00	6.77	10.18	28.37	51.31	1.02	−0.38
Wellhead back pressure	15.47	2.43	−30.64	13.68	16.04	17.26	19.70	−4.42	67.62
Wellhead temperature	53.54	9.30	0.00	49.70	53.70	60.00	74.70	−1.70	6.08
Post-choke temperature	41.17	14.01	0.00	28.50	42.00	52.30	73.30	−0.00	−0.90
Total gas production	4.67	2.62	0.00	2.43	4.59	6.94	9.20	−0.03	−1.15
Total oil production	7.94	4.42	0.01	4.28	7.89	11.73	15.48	−0.08	−1.12
Total water production	0.21	0.08	0.00	0.18	0.27	0.27	0.27	−1.30	0.20
Gas-oil ratio	5999.60	667.95	1975.07	6099.67	6100.10	6174.67	10,742.57	−1.68	7.52
Bottomhole pressure	93.63	22.13	0.68	77.79	88.85	120.71	128.72	−0.53	1.50

Some data exhibited significant outliers (the minimum Wellhead back pressure value of −30.64 MPa is inconsistent with actual conditions), necessitating the handling of these outliers.

2.3.3. Box Plot Analysis and Outlier Removal

The statistical characteristics table indicates that some variables contain significant outliers (e.g., minimum Wellhead back pressure of −70.44, extremely skewed Water cut), which may interfere with model training.

In this study, the data were first subjected to outlier removal using box plots (IQR) combined with logical judgment (Figure 3). Following this, a filtering process was performed to eliminate extreme values in key parameters such as Wellhead back pressure, Water cut, and Daily water production, thereby ensuring that only physically consistent samples were retained.

https://doi.org/10.3390/pr14081194

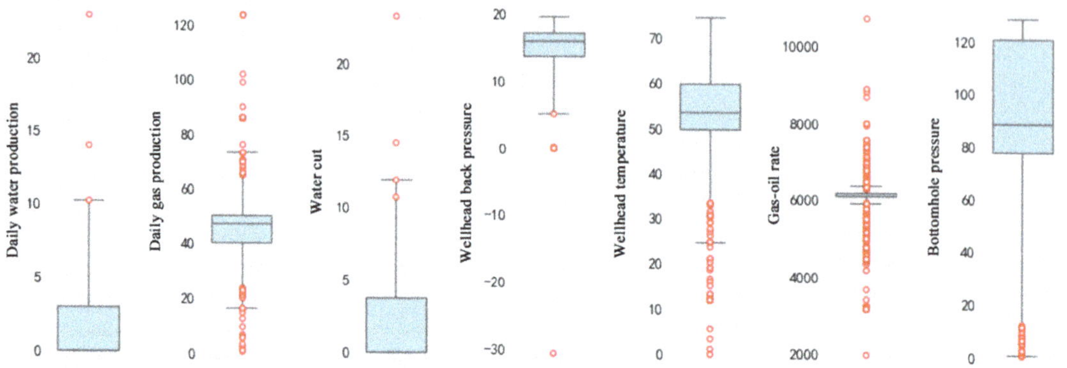

Figure 3. Box plot of certain features.

2.3.4. Feature Standardization

Standardization facilitates the model's accelerated search for optimal solutions. To eliminate the influence of feature dimensions, continuous variables in this study were standardized using Z-score normalization, resulting in distributions with a mean of 0 and a standard deviation of 1. This ensures that each independent variable exhibits values on a uniform scale.

$$z = \frac{x - \mu}{\sigma}, \tag{9}$$

In the equation, z is the standard score; x is the raw data value; μ is the mean; and σ is the population standard deviation.

2.3.5. Correlation Analysis and Feature Selection

The predictive performance of statistical analysis and machine learning models heavily depends on data quality. While the entire dataset underwent cleaning as described above, the retention of all fourteen independent variables was excessive for predicting bottomhole pressure. A reduction in the number of input parameters can effectively enhance model efficiency by mitigating the negative impacts of excessive data on training, such as prolonged training times and reduced prediction accuracy. Therefore, selecting an appropriate number of independent variables is crucial. To this end, the independent variables were screened using Spearman's correlation coefficient in this study.

The Spearman correlation diagram revealed strong positive or negative correlations among certain independent variables (Figure 4), as well as between these variables and bottomhole pressure. Among them, variables such as daily water production, water cut, oil pressure, tubing pressure, post-choke temperature, total gas production, total oil production, and total water production exhibited high correlations with the predicted variable (bottomhole pressure). These variables were therefore identified as preliminary candidates for feature selection.

However, strong correlations among independent variables can lead to collinearity and multicollinearity issues in linear machine learning models. If two variables are highly correlated, one variable can predict the other. Highly correlated independent variables provide redundant information for predicting the target output, thereby compromising the predictive capability of the developed model. Therefore, removing some highly correlated independent variables helps reduce dimensionality and improves prediction speed and accuracy.

 https://doi.org/10.3390/pr14081194

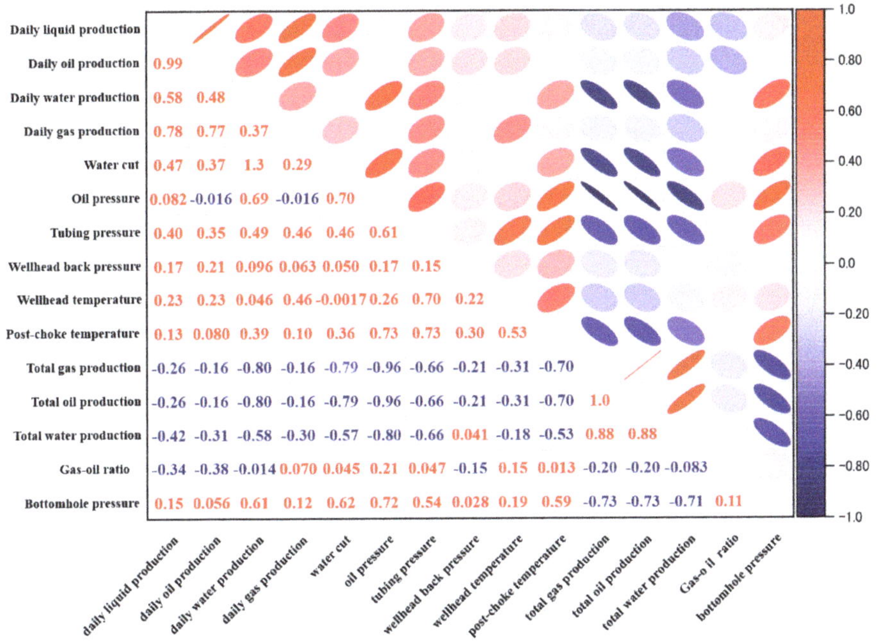

Figure 4. Spiegelman thermal coefficient for all features.

To further refine the number of independent variables required for predicting Bottomhole pressure and reduce the impact of multicollinearity on predictions, this study continued to refine feature selection based on Spearman's correlation coefficient. Information redundancy existed among the three features: Total oil production, Total gas production, and Total water production. To ensure the statistical effectiveness of the model, only cumulative gas production was ultimately selected from these three. The final features chosen for this study were "Total gas production", "Oil pressure", "Water cut", "Daily water production", "Post-choke temperature", and "Tubing pressure".

2.3.6. Data Expansion

To address the challenges of sparse effective data and high variability across different production periods within the complex context of non-steady-state flow and variable operating conditions, a data augmentation method based on the principles of sliding window replication with translation and noise augmentation is introduced in this study. The "sliding window replication with translation" technique was systematically applied by moving a fixed-size window to traverse and replicate data segments, thereby effectively expanding the coverage of the training set. The core parameters of the adopted translation sliding window method in this study are set as window size 1200 and an overlap step 200. This process resulted in a significant increase in the size of the sample set and a marked improvement in the density of the training data distribution. The workflow of the sliding window is shown in Figure 5.

https://doi.org/10.3390/pr14081194

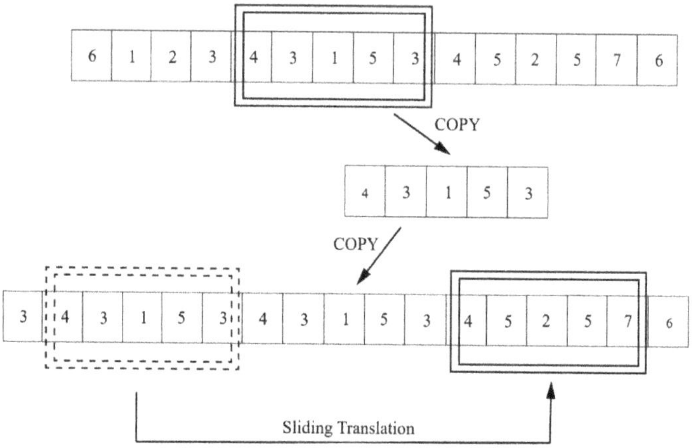

Figure 5. The workflow of Sliding window Data Extension.

Small Gaussian noise was added to each input sample as an additive perturbation to enhance model robustness.

$$x_{aug} = x + \mathcal{N}\left(0, \sigma^2\right), \tag{10}$$

In the equation, x_{aug} denotes noisy data, i.e., data with added noise; x represents the raw data, which are unperturbed input samples; $\mathcal{N}$ is a Gaussian normal distribution function; σ represents the noise standard deviation, used to control noise intensity.

During the data augmentation process, the noise intensity is set to 5%. This ratio is determined based on the actual noise level of oilfield data, which is particularly consistent with the measurement error characteristics of downhole and surface sensors and conforms to the actual working conditions of data acquisition in petroleum engineering. In this study, Gaussian noise perturbation with a mean (loc) of 0.0 and a standard deviation (scale) of 0.05 is employed to implement data augmentation. These parameters are optimized and selected through multiple experiments. By applying small-scale perturbations, only weak noise is added to the normalized bottomhole pressure labels. This method can not only effectively expand the dataset and enhance the generalization ability of the model, but also avoid data distortion, excessive augmentation and artifacts caused by an excessively high noise ratio or amplitude. It preserves the temporal distribution characteristics of the raw data to the maximum extent, thereby ensuring the physical authenticity and engineering rationality of the augmented data.

3. Results and Discussion

A unified dataset was utilized to train both the aforementioned novel model and conventional machine learning models (CNN, LSTM, GRU, BiGRU) along with their hybrid variants, and their performance was subsequently compared in this study. A reliable prediction evaluation should rely on test data rather than training data. Most deep learning models perform well on training sets but exhibit error accumulation on test sets. Therefore, the prediction results of each model on the prediction set were presented in this section to evaluate their generalization capabilities.

Among the basic models, both traditional time series models and convolutional models demonstrated certain convergence capabilities, though with noticeable differences. In the comparison of fusion models, performance disparities between different architectures became even more pronounced. Among the fusion models, the LSTM-GRU and LSTM-BiGRU models exhibited slower loss reduction during training, with higher convergence

https://doi.org/10.3390/pr14081194

values in the later stages, accompanied by some degree of fluctuation. This indicated that their modeling capabilities for key features remained insufficient in the highly nonlinear and strongly time-correlated task of wellbore pressure estimation. In contrast, a faster and more stable decline in the loss function was observed for the CNN-BiGRU model, which ultimately converged to a lower level. An effective complementarity between the convolutional network and bidirectional temporal modeling was evident in the extraction of local operating condition features, as demonstrated by this result.

The introduction of the multi-head attention mechanism further enhanced the model's performance. The CNN-BiGRU-Multi-Head Attention model consistently maintained the lowest loss values throughout training and exhibited the smoothest convergence curve. This demonstrated that the multi-head attention mechanism adaptively enhanced focus on critical operating condition features and significant time steps, which effectively suppressed the interference of redundant information during model training. The comparison of loss functions for each model is shown in Figure 6. The horizontal axis represents the training epoch, and the vertical axis represents the loss function value, showing the loss convergence of each model during the training process.

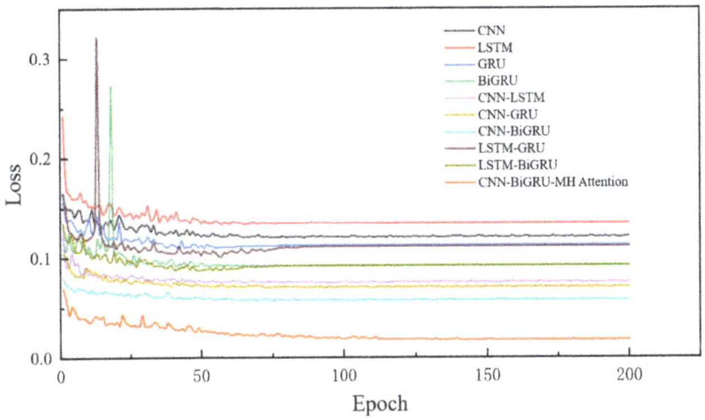

Figure 6. Loss plots for different models.

The scatter plots of predicted versus actual bottomhole pressure (BHP) for different models is shown in Figure 7. The horizontal axis represents the actual BHP, and the vertical axis represents the predicted BHP. The blue dashed line is the ideal prediction line (True Line), and the closer the scatter points are to this line, the higher the prediction accuracy of the model.

In the basic model, the scatter plot generally followed a diagonal distribution, but the point cloud remained relatively dispersed, with more pronounced deviations in extreme pressure ranges such as high/low pressure. This indicated that a single structural model still has limitations in fitting highly nonlinear and strongly time-correlated signals like bottomhole pressure. In the fusion models, the scatter points were generally distributed more closely along the diagonal, and exhibited a higher degree of concentration. This demonstrated that the fusion architecture exhibited stronger complementarity in feature representation and temporal modeling, which effectively enhanced the consistency and stability of bottomhole pressure predictions. However, differences persisted among the fusion architectures. Some models still exhibited notable dispersion in extreme pressure ranges, indicating room for improvement in capturing complex operational variations. The superior capability of the CNN-BiGRU model in modeling bottomhole pressure dynamics was evidenced by its more concentrated scatter plot distribution, where most data points aligned closely along the diagonal, as revealed in a further comparison of the key models.

https://doi.org/10.3390/pr14081194

With the incorporation of the multi-head attention mechanism, the scatter plot of the CNN-BiGRU-Multi-Head Attention model showed further convergence. Points near the diagonal became denser, and outliers decreased. These changes demonstrated that the multi-head attention mechanism could adaptively amplify key information while suppressing redundant noise, thereby enhancing the model's overall fitting accuracy and robustness.

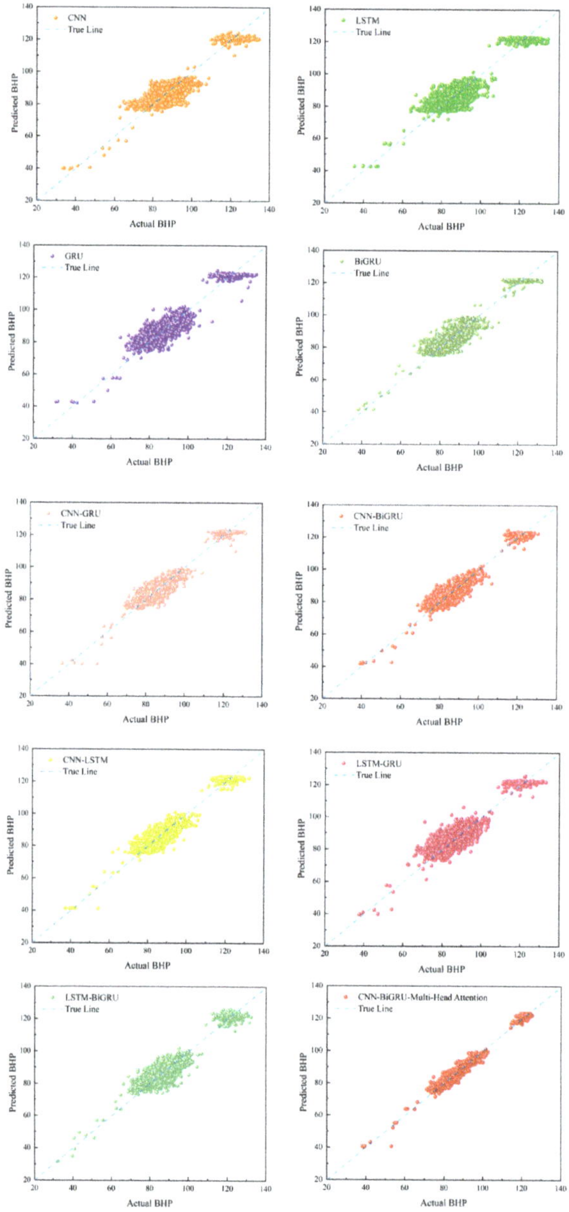

Figure 7. Cross plot of the actual and predicted values of the ten models.

To comprehensively evaluate the predictive performance of each model, radar charts (Figure 8) are constructed based on five metrics: coefficient of determination (R^2, measuring

https://doi.org/10.3390/pr14081194

the goodness of fit of the model; the closer to 1 the better the performance), root mean square error (RMSE, measuring the deviation between predicted and actual values), mean absolute error (MAE, measuring the average absolute value of prediction errors), symmetric mean absolute percentage error (SMAPE, a symmetric indicator of relative errors), and mean directional accuracy (MDA, measuring the accuracy of prediction direction), which are used to comprehensively compare the bottomhole pressure prediction performance of each model. The specific values of each evaluation metric were presented in Table 2.

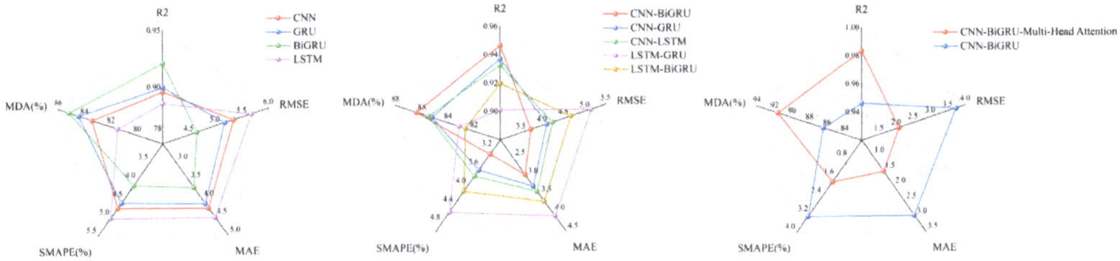

Figure 8. The comparison of model performance on the test set.

Table 2. Model performance data on the test set.

Model	RMSE	R^2	MAE	SMAPE(%)	MDA(%)
CNN	5.34595	0.8954	4.257	4.75	83.38
LSTM	5.67523	0.8853	4.510	5.03	81.45
GRU	5.18282	0.8993	4.132	4.61	84.38
BiGRU	4.64768	0.9204	3.689	4.13	85.13
CNN-LSTM	4.23495	0.9324	3.399	3.78	85.46
CNN-GRU	4.10117	0.9369	3.260	3.65	85.21
CNN-BiGRU	3.72384	0.9463	2.948	3.31	86.38
LSTM-GRU	5.14049	0.9011	4.078	4.62	83.12
LSTM-BiGRU	4.66329	0.9197	3.669	4.11	82.62
CNN-BiGRU -MH Attention	2.07094	0.9831	1.580	1.80	91.56

Among the base models, BiGRU demonstrated the best overall performance. Compared to CNN/LSTM/GRU, BiGRU showed superiority in both error and goodness-of-fit, indicating that bidirectional temporal dependency modeling is more effective for signals with strong temporal correlations. In the corresponding radar chart of base models, BiGRU also covered a larger area, reflecting its more balanced comprehensive performance.

The fusion model generally outperformed the base model, indicating that the combined "local feature extraction + temporal modeling" approach could more fully characterize the dynamic variation patterns of bottomhole pressure. Among these, CNN-BiGRU demonstrated the best performance, exhibiting a larger coverage area and a more balanced shape in the radar chart among the fusion models.

Compared to BiGRU, which was the best basic model before fusion, the improvement of CNN-BiGRU was quite evident. This indicates that CNN's extraction of local patterns within input sequences effectively complements BiGRU's bidirectional temporal modeling. It should be noted that the fusion effect of LSTM-GRU and LSTM-BiGRU was not ideal. For instance, LSTM-GRU performed close to or even fell short of some baseline models. Although LSTM-BiGRU outperformed LSTM-GRU, it lagged significantly behind CNN-based fusion models. This suggested that merely stacking or fusing recurrent structures yields relatively limited gains for sequences like bottomhole pressure, which exhibit both local fluctuation characteristics and coexisting short- and long-term dependencies.

https://doi.org/10.3390/pr14081194

By introducing multi-head attention to the optimal fusion architecture CNN-BiGRU, the CNN-BiGRU-Multi-Head Attention model achieved the best results across the entire table. Compared to CNN-BiGRU, it delivered significant improvements: RMSE decreased by approximately 55.6%, MAE further decreased by about 46.4%, SMAPE decreased by approximately 45.6%, R^2 increased by 0.0368, and MDA increased by 5.18%. In the radar chart of attention models, it exhibited the largest coverage area, with more "extended" and balanced dimensions across all metrics. This demonstrated that multi-head attention can adaptively enhance the contribution of critical time steps/key features, significantly improving trend judgment capability (MDA) while reducing errors. Consequently, it proved more suitable for predicting bottomhole pressure under complex operating conditions.

The efficacy of the introduced attention mechanism in capturing long-term dependencies and complex features within the data was demonstrated by the above findings, leading to an enhancement of the model's predictive capability.

Despite the advantages of the proposed CNN-BiGRU-Multi-Head Attention mechanism in this study, it should be acknowledged that the present work is subject to certain limitations. Firstly, the model was trained and validated on a single-well dataset comprising only 2027 samples. While this dataset is representative of the target well's operating conditions, its relatively small size may limit the robustness and generalization capability of the proposed model, particularly given the complexity of the deep learning architecture employed, which carries an inherent risk of overfitting. Additionally, the lack of validation across multiple wells, different reservoir types, and varying operating conditions restricts the broader applicability of the current findings. Future work will focus on expanding the dataset to include multi-well, multi-reservoir, and multi-condition data to further verify and enhance the generalization performance of the model.

4. Conclusions

This paper addresses challenges in bottomhole pressure prediction, including sparse samples, strong temporal fluctuations, and high modeling complexity, by proposing a deep prediction method based on data augmentation and multi-scale sliding windows. Key findings are as follows:

1. Data augmentation strategies, including Gaussian perturbation and sliding window translation, significantly enhanced model training sample diversity and generalization capabilities without introducing physical anomalies.
2. The sliding window mechanism effectively balanced short-term disturbances and long-term trends, improving the model's sensitivity to complex pressure variations.
3. The CNN-BiGRU-Multi-Head Attention model architecture fully integrated local feature extraction, sequence modeling, and attention-based focus capabilities, enabling more precise modeling of pressure sequences.
4. Compared to traditional models, the proposed method outperformed them across all metrics, demonstrating particularly significant advantages in RMSE and MAE.

Author Contributions: Conceptualization: X.X. and G.Y.; Methodology: X.X. and X.J. (Xuecheng Jiang); Software: X.X., X.J. (Xuecheng Jiang) and G.Y.; Validation: X.J. (Xuecheng Jiang); Formal Analysis: X.X., X.J. (Xuecheng Jiang) and S.L.; Investigation: X.J. (Xuecheng Jiang); Data Curation: X.J. (Xuecheng Jiang) and S.L.; Resources: G.Y. and S.L.; Writing—Original Draft: X.X., X.J. (Xuecheng Jiang) and S.L.; Writing—Review and Editing: X.X., X.J. (Xuecheng Jiang) and S.L.; Visualization: X.J. (Xuecheng Jiang) and S.L.; Supervision: X.X., G.Y. and X.J. (Xujian Jiang); Project Administration: G.Y.; Funding Acquisition: G.Y. All authors have read and agreed to the published version of the manuscript.

https://doi.org/10.3390/pr14081194

Funding: This research was funded by the National Oil & Gas Major Project, grant number 2025ZD1401502.

Data Availability Statement: With the consent of the corresponding author, data will be provided as needed.

Conflicts of Interest: Author Xujian Jiang was employed by the Research Institute of Petroleum Exploration and Development, Tarim Oilfield Company. The remaining authors declare that the research was conducted in the absence of any commercial or financial relationships that could be construed as a potential conflict of interest.

References

1. Campos, D.; Wayo, D.D.K.; De Santis, R.B.; Martyushev, D.A.; Yaseen, Z.M.; Duru, U.I.; Saporetti, C.M.; Goliatt, L. Evolutionary automated radial basis function neural network for multiphase flowing bottom-hole pressure prediction. *Fuel* **2024**, *377*, 132666. [CrossRef]
2. Hailong, L. Semi-analytical solution for bottomhole pressure transient analysis of a hydraulically fractured horizontal well in a fracture-cavity reservoir. *Sci. Rep.* **2022**, *12*, 22095. [CrossRef] [PubMed]
3. Adeyinka, A.; Oriola, O.; Tomomewo, O.S. An innovative approach for oil well bottomhole pressure forecasting using Kolmogorov-Arnold Neural Networks (KANs): A case study in an offshore oilfield. *Deep Resour. Eng.* **2025**, *3*, 100233. [CrossRef]
4. Ahmadi, M.A.; Chen, Z. Machine learning models to predict bottom hole pressure in multi-phase flow in vertical oil production wells. *Can. J. Chem. Eng.* **2019**, *97*, 2928–2940. [CrossRef]
5. Maraggi, L.M.R.; Walsh, M.P.; Lake, L.W.; Male, F.R. Bayesian variable pressure decline-curve analysis for shale gas wells. *Unconv. Resour.* **2024**, *4*, 100103. [CrossRef]
6. Zhang, C.; Zhang, R.; Zhu, Z.; Song, X.; Su, Y.; Li, G.; Han, L. Bottom hole pressure prediction based on hybrid neural networks and Bayesian optimization. *Pet. Sci.* **2023**, *20*, 3712–3722. [CrossRef]
7. Makarov, N.; Al-Shargabi, M.; Wood, D.A.; Burnaev, E.; Davoodi, S. Prediction of oil production rate in multiple wells of a producing field applying combined deep–learning and optimization techniques. *Fuel* **2026**, *406*, 136847. [CrossRef]
8. Al Shehri, F.H.; Gryzlov, A.; Al Tayyar, T.; Arsalan, M. Utilizing machine learning methods to estimate flowing bottom-hole pressure in unconventional gas condensate tight sand fractured wells in Saudi Arabia. In Proceedings of the SPE Russian Petroleum Technology Conference, SPE, Virtual, 26–28 October 2020; D043S032R002. [CrossRef]
9. Chen, W.; Di, Q.; Ye, F.; Zhang, J.; Wang, W. Flowing bottomhole pressure prediction for gas wells based on support vector machine and random samples selection. *Int. J. Hydrogen Energy* **2017**, *42*, 18333–18342. [CrossRef]
10. Mohammadpoor, M.; Shahbazi, K.; Torabi, F.; Qazvini, A. A new methodology for prediction of bottomhole flowing pressure in vertical multiphase flow in Iranian oil fields using artificial neural networks (ANNs). In Proceedings of the SPE Latin America and Caribbean Petroleum Engineering Conference, SPE, Lima, Peru, 1–3 December 2010; SPE-139147-MS. [CrossRef]
11. Rathnayake, S.; Rajora, A.; Firouzi, M. A machine learning-based predictive model for real-time monitoring of flowing bottom-hole pressure of gas wells. *Fuel* **2022**, *317*, 123524. [CrossRef]
12. Okoro, E.E.; Sanni, S.E.; Obomanu, T.; Igbinedion, P. Predicting the effects of selected reservoir petrophysical properties on bottomhole pressure via three computational intelligence techniques. *Pet. Res.* **2023**, *8*, 118–129. [CrossRef]
13. Liu, X.Q.; Li, Y.Z. Prediction of flowing bottomhole pressures for two-phase coalbed methane wells. *Acta Pet. Sin.* **2010**, *31*, 998. Available online: https://kns.cnki.net/kcms2/article/abstract?v=Jz5IuRg0t03LZIy-NkmJv4JSZwEDPtXs2ldCSVwMFphlny1hvFQ5_waY_fZgGXUnazMoKCVR-thX23zKYrNHQ3ZCI3ogKZIaTa5lOiDQ_ZHhKfoTWX1TkOlcEKM7GeEwz7PAgkIVHPSLYaHHuhR44Hay9NNfk8XO-WG25c-Vxfi5_gSDBQCrcQ==&uniplatform=NZKPT&language=CHS (accessed on 5 April 2026).
14. Al-Otaibi, M.B.; Elkamel, A.; Nassehi, V.; Abdul-Wahab, S.A. A computational intelligence based approach for the analysis and optimization of a crude oil desalting and dehydration process. *Energy Fuels* **2005**, *19*, 2526–2534. [CrossRef]
15. Gladchenko, E.; Orlov, D.; Cheremisin, A.; Chernov, A.; Sharonov, M.; Samsonov, I.; Savchuk, D.; Novikov, A.; Koroteev, D. Gradient Boosting Justification for Predicting Flow Assurance Issues in the Bottomhole Zone of Gas-Condensate Wells. *Petroleum* 2025, *in press*. [CrossRef]
16. Jinming, L.; Xuesen, Q.; Jian, D.; Zhi, Z.; Yue, Z.; Jiangping, D. Dynamic Prediction and Control Method for Annular Pressure in Offshore Gas Wells Based on Real-Time Monitoring Data. *Energy Sci. Eng.* **2025**, *13*, 6040–6049. [CrossRef]
17. Liu, X.; Zhou, S.; Zhang, W.; Cheng, Q.; Liu, D.; Yan, D.; Wang, H. Machine learning method for lacustrine shale oil reservoirs: Improving movable fluid porosity prediction. *Unconv. Resour.* **2026**, *9*, 100263. [CrossRef]
18. Almudhhi, S.; Lababidi, H.M.S.; Garrouch, A.A. Application of Machine Learning for Modeling Heavy Oil Viscosity. *J. Eng. Res.* **2025**, *13*, 6040–6049. [CrossRef]

 https://doi.org/10.3390/pr14081194

19. Liu, Z.; Dong, L.; Li, C.; Xu, H. Spatio-temporal feature perception and fusion for multi-frame maritime infrared small target detection. *Adv. Eng. Inform.* **2026**, *74*, 104641. [CrossRef]
20. Li, X.; Miskimins, J.L.; Hoffman, B.T. A combined bottom-hole pressure calculation procedure using multiphase correlations and artificial neural network models. In Proceedings of the SPE Annual Technical Conference and Exhibition, SPE, Amsterdam, The Netherlands, 27–29 October 2014; SPE-170683-MS. [CrossRef]
21. Antonelo, E.A.; Camponogara, E.; Foss, B. Echo state networks for data-driven downhole pressure estimation in gas-lift oil wells. *Neural Netw.* **2017**, *85*, 106–117. [CrossRef]
22. Jiang, H.; Ceng, J.; Li, L.; Wen, T.; Zhou, J.; Chen, X.; Wang, J. Data—driven bottomhole pressure prediction for gas storage reservoirs. *Spec. Oil Gas* **2025**, *32*, 122–129. Available online: https://kns.cnki.net/kcms2/article/abstract?v=Jz5IuRg0t0381oN02D54u01__-MxPjr4saohBieEkkzZRBvGTRAr7wBWGnB8tWg1A76sv7a44ZfLN6psH2iqLaFPb88V6QerXumiewdZ8tOVG9CDpPdqOYihEluaE2VHD1zEI_lAFrYiLz_JvKifeHPMtINWRlNSMcH5d5mFUdUHcJ6P1I65sg==&uniplatform=NZKPT&language=CHS (accessed on 5 April 2026).
23. Guo, H. Rapid calculation method of bottom hole pressure for gas storage based on incremental learning. *Fault-Block Oil Gas Field* **2025**, *32*, 292–299. Available online: https://kns.cnki.net/kcms2/article/abstract?v=Jz5IuRg0t02jZYHf_UXxUjhZZ3pG-rhQ7Mcivvp7VgCwuU3obflTF1B7BmK3ZYTmK7GCrjThjjmLeKz3r3njwVVVWdCZlRFBd5UnAjISJ48GZe8Mt-nWLOQMeNx8UlYisAryOuY21-ho7WJItunLQA-1-9bQ1-upVWkhY6nBJ0SM9mFZe6u5TQ==&uniplatform=NZKPT&language=CHS (accessed on 5 April 2026).
24. Spesivtsev, P.; Sinkov, K.; Sofronov, I.; Zimina, A.; Umnov, A.; Yarullin, R.; Vetrov, D. Predictive model for bottomhole pressure based on machine learning. *J. Pet. Sci. Eng.* **2018**, *166*, 825–841. [CrossRef]
25. Tian, C. Machine Learning Approaches for Permanent Downhole Gauge Data Interpretation. Ph.D. Thesis, Stanford University, Stanford, CA, USA, 2018; 199p. Available online: https://pangea.stanford.edu/ERE/pdf/pereports/PhD/Tian_Chuan2018.pdf?utm_source=chatgpt.com (accessed on 5 April 2026).
26. Ignatov, D.I.; Sinkov, K.; Spesivtsev, P.; Vrabie, I.; Zyuzin, V. Tree-based ensembles for predicting the bottomhole pressure of oil and gas well flows. In *Proceedings of the International Conference on Analysis of Images, Social Networks and Texts*; Springer International Publishing: Cham, Switzerland, 2018; pp. 221–233. [CrossRef]
27. Firouzi, M.; Rathnayake, S. Prediction of the flowing bottom-hole pressure using advanced data analytics. In Proceedings of the //SPE/AAPG/SEG Asia Pacific Unconventional Resources Technology Conference, URTEC, Brisbane, Australia, 18–19 November 2019; D021S013R001. [CrossRef]
28. Ternyik, I.V.J.; Bilgesu, H.I.; Mohaghegh, S. Virtual measurement in pipes: Part 2-liquid holdup and flow pattern correlations. In Proceedings of the SPE Eastern Regional Meeting, SPE, Morgantown, WV, USA, 17–19 October 1995; SPE-30976-MS. [CrossRef]
29. Osman, E.A.; Ayoub, M.A.; Aggour, M.A. Artificial neural network model for predicting bottomhole flowing pressure in vertical multiphase flow. In Proceedings of the SPE Middle East Oil and Gas Show and Conference, SPE, Manama, Bahrain, 12–15 March 2005; SPE-93632-MS. [CrossRef]
30. Sami, N.A.; Ibrahim, D.S. Forecasting multiphase flowing bottom-hole pressure of vertical oil wells using three machine learning techniques. *Pet. Res.* **2021**, *6*, 417–422. [CrossRef]
31. Al-Shammari, A. Accurate prediction of pressure drop in two-phase vertical flow systems using artificial intelligence. In Proceedings of the SPE Kingdom of Saudi Arabia Annual Technical Symposium and Exhibition, SPE, Dhahran, Saudi Arabia, 22–24 May 2011; SPE-149035-MS. [CrossRef]
32. Sun, H.; Luo, Q.; Xia, Z.; Li, Y.; Yu, Y. Bottomhole pressure prediction of carbonate reservoirs using XGBoost. *Processes* **2024**, *12*, 125. [CrossRef]
33. Li, X.; Li, X.; Xie, H.; Feng, C.; Cai, J.; He, Y. Enhanced coalbed methane well production prediction framework utilizing the CNN-BL-MHA approach. *Sci. Rep.* **2024**, *14*, 14689. [CrossRef]
34. Pang, Z.; Zhang, R.; Ma, M.; Wang, H.; Li, Q.; Wang, C. Real-Time Prediction of Bottom Hole Pressure via Graph Neural Network. *Processes* **2025**, *13*, 4081. [CrossRef]

 https://doi.org/10.3390/pr14081194

processes

MDPI

Article

Smart Oil Production Forecasting Process Using Deep Learning and African Vulture Optimization Algorithm

Xiankang Xin [1,2,3,4], Zhao Xie [5], Saijun Liu [1,2,3,4,*], Gaoming Yu [1,2,3,4,*] and Jing Cao [5]

1 School of Petroleum Engineering, Yangtze University, Wuhan 430100, China; xiankang.xin@hotmail.com
2 State Key Laboratory of Low Carbon Catalysis and Carbon Dioxide Utilization, Yangtze University, Wuhan 430100, China
3 Hubei Key Laboratory of Oil and Gas Drilling and Production Engineering, Yangtze University, Wuhan 430100, China
4 National Engineering Research Center for Oil & Gas Drilling and Completion Technology, School of Petroleum Engineering, Yangtze University, Wuhan 430100, China
5 College of Information and Mathematics, Yangtze University, Jingzhou 434023, China; 2022710163@yangtzeu.edu.cn (Z.X.); 100521@yangtzeu.edu.cn (J.C.)
* Correspondence: liusj@yangtzeu.edu.cn (S.L.); ygm@yangtzeu.edu.cn (G.Y.)

Abstract

Accurate prediction of reservoir production dynamics remains a key challenge in the oil and gas industry, especially for complex, high-dimensional time-series data. Conventional models fail to capture temporal dependencies, while existing hybrid models suffer from high parameter complexity and lack automated hyperparameter tuning, increasing training difficulty. To address these issues, this study proposes a novel hybrid model, TCN-LSTM-AVOA, combining a temporal convolutional network (TCN) with a long short-term memory network (LSTM) and incorporating the African Vulture Optimization Algorithm (AVOA) to enhance forecasting accuracy. The model not only captures complex temporal relationships and nonlinear features in reservoir data but also facilitates automated tuning of critical hyperparameters (e.g., the number of TCN kernels, LSTM units, batch size, and learning rate), which significantly enhances its robustness. Compared to eight benchmark models (back propagation neural network (BPNN), LSTM, convolutional neural network(CNN)-LSTM, TCN-LSTM, LSTM-AVOA, CNN-AVOA, TCN-AVOA), TCN-LSTM-AVOA achieves superior performance on a two-dimensional, three-phase heterogeneous reservoir, yielding a root mean square error (RMSE) of 7.0806, mean absolute error (MAE) of 3.4780, coefficient of determination (R^2) of 0.9975, and mean absolute percentage error (MAPE) of 1.81%. This work demonstrates a more accurate and efficient methodology for reservoir production prediction, with significant potential for oilfield production optimization and resource management.

Keywords: temporal convolutional network; long short-term memory network; African vulture optimization algorithm; production forecasting

1. Introduction

1.1. Background and Motivation

With the continuous growth of global energy demand and the gradual depletion of conventional oil and gas resources, the petroleum industry faces significant challenges in transitioning from easily developable homogeneous reservoirs to complex heterogeneous and unconventional reservoirs, such as shale oil, tight oil, and carbonate fractured reservoirs [1–3]. These unconventional reservoirs typically exhibit strong spatial heterogeneity, complex pore structures, and multi-scale flow mechanisms, making their production behavior far more difficult to predict than that of conventional ones. Reservoir production forecasting, as a core technology in reservoir engineering, aims to quantitatively predict the production rate, pressure changes, and residual oil distribution patterns of individual wells or entire reservoirs, thereby providing a scientific basis for decision-making in field development planning and enhanced oil recovery (EOR) [4]. However, traditional physical models and empirical decline curve methods often fail to capture the nonlinear and high-dimensional temporal dependencies inherent in production data from heterogeneous reservoirs. Moreover, conventional neural networks, as a typical data-driven approach, either fail to adequately model temporal dependencies or rely heavily on manual parameter tuning. Therefore, developing more accurate and automated forecasting models that can effectively learn from complex production time-series data has become an urgent research priority. Addressing these challenges is essential for optimizing well placement, designing stimulation treatments, and improving ultimate recovery factors in unconventional resources [5].

1.2. Current Challenges and Limitations

Early oilfield development primarily relied on the field experience and intuitive judgment of geologists, e.g., plotting the production rate versus time and cumulative production curves, combined with decline curve models for prediction [6,7]. However, traditional physical models and mathematical formulas face significant limitations in accuracy and applicability when dealing with the high complexity and nonlinear characteristics of reservoir systems. With the rapid development of machine learning and deep learning technologies, their application in reservoir production forecasting has demonstrated strong potential [8]. The reservoir system involves multiple complex factors, including geological structures, physical properties, production dynamics, and fluid characteristics, which are interrelated through highly nonlinear relationships. Traditional linear models often fail to achieve sufficient prediction accuracy due to the simplification of these complex interactions. In contrast, deep learning models can automatically learn multidimensional temporal features from massive historical production data, significantly improving prediction efficiency and accuracy, such as the long short-term memory (LSTM) network and temporal convolutional network (TCN) [9,10]. However, existing research still faces several challenges: On one hand, single neural network models struggle with insufficient local feature extraction or long-term temporal relationship modeling when handling high-dimensional, nonlinear reservoir data. On the other hand, hybrid models exhibit high parameter complexity, and their hyperparameter optimization (e.g., learning rate, batch size, and network units) still relies on manual adjustment, while traditional optimization algorithms are prone to falling into local optima in high-dimensional parameter space, making it difficult to balance convergence speed and global search capability.

1.3. Literature Review

Oilfield production forecasting plays an extremely important role in oil and gas field development, helping engineers make effective development decisions. Traditional methods use production decline models, water flooding characteristic curves, and multivariate linear regression models to predict oilfield production [11–13]. The production decline models include the Arps decline model, the stretched exponential production decline (SEPD) model, the Duong decline model, etc. The Arps model is suitable for oil and gas wells that have reached a quasi-steady state, and its decline rate remains constant. The SEPD model, similar to the Arps model in its exponential decline, is more suitable for scenarios where the decline rate varies. When gas wells are in the phase of slow production

https://doi.org/10.3390/pr14101558

decline, the Duong model provides more accurate production forecasts. Wang et al. [14] integrated the Arps, SEPD, and Duong models and proposed a new decline model, which serves as a reliable basis for analyzing the production decline patterns of oil and gas wells. Gao et al. [15] proposed the generalized water flooding characteristic curve. The new water flooding characteristic curve incorporates the characteristics of the Nazaroff curve ($n = 0$) and the Sipachev curve ($n = 1$) and is capable of describing concave, convex, and S-shaped water cut variation patterns. It is of great significance for enhancing recovery efficiency and optimizing water injection development. Oil production dynamics are governed by multivariate geological and engineering factors. However, operational constraints often limit the comprehensive quantification of these interdependencies in field-scale models. Ren et al. [16] used SPSS software to select the main influencing factors and establish a prediction model of oil production based on multiple linear regression. Based on single-well production dynamic indicators, Bing [17] proposed 19 single-well recoverable reserve prediction models that apply water flooding, heavy oil thermal recovery, and other development methods to low-permeability and medium-to-high permeability reservoirs. These models can adapt to the complex and diverse practical conditions of oilfields and effectively predict single-well recoverable reserves.

With the rapid development of artificial intelligence and machine learning technologies, researchers have begun exploring the application of these technologies in production prediction, aiming to construct simpler and more effective reservoir prediction models. AI technologies address the issues of large data requirements, computational complexity, and limited applicability in traditional methods. Zhang and Hu [18] proposed a method based on multivariate time series (MTS) and vector autoregression (VAR) machine learning models for predicting well production in a water flooding reservoir. It optimizes injection–production well data through multivariate time-series analysis and uses the VAR model to determine the linear relationship between different injection–production wells, achieving accurate oil production prediction. Fang et al. [19] applied a random forest (RF) model to predict productivity in a tight gas reservoir. Through experimental comparison, the RF model performed best in high-production and medium-production wells, with an RMSE of 3.98 and an R^2 of 0.91. Du et al. [20] applied three machine learning algorithms to predict hydrogen injection mass, production mass, and production rate under different geological parameters. Yahya et al. [21] proposed an integrated machine learning framework that combines various machine learning algorithms to predict oil recovery based on field and experimental data. Among these models, the RF model was proven to be the most effective, achieving an R^2 value of 0.99 on both field Dataset A and experimental Dataset B. A hybrid model based on particle swarm optimization (PSO) and support vector machine (SVM) is proposed by Han and Bian [22] to predict the production rate in low-permeability reservoirs. The literature review on machine learning-based reservoir production forecasting methods is summarized in Table 1. However, the machine learning-based reservoir production prediction model takes a long time to train, requires a large amount of computational resources, and performs poorly when dealing with high-dimensional reservoir engineering data, which may lead to low prediction accuracy of reservoir production data.

https://doi.org/10.3390/pr14101558

Table 1. The literature review on machine learning-based reservoir production forecasting methods.

Literature	Methods	Contributions	Limitations of the Study
Zhang and Hu [18]	MTS, VAR	A hybrid MTS-VAR machine learning model for interaction law extraction in multivariate time series and oil well production prediction in waterflooding reservoirs.	The complexity of subsurface fluid flow patterns renders the method less effective; meanwhile, well groups with weak inter-well correlation also exhibit relatively limited prediction accuracy.
Du et al. [20]	XGBoost, RF, SVR	These ML algorithms can achieve high-precision predictions of hydrogen injection mass, production mass, and production rate while significantly enhancing computational efficiency. The XGBoost model demonstrated superior performance over the other two models (RF and SVR) by integrating hyperparameter optimization methods.	The initial training set construction requires 110.4 h (involving 1000 simulations). Although surrogate models significantly reduce subsequent analysis time, the upfront computational cost remained substantial.
Yahya et al. [21]	DT, RF, GBR, SVR, MLP, and KNN	Six ML techniques were applied to predict oil recovery. Among them, the RF model achieved an R^2 value as high as 0.99, demonstrating its capability to provide significantly more accurate recovery prediction.	Complex models such as MLP and GBR require significant computational resources and face challenges in memory consumption and training time, especially when processing large-scale datasets.
Han and Bian [22]	PSO-SVM	This study presents a hybrid PSO-SVM model for predicting the recovery factor of low-permeability reservoirs. The PSO-SVM model demonstrated superior performance over both the PSO-optimized back propagation (PSO-BP) neural network and empirical methods in recovery factor prediction.	The model validation is insufficient, as PSO-SVM is only compared with PSO-BP and empirical formulas.

As an important branch of machine learning, deep learning focuses on solving complex problems by constructing multilayer neural networks. Compared with traditional machine learning methods, deep learning performs well when handling big data and high-dimensional features, and it has been widely applied in the field of reservoir engineering [23]. Researchers [24–26] have trained a BP neural network based on existing production data to analyze dynamic production changes in an oilfield, enabling reservoir production prediction. However, a BP neural network cannot capture the temporal relationship in reservoir production data. LSTMs [27,28] have the ability to process long time-series data. Researchers [29–32] have applied LSTM to production forecasting, effectively predicting future production data based on historical records. Although LSTM performs well in handling time-series data, its capability for feature extraction is limited, especially when dealing with high-dimensional and complex datasets. To address this issue and further improve

https://doi.org/10.3390/pr14101558

prediction accuracy, various hybrid models that fully leverage the strengths of different algorithms have been proposed by recent studies to tackle the challenges of complexity and randomness in oil well production forecasting. For instance, the integration of multilevel discrete wavelet decomposition (MDWD) with a Transformer neural network effectively reduced the instability of time series and enhanced long-term prediction capability [33]; the fusion of a CNN-LSTM model with the partial unknown recursive prediction strategy (PURPS) demonstrated significant advantages in monthly gas production forecasting [34]. Additionally, the BiGRU-MTL framework employed a multi-task learning mechanism to enable knowledge sharing across multi-phase production forecasting tasks, thereby improving the robustness of multi-step-ahead predictions [35]. The VMD-blending model further optimized prediction performance through a multi-model ensemble strategy [36]. Meanwhile, the CNN-GRU model combined the feature extraction capability of convolutional neural networks with the temporal modeling strengths of gated recurrent units, achieving outstanding performance in enhanced oil recovery (EOR) efficiency prediction [37]. These hybrid methods, by incorporating techniques such as multi-scale decomposition, multi-task learning, and model ensemble strategies, significantly overcome the shortcomings of traditional single-model approaches in complex feature extraction and long-term dependency modeling, providing more reliable technical support for dynamic production optimization in oil and gas fields. TCNs capture temporal features in data through causal convolution and dilated convolution while expanding the receptive field of convolutional layers [9], making it more suitable than CNNs for extracting temporal features from reservoir production data. The performance of neural network models in production prediction depends on the choice of hyperparameters. The incorrect setting of hyperparameters can negatively impact the model's prediction accuracy and stability. Therefore, optimization algorithms can be introduced to further improve model performance [38–40]. Li et al. [38] proposed a PSO-CNN-LSTM model for reservoir production prediction. Compared to single neural network models, the PSO-CNN-LSTM model performs better in time-series prediction, reducing the time required for hyperparameter tuning and improving prediction accuracy. An overview of deep learning-based reservoir production forecasting methods identified from the literature is provided in Table 2. Abdollahzadeh et al. [41] proposed AVOA, which outperformed various algorithms, including the PSO and Sparrow Search Algorithm (SSA), on 30 standard test functions. Compared with other optimization algorithms, this study employs AVOA as the optimization algorithm for the neural network model to obtain the optimal solution for the model parameters. It maintains a better balance between exploration and exploitation through a starvation rate mechanism and multiple search strategies, which effectively prevent premature convergence and help avoid local optima [41].

To address these issues, integrating deep learning models with optimization algorithms to create hybrid models capable of automatically adjusting hyperparameters has become the mainstream approach for reservoir production forecasting [38,39]. Based on the above considerations regarding the development of reservoir production forecasting, this paper proposes the TCN-LSTM-AVOA model, which integrates the long short-term memory (LSTM) network, temporal convolutional network (TCN), and African Vulture Optimization Algorithm (AVOA). This model is designed to predict the bottom-hole flowing pressure (BHP), gas production rate (GPR), and water production rate (WPR) from five production wells and the BHP from one injection well in a heterogeneous reservoir. The model leverages TCN to efficiently extract local temporal features from reservoir production data and combines LSTM's capability to predict long-term dynamic evolution trends. Additionally, the AVOA algorithm is introduced to globally optimize hyperparameters such as the number of TCN kernels, LSTM units, batch size, and learning rate, achieving precise production data prediction while reducing manual hyperparameter tuning efforts. This

 https://doi.org/10.3390/pr14101558

innovative approach not only improves the accuracy of reservoir production forecasting but also reduces manual hyperparameter tuning efforts, providing a reliable data-driven tool for dynamic production management and operational risk mitigation in the petroleum industry. The research roadmap of this study is illustrated in Figure 1.

Table 2. The literature review on deep learning-based reservoir production forecasting methods.

Literature	Methods	Contributions	Limitations of the Study
Zha et al. [34]	CNN-LSTM	This study proposed a CNN-LSTM model to predict gas field production. The CNN extracted data features, while the LSTM performed time-series prediction. The CNN-LSTM model outperformed seven models, including CNN and LSTM, in terms of accuracy.	Since geological factors were not considered in the current study, the proposed method is not applicable to newly developed gas reservoirs.
Li et al. [35]	BiGRU-MTL	This study proposed a novel multistep-ahead production prediction model based on BiGRU and MTL. The BiGRU comprehensively captured bidirectional temporal patterns from production sequences, while the MTL framework, incorporating a common subnetwork and a multitask loss function, achieved complementary information sharing between multiphase production tasks. Compared to RNN, LSTM, GRU, and STL models, the BiGRU-MTL model demonstrated superior accuracy in aligning its predictions with actual production data.	There are aspects of the proposed model that could be further optimized.
Chen et al. [37]	CNN-GRU	This study proposed the CNN-GRU model to accurately predict oil production through CO_2 flooding or water flooding methods, effectively capturing the nonlinear relationship between production constraints and oil production.	The CNN-GRU model requires a longer runtime compared to the GRU model. This research only focused on the impact of injection rate (CO_2 or water) on oil production.
Zhang et al. [40]	Improved TCN models	This study proposed a single-well production prediction method based on the temporal convolutional network (TCN) model, optimized using the Sparrow Search Algorithm (SSA) to tune hyperparameters. Data dimensionality reduction and imputation were performed with principal component analysis (PCA) and Random Forest, ensuring dataset authenticity and completeness. Experimental results showed that the improved TCN model demonstrated superior predictive performance compared to 11 existing time-series models, particularly for sequences with significant fluctuation and periodic patterns.	The improved TCN model has a complex structure, which may limit its practical applications.

https://doi.org/10.3390/pr14101558

Table 2. *Cont.*

Literature	Methods	Contributions	Limitations of the Study
Li et al. [38]	PSO-CNN-LSTM	This study proposed the PSO-CNN-LSTM model, where the CNN extracted spatial features from data, the LSTM predicted time-series trends, and the PSO algorithm optimized the hyperparameters of the CNN-LSTM framework.	Current applications have focused on single-well production forecasting, while future work could explore the model's performance in multi-well production prediction within the same reservoir, considering well interference effects.
Huo et al. [39]	PSO-TCN-LSTM	This study introduced the PSO-TCN-LSTM model for accurate sweet-zone evaluation in reservoir characterization. The TCN-LSTM architecture performed feature screening and prediction through causal convolutions with progressively expanded dilation rates across network layers, enabling long-term temporal memory capabilities. The PSO algorithm optimized critical hyperparameters of the TCN-LSTM, eliminating biases caused by manual parameter tuning.	The combination of TCN and LSTM enhances the ability to extract sequential features, but the necessity of the hybrid model has not been demonstrated through ablation experiments. And there may be issues of redundancy or high computational complexity.

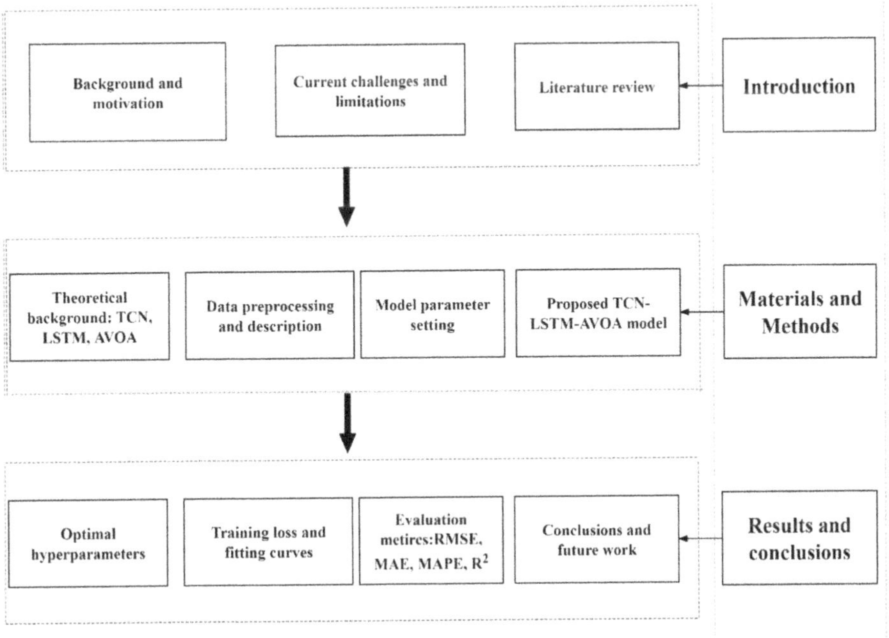

Figure 1. Research roadmap of the proposed study.

It should be noted that the scope of this work is focused on a data-driven approach utilizing historical production and operational data. While we recognize the fundamental influence of static geological properties (e.g., permeability, porosity) on reservoir dynamics,

https://doi.org/10.3390/pr14101558

these parameters are not explicitly included as varying inputs in the current model framework. This simplification allows us to first establish the core forecasting capability, with the integration of geological heterogeneity reserved for future work. The main structure of this paper is as follows. The theoretical background of the TCN-LSTM-AVOA, as well as the dataset description and model parameter configuration, is provided in Section 2. The performance of the model in the experiment is discussed in Section 3. The conclusions are presented in Section 4.

2. Materials and Methods

2.1. Theoretical Background

2.1.1. Temporal Convolutional Network

While CNN is commonly employed for image-processing tasks, its architecture can be adapted to utilize convolutional layers for scanning along the temporal dimension, enabling the extraction of local temporal patterns [42]. Building upon this adaptability, TCN, a neural network model improved from CNN, enhances temporal modeling by incorporating three key components: causal convolution, dilated convolution, and residual connection [43]. In time-series modeling, causal convolution ensures that the output at the current time step depends only on the current and previous input information, without being influenced by future inputs. So causal convolution can preserve the sequential order and causality of the data series. The network structure of TCN is shown in Figure 2, where the model processes an input sequence of $\{x_0, x_1, \ldots, x_t\}$ into an output sequence of $\{y_0, y_1, \ldots, y_t\}$. The convolution kernel size is 3×1, and each layer has three inputs at the current time step. Dilated convolution introduces an interval (dilation rate) between the convolution kernel elements, effectively expanding the receptive field of units in the convolutional layer. Dilated convolution helps the network capture a broader range of production data time-series features without increasing the number of model parameters. The dilation rate d determines the interval size of the convolution kernel, enabling the entire network to extract features at different temporal or spatial scales through the convolution kernel. As shown in Figure 2, the dilation rate d is configured as 1, 2, and 4 for successive layers in the TCN architecture. TCN also employs residual connections to prevent issues like gradient vanishing or overfitting. Residual connection skips over convolutional layers, adding the input to the output [44]. Because of the residual connection, information can be propagated to deeper layers of the network, which promotes network training and convergence.

2.1.2. Long Short-Term Memory Network

LSTM incorporates a gating mechanism with an input gate, forget gate, and output gate (Figure 3). Compared to a traditional recurrent neural network (RNN), LSTM combines long-term memory and short-term memory, enabling better handling of long sequences of data and partially addressing the gradient-vanishing problem inherent in RNNs [45]. The calculation process is as follows:

$$g_t = \varphi(w_g[h_{t-1}, x_t] + b_g), \quad (1)$$

$$i_t = \sigma(w_i[h_{t-1}, x_t] + b_i), \quad (2)$$

$$f_t = \sigma(w_f[h_{t-1}, x_t] + b_f), \quad (3)$$

$$o_t = \sigma(w_o[h_{t-1}, x_t] + b_o), \quad (4)$$

$$s_t = g_t * i_t + s_{t-1} * f_t, \quad (5)$$

$$h_t = \varphi(s_t) * o_t, \quad (6)$$

https://doi.org/10.3390/pr14101558

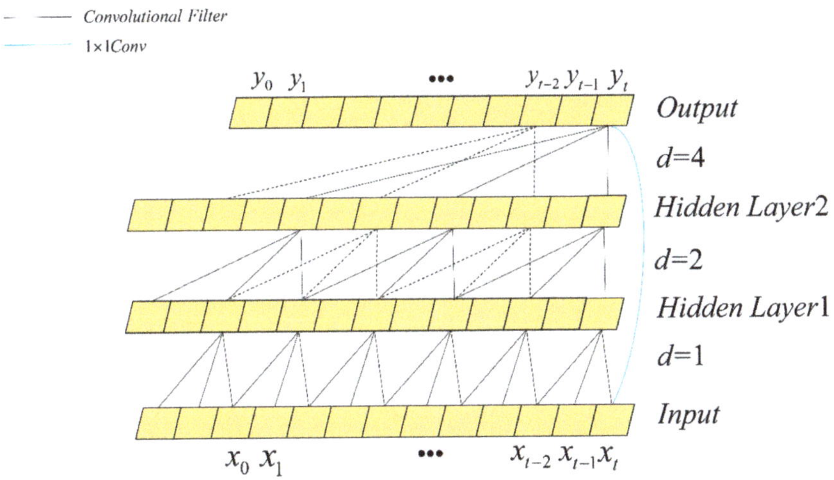

Figure 2. Structure diagram of TCN.

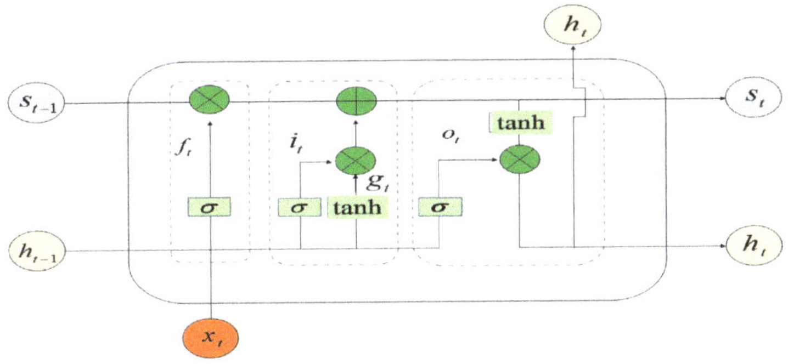

Figure 3. Structure diagram of LSTM.

The input node g_t, along with the input gate i_t, forget gate f_t, and output gate o_t, serves as the key components of LSTM. w_g is the weight matrices for the input node. b_g is the bias matrix for the input node. w_i is the weight matrix for the input gate, and b_i is the bias matrix for the input gate. w_f is the weight matrix for the forget gate, and b_f is the bias matrix for the forget gate. w_o is the weight matrix for the output gate, and b_o is the bias matrix for the output gate. h_{t-1}, h_t represents the output information at time $t-1$, t. x_t represents the input information at time t. s_{t-1} and s_t are the LSTM cell states at times $t-1$ and t, respectively. φ is the tanh activation function, and σ is the sigmoid activation function.

Based on the current input and the previous hidden state, the input gate determines whether the information should be updated into the memory cell. The input gate produces a value between 0 and 1 through the sigmoid activation function, where 1 indicates that the model chooses to accept the new information. Based on the current input and the previous hidden state, the forget gate determines whether the information in the memory cell should be forgotten. Similarly, it produces a value between 0 and 1 through the sigmoid activation function. When the value is close to 1, the information is retained. The output gate determines the output at the current moment based on the current input and the hidden state of the previous moment.

https://doi.org/10.3390/pr14101558

2.1.3. African Vulture Optimization Algorithm

AVOA is a swarm optimization algorithm based on the foraging and navigation abilities of African vultures. This algorithm mimics the intelligent strategies vultures use when widely searching for food, including exploring vast space and quickly locking onto targets, to find optimal or near-optimal solutions in a complex search space. Using the AVOA algorithm, automatic hyperparameter optimization in the TCN-LSTM-AVOA model can be achieved to obtain the best-performing model [41,46,47]. The AVOA algorithm process is as follows:

(1) Identify the optimal vultures in the population. By setting parameters such as population size, the number of iterations, and search-space boundaries, an initial population is formed. MSE is used as the fitness function of the algorithm, and the fitness values of all solutions are calculated. The two solutions with the lowest fitness values are selected as the optimal vulture and sub-optimal vulture (vultures with existing food), while the remaining individuals form a new population;

(2) The starvation rate of vultures. Vultures are in one of two states—satiated or hungry—which correspond to global search and local search behaviors, respectively. To balance global exploration and local exploitation, the algorithm calculates a starvation rate for the vultures. This rate is dynamically adjusted based on the current iteration number and the total number of iterations. When satiated, vultures engage in extensive foraging over a wide area. In a hungry state, they exhibit heightened aggression and tend to approach the optimal and sub-optimal vultures.

$$F = (2 \times r1 + 1) \times z \times (1 - \frac{iter_i}{\text{max\_}iter}) + a, \tag{7}$$

$$a = h \times \left(\sin(\omega \times \frac{\pi}{2} \times \frac{iter_i}{\text{max\_}iter}) + \cos(\omega \times \frac{\pi}{2} \times \frac{iter_i}{\text{max\_}iter}) - 1 \right), \tag{8}$$

In Equation (7), F represents the starvation rate; $r1$ is a random number in the interval $[0, 1]$; $iter_i$ denotes the current iteration number; max_iter is the maximum number of iterations; h is a random number in the interval $[-2, 2]$; ω is a pre-set parameter that determines the probability of the vulture entering the global exploration phase; z is a random number in the interval $[-1, 1]$—when $z > 0$, the vulture is in a satiated state, and when $z < 0$, the vulture is in a hungry state;

(3) Global exploration phase. When the vultures are less hungry, individuals have sufficient energy to perform a global search. At this stage, the vultures randomly adjust their positions and search for food over a wider area, effectively avoiding getting trapped in a local optimal solution;

(4) Local exploitation phase. As the number of iterations increases, the vultures' hunger level rises, and the algorithm transitions into different exploitation phases. The algorithm simulates the vultures' circling behavior around food. By incorporating the Levy flight strategy, the vultures move toward the optimal or sub-optimal individuals, accelerating the convergence of the algorithm to the optimal solution.

2.2. Proposed Method

In the TCN-LSTM-AVOA model, the number of TCN kernels determines each convolutional layer's ability in the TCN to learn features, while kernel size determines the range of temporal features that the TCN captures in the input data. The model enhances its ability to capture long-term dependencies by stacking multiple LSTM layers [48]. Additionally, the number of units in each layer determines the network's capacity to process complex temporal patterns in the data. By optimizing the number of TCN kernels, LSTM units, learning

https://doi.org/10.3390/pr14101558

rate, and batch size via the AVOA, the model complexity can be effectively reduced, while its generalization performance is simultaneously enhanced. The TCN-LSTM-AVOA model workflow is shown in Figure 4, with the detailed process as follows:

(1) Data import and preprocessing. Load the dataset, preprocess the data, and split it into training and testing sets;
(2) Build the TCN-LSTM model structure. Construct the TCN-LSTM model, which includes an input layer, two convolutional layers, two LSTM layers, and an output layer. To maintain gradient flow and preserve important features in deep networks, TCN employs residual connections. Each residual block typically consists of two dilated causal convolution layers, each followed by weight normalization and a ReLU activation function. Through such shortcut connections, the model can be trained more effectively in deep architectures and mitigate the gradient-vanishing problem;
(3) Initialize AVOA parameters. Initialize the parameters for the AVOA algorithm, including population size, maximum number of iterations, and the search-space range for the hyperparameters. For each parameter combination in the initial population, calculate the model's fitness function value to identify the optimal vulture and the sub-optimal vulture. Based on the positions of the optimal and sub-optimal vultures, fitness function values, hunger rate, and vulture flight strategies, vultures adjust their positions to gradually approach the optimal and sub-optimal vultures. The fitness function values are recalculated, and the vulture positions are updated iteratively until the algorithm reaches the maximum number of iterations (max_iter = 20);
(4) Model evaluation. After obtaining the optimal hyperparameter combination, the TCN-LSTM-AVOA minimizes the error between the predicted values and the actual values. The model is evaluated on the testing set, and the prediction results for production data are output.

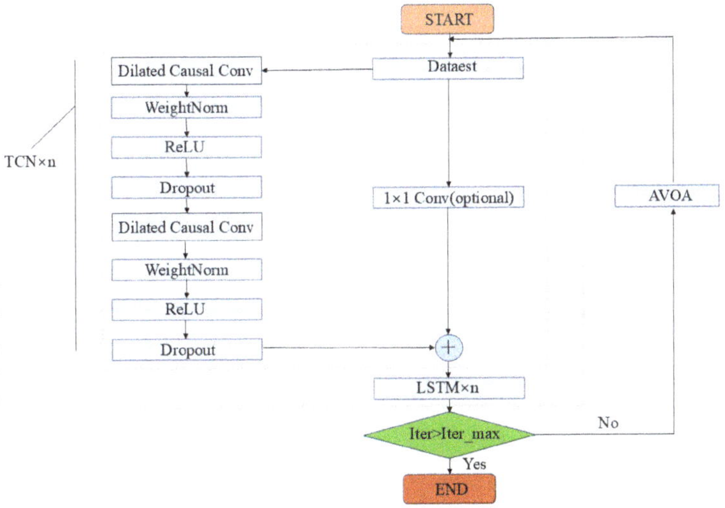

Figure 4. The TCN-LSTM-AVOA modeling process.

2.3. Data Preprocessing

2.3.1. Data Standardization

The reservoir dynamic production data is standardized using Equation (9) to eliminate the impact of dimensionality and facilitate the model's more accurate convergence to the optimal solution.

https://doi.org/10.3390/pr14101558

$$X = \frac{x - \mu}{\sigma}, \tag{9}$$

In Equation (9), X is the standardized sample data, x is the original data before standardization, and μ and σ represent the mean and standard deviation of x, respectively.

2.3.2. Correlation Analysis

The Pearson correlation coefficient (r) quantifies linear relationships between variables [49], with a value range from −1 to 1. It is given by Equation (10):

$$r = \frac{\sum_{i=1}^{n} (G_i - \overline{G})(H_i - \overline{H})}{\sqrt{\sum_{i=1}^{n} (G_i - \overline{G})^2} \sqrt{\sum_{i=1}^{n} (H_i - \overline{H})^2}}, \tag{10}$$

where r represents the Pearson correlation coefficient between variables G and H, and $\overline{G}$, $\overline{H}$ are the mean values of G, H, respectively. The value n represents the sample size. When $|r| \geq 0.6$, it indicates a strong correlation between G and H. When $|r| \in [0.4, 0.6)$, it shows a moderate correlation.

2.4. Data Description

The case study focuses on a two-dimensional, three-phase heterogeneous reservoir model. The reservoir is discretized into a 20 × 30 × 1 grid (grid size = 300 ft), exhibiting strong heterogeneity and containing a high-permeability channel, with the planar permeability distribution shown in Figure 5. The reservoir includes one injection well and five production wells. The injection well is labeled as I1, while production wells P2 and P5 are located in the high-permeability zones. Production wells P3, P4, and P6 are situated in areas with relatively low permeability. As shown in Figure 6 and Table 3, the dataset covers six wells with three metrics: BHP varies between 1875 psi and 4004 psi (mean 2723–3667 psi), while GPR spans 106–802 MSCF/D (mean 146–247 MSCF/D). WPR is near-zero for five wells (mean <0.03 STB/D) but reaches a maximum of 1465 STB/D for well P5 (mean 239 STB/D), highlighting strong heterogeneity in water production.

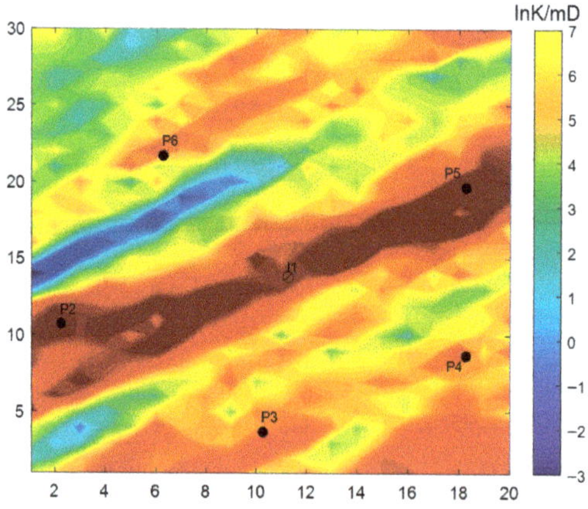

Figure 5. Distribution of the horizontal permeability field in the reservoir (logarithmic scale).

https://doi.org/10.3390/pr14101558

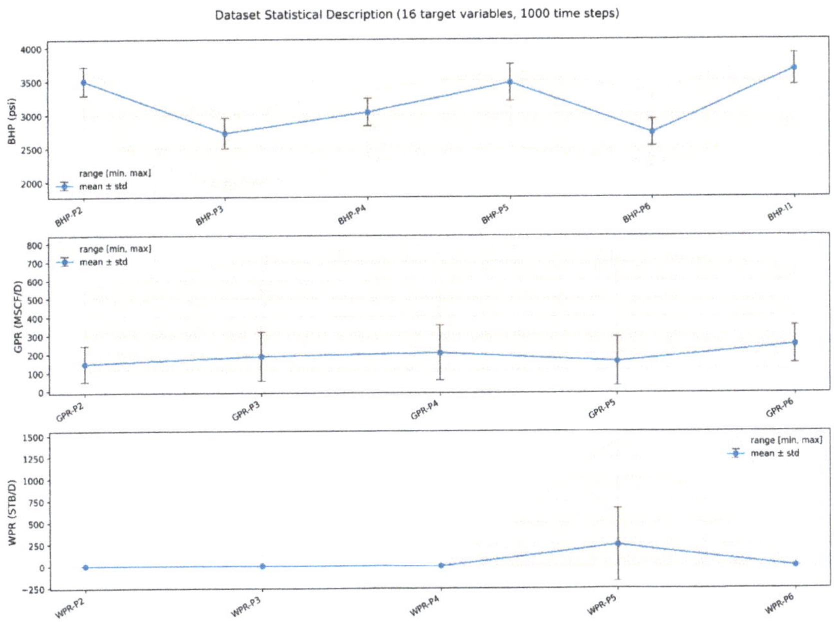

Figure 6. Descriptive statistics (mean ± std and range) of the production metrics across all wells.

Table 3. Descriptive statistics of BHP, GPR, and WPR for each well (P2–P6 and I1).

Target Variables	Mean	Std	Min	Max	Range
BHP-P2 (psi)	3500.31152	211.3979818	2707.6411	3655.7378	948.0967
BHP-P3 (psi)	2723.273458	228.39941	1875.4287	3034.7554	1159.3267
BHP-P4 (psi)	3035.01072	202.5547866	2290.2671	3368.4167	1078.1496
BHP-P5 (psi)	3471.445814	273.9156037	2511.1179	3669.9773	1158.8594
BHP-P6 (psi)	2727.442828	199.6633177	2220.3892	3354.2273	1133.8381
BHP-I1 (psi)	3666.63571	233.8714812	2808.2766	4003.729	1195.4524
GPR-P2 (MSCF/D)	145.7848195	97.82810689	108.0326	556.09637	448.06377
GPR-P3 (MSCF/D)	185.8139902	131.9221866	105.88278	732.58704	626.70426
GPR-P4 (MSCF/D)	204.1849204	148.7672503	108.0326	801.59125	693.55865
GPR-P5 (MSCF/D)	157.281227	131.3260838	108.0326	751.09747	643.06487
GPR-P6 (MSCF/D)	247.1015404	101.5332439	108.0326	683.77246	575.73986
WPR-P2 (STB/D)	0.021353498	0.089164092	0	1.1563998	1.1563998
WPR-P3 (STB/D)	0.006680595	0.004732047	0.00118587	0.026342342	0.025156472
WPR-P4 (STB/D)	0.008124899	0.005235236	0.001367188	0.030019673	0.028652485
WPR-P5 (STB/D)	238.5192479	418.1289953	0	1464.9373	1464.9373
WPR-P6 (STB/D)	0.015301256	0.005327041	0.001521307	0.031797703	0.030276396

Using Eclipse 2011 software to calculate a real reservoir model, data spanning 5000 days were generated, including BHP, GPR, and WPR of 5 oil production wells, as well as a BHP of 1 water injection well. During the model training phase, this study employed data augmentation techniques to enhance the model's generalization capability. Specifically, the Gaussian noise, following a normal distribution with a mean of 0 and a standard deviation of 0.05, was dynamically added only to the input features of the training set. It is noteworthy that, to evaluate the model's generalization performance under clean conditions and to improve its robustness, the noise-augmented dataset was used exclusively for training, while the validation and test sets remained free of any noise

https://doi.org/10.3390/pr14101558

injection. A noise robustness analysis was performed using the TCN-LSTM-AVOA, where σ was varied over {0, 0.02, 0.04, 0.06, 0.08, 0.10, 0.15} with three independent random seeds per level (21 training runs total). The R^2, MAE, and RMSE on the clean test set, reported as mean ± standard deviation, confirm that the model remains stable across the tested σ range.

As shown in Figure 7, it displays the Pearson correlation coefficients between the 16 injection and production data, including BHP-I1, BHP-P2, BHP-P3, BHP-P4, BHP-P5, BHP-P6, GPR-P2, GPR-P3, GPR-P4, GPR-P5, GPR-P6, WPR-P2, WPR-P3, WPR-P4, WPR-P5, and WPR-P6. Except for the correlation coefficient $|r| = 0.49$ between the bottom-hole flowing pressure of the production wells P2 and P6, which has a moderate correlation, the absolute value of the Pearson correlation coefficients between the other features is greater than 0.6, indicating a strong linear correlation between each pair of injection and production data. Therefore, in the model's input, the 16 bottom-hole flowing pressure, gas production rate, and water production rate data at the same time step are combined into a single sample. The model can simultaneously consider dynamic production data from multiple wells. This enables it to better capture the dynamic changes in the injection and production data.

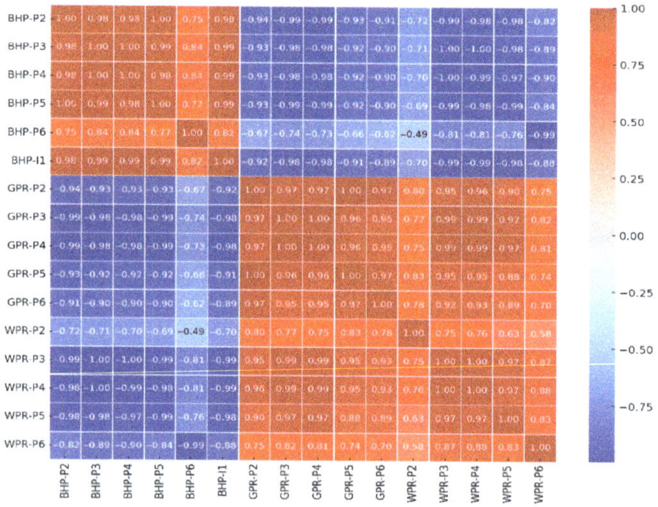

Figure 7. Correlation analysis of injection and production data.

2.5. Model Parameter Setting

Based on the correlation analysis (Section 2.4), each input sample combines BHP, GPR, and WPR measurements at synchronized timesteps. The dataset is partitioned into training, validation, and testing sets at a 7:2:1 ratio, ensuring data isolation throughout the model training and evaluation processes. The proposed TCN-LSTM-AVOA model is benchmarked against seven architectures: BPNN, LSTM, CNN-LSTM, TCN-LSTM, LSTM-AVOA, CNN-AVOA, and TCN-AVOA. In the study, the model training process employs a maximum of 100 epochs, with hyperparameter optimization constrained to 20 iterations. After chronologically splitting the dataset into training, validation, and test sets, a sliding window of 32 time steps is applied independently within each set. This allows the model to leverage historical data from the preceding 32 time steps to forecast the production metric at the subsequent time step, while guaranteeing that no window crosses the segment boundaries. The chronological data split and an example of sliding windows generated within the training segment are shown in Figure 8.

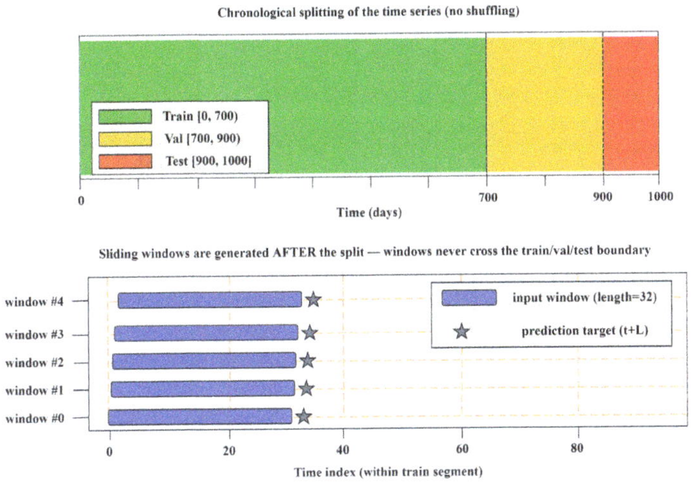

Figure 8. Chronological splitting of the time series and illustration of sliding windows within the training segment.

Learning rate is a critical hyperparameter in deep learning optimization algorithms, determining the step size by which model parameters are updated during training [50]. The magnitude of the learning rate directly affects the training speed, convergence, and final performance of the model. Batch size refers to the number of samples processed in each parameter update step, which influences the efficiency of gradient calculation and the training speed. The number of convolutional kernels determines the quantity and complexity of features extracted by the convolutional layer. More kernels enhance the model's expressive power but also increase the computational load and the number of parameters. Li et al. [38] employed the PSO algorithm to optimize the CNN-LSTM model, exploring optimal values for hyperparameters such as the number of convolutional kernels, training epochs, learning rate, and batch size, thereby achieving accurate predictions of oil production. Building on this study, AVOA is selected to optimize the number of LSTM units, the number of convolutional kernels, the learning rate, and the batch size for the TCN-LSTM-AVOA model. The key meta-parameters for AVOA used in our study are population size = 10 and maximum iterations = 20. Below are the specific parameter settings for each model:

- BPNN: Comprises three fully connected layers (64-128-64 neurons) with ReLU activation, trained at a learning rate of 0.0001;
- LSTM: Integrates two LSTM layers (64 units each), followed by a dense layer (64 neurons), optimized with a learning rate of 0.0001;
- CNN-LSTM: Combines two convolutional layers (64 kernels, kernel size = 2) with two LSTM layers (64 units each);
- TCN-LSTM: Features two LSTM layers (64 units) and two dilated convolutional layers (64 kernels, kernel size = 2, dilation rates = [1, 2]);
- LSTM-AVOA: Optimizes hyperparameters including LSTM layer units, learning rate, and batch size;
- CNN-AVOA: Optimizes the number of convolutional kernels, learning rate, and batch size, with fixed kernel size = 2;

https://doi.org/10.3390/pr14101558

- TCN-AVOA: Optimizes the number of convolutional kernels across three layers, with fixed kernel size = 2 and dilation rates = [1, 2, 4];
- TCN-LSTM-AVOA: Jointly optimizes the number of convolutional kernels, LSTM units, learning rate, and batch size, using fixed kernel size = 2 and dilation rates = [1, 2].

2.6. Model Evaluation Metrics

To reflect the predictive performance comparison of the TCN-LSTM-AVOA model with the BPNN, LSTM, CNN-LSTM, TCN-LSTM, LSTM-AVOA, CNN-AVOA, and TCN-AVOA models, RMSE, MAE, and R^2 are used as model evaluation metrics. The calculation formulas are shown in Equations (11)–(14). The closer the *RMSE* and *MAE* are to 0, and the closer the R^2 is to 1, the better the model's predictive performance.

$$RMSE = \sqrt{\frac{1}{n}\sum_{i=1}^{n}\left(y_i - y_{pred,i}\right)^2}, \tag{11}$$

$$MAE = \frac{1}{n}\sum_{i=1}^{n}\left|y_i - y_{pred,i}\right|, \tag{12}$$

$$MAPE = \frac{1}{n}\sum_{i=1}^{n}\left(\left|\frac{y_i - y_{pred,i}}{y_i}\right| \times 100\%\right), \tag{13}$$

$$R^2 = 1 - \frac{\sum_{i=1}^{n}\left(y_{pred,i} - \bar{y}\right)^2}{\sum_{i=1}^{n}\left(y_i - \bar{y}\right)^2}, \tag{14}$$

where y_i is the actual value of the i-th sample, $\bar{y}$ is the mean of the actual values of the samples, and $y_{pred,i}$ is the predicted value of the i-th sample.

3. Results and Discussion

3.1. Optimal Hyperparameter Values

The hyperparameter optimization ranges and corresponding optimal values for the models LSTM-AVOA, CNN-AVOA, TCN-AVOA, and TCN-LSTM-AVOA are presented in Tables 4–7.

Table 4. Hyperparameter search range and values in LSTM-AVOA.

Optimized Parameters	Search Range	Best Parameters
units	[32, 100]	33
batch size	[128, 256]	136
learning rate	[0.00001, 0.001]	0.0003

Table 5. Hyperparameter search range and values in CNN-AVOA.

Optimized Parameters	Search Range	Best Parameters
kernels	[32, 100]	32
batch size	[128, 256]	128
learning rate	[0.00001, 0.001]	0.0002

Table 6. Hyperparameter search range and values in TCN-AVOA.

Optimized Parameters	Search Range	Best Parameters
kernels	[32, 100]	57
batch size	[128, 256]	215
learning rate	[0.00001, 0.001]	0.0039

https://doi.org/10.3390/pr14101558

Table 7. Hyperparameter search range and values in TCN-LSTM-AVOA.

Optimized Parameters	Search Range	Best Parameters
units	[32, 100]	90
kernels	[32, 100]	91
batch size	[128, 256]	139
learning rate	[0.00001, 0.001]	0.0018

Under the optimal parameter combinations, the respective models' best achieved performance is enabled by these configurations. Training loss variation curves, fitting curves, and model evaluation metrics are obtained by training and testing these models on heterogeneous reservoirs.

3.2. *Training Loss Analysis*

The loss curves shown in Figure 9 intuitively illustrate the trend of the model's loss during training, helping to determine whether the model has converged.

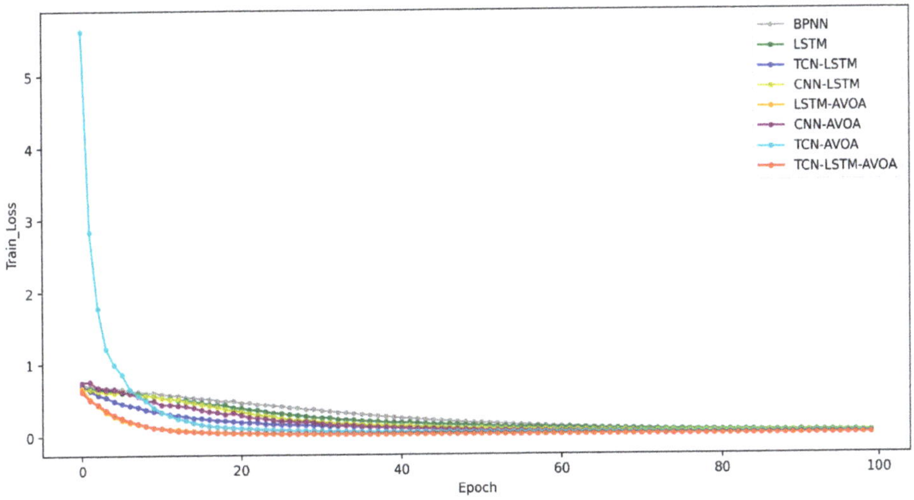

Figure 9. Comparison of training loss curves.

Over the 100 training epochs, all models exhibited declining training losses with increasing epochs, though significant differences emerged in convergence speed and final performance. Regarding training loss, the TCN-LSTM-AVOA model achieved the fastest convergence to minimal loss, whereas the conventional models, BPNN and LSTM, maintained persistently high loss values. In terms of validation loss, TCN-LSTM-AVOA similarly maintained the lowest loss with a stable decreasing trend. Notably, all models incorporating the AVOA demonstrated stable generalization performance on the validation set, highlighting the algorithm's effectiveness in mitigating overfitting risks and enhancing model robustness. Collectively, these results indicate that the TCN-LSTM-AVOA model, which integrates temporal feature extraction, long-sequence modeling, and collective optimization strategies, delivers optimal performance with a training efficiency and generalization capabilities that are significantly superior to those of other benchmark models.

https://doi.org/10.3390/pr14101558

3.3. Fitting Curve Analysis

The matching results of eight models for the bottom-hole flowing pressure, gas production rate, and water production rate of some injection–production wells are, respectively, presented in Figures 10–16. In these figures:

- Black scatter points indicate the actual values of the corresponding production data;
- The red curve represents the matched values of the TCN-LSTM-AVOA model for the production data;
- Two vertical dashed lines indicate the divisions between the training set and validation set, and between the validation set and testing set, respectively.

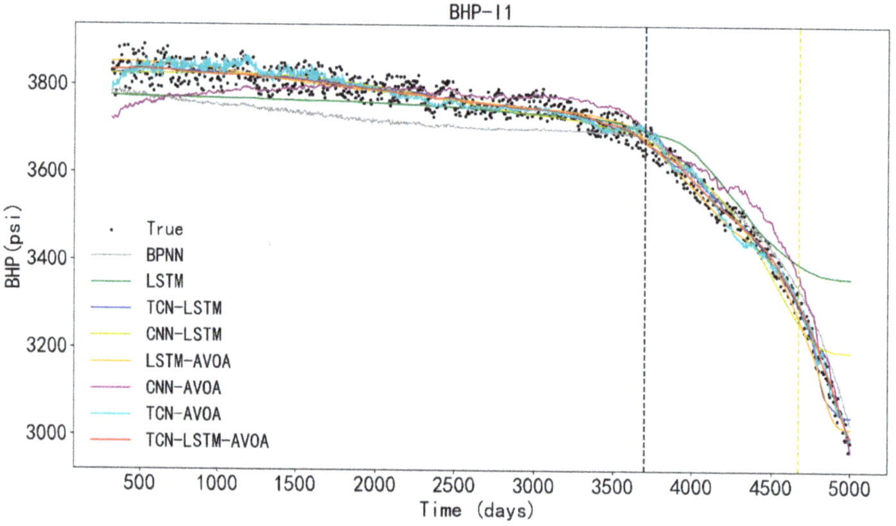

Figure 10. Matching results of bottom-hole flowing pressure for well I1.

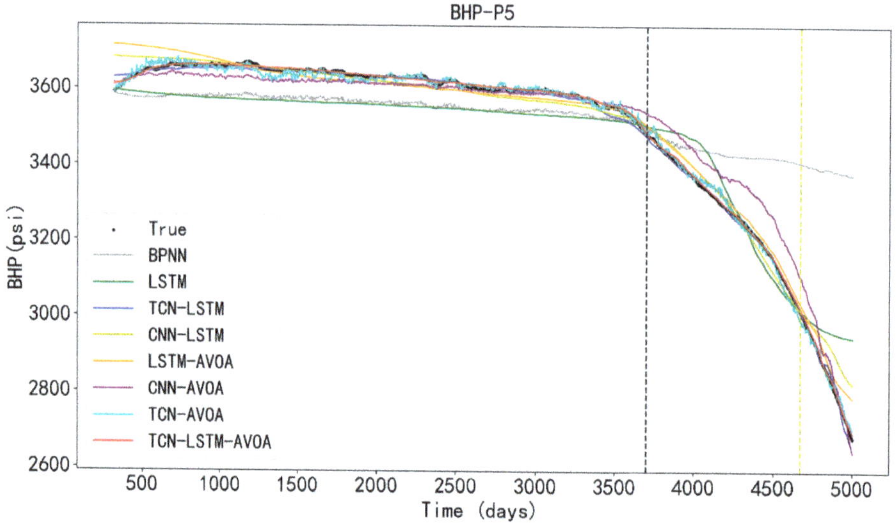

Figure 11. Matching results of bottom-hole flowing pressure for well P5.

https://doi.org/10.3390/pr14101558

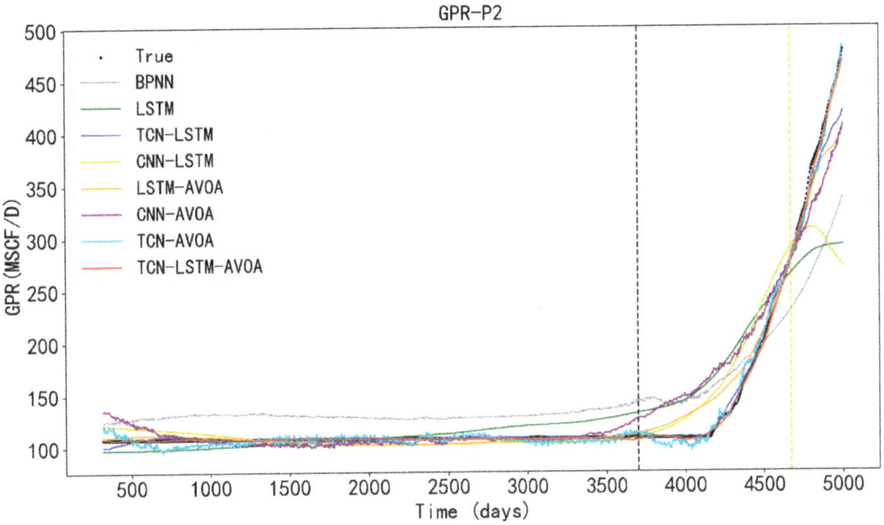

Figure 12. Matching results of gas production rate for well P2.

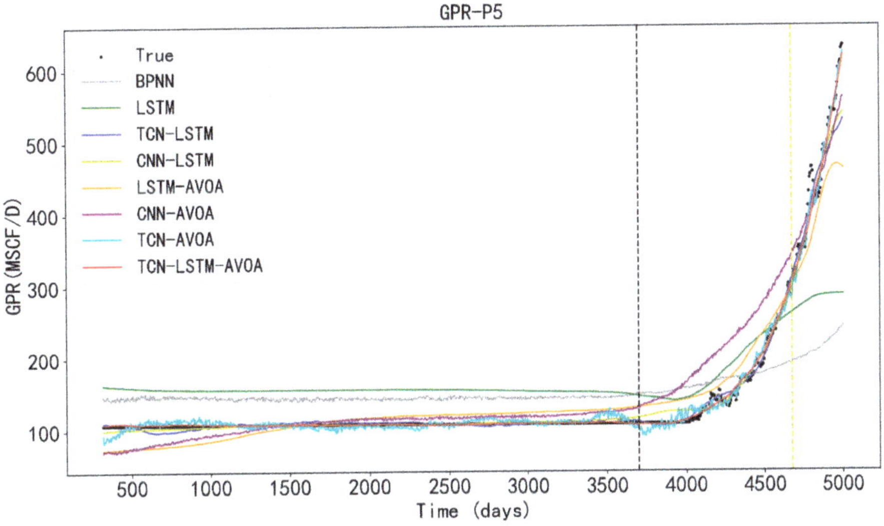

Figure 13. Matching results of gas production rate for well P5.

The horizontal axis denotes time (days), while the vertical axes display values of BHP (psi), GPR (MSCF/D), and WPR (STB/D).

As shown in Figures 10 and 11, well I1 is a water injection well located in a high-permeability zone, and adjustments in water injection operations cause significant fluctuations in the actual values of BHP-I1. The BHP-I1 generally decreases over time, while well P5 exhibits a similar trend in BHP variation, though its initial pressure is lower than that of well I1. That the gas production rates of well P2 and well P5 remain stable within the training set range but increase significantly within the validation set and testing set range is revealed by Figures 12 and 13. The actual gas production rate of well P5 has fluctuated due to the influence of water injection well I1, reservoir heterogeneity, and other factors.

https://doi.org/10.3390/pr14101558

From Figure 14 to Figure 16, it is observed that the water production rates of well P3, well P4, and well P6 gradually increase over time.

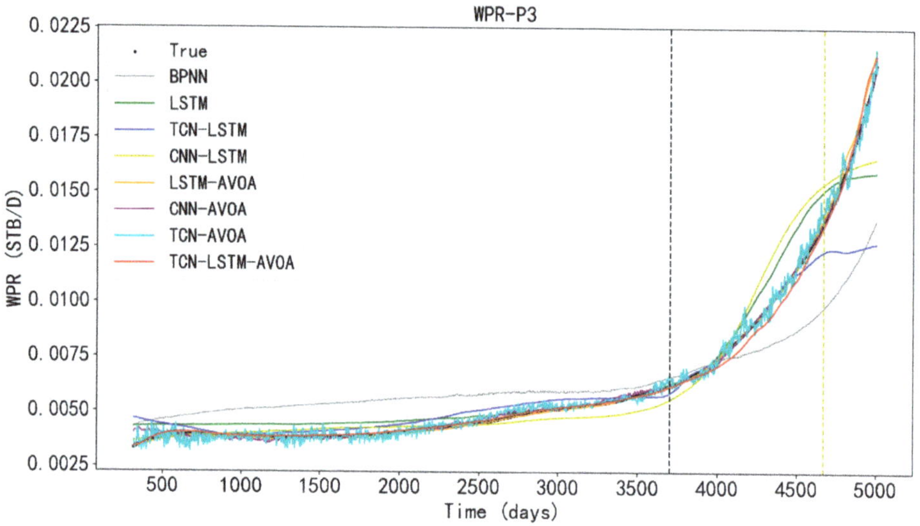

Figure 14. Matching results of water production rate for well P3.

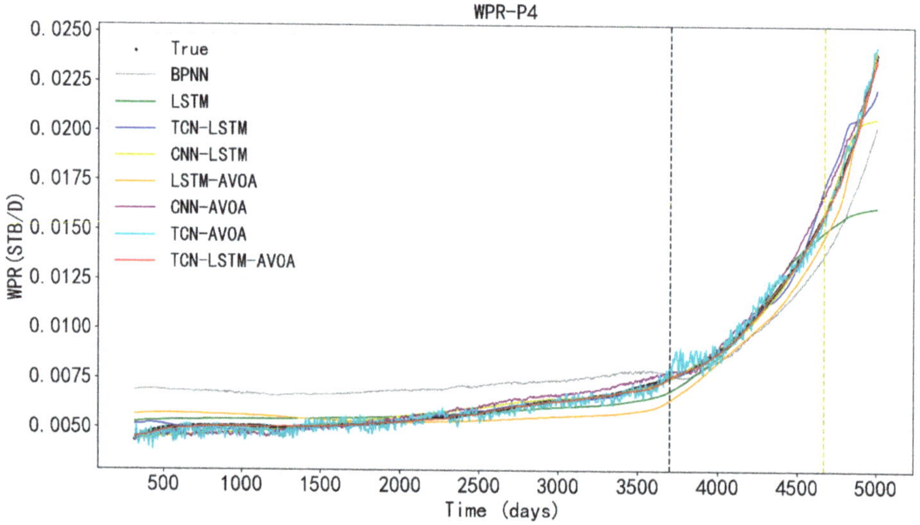

Figure 15. Matching results of water production rate for well P4.

An analysis of Figures 10–16 reveals that CNN and LSTM exhibit weak capabilities in matching the production data from all wells, showing significant deviations between their predictions and the actual values. Although TCN-LSTM underperforms in WPR-P6 prediction, its fitting curves for other features (BHP, GPR) closely align with the actual values, indicating enhanced temporal feature extraction through the fusion of different network architectures. As the training iterations progress, the LSTM-AVOA, CNN-AVOA, and TCN-AVOA models approach the true values more closely compared to the single models, validating the optimization effects of the AVOA algorithm on model convergence

 https://doi.org/10.3390/pr14101558

and accuracy. Notably, optimal performance across all well-matching tasks is demonstrated by the TCN-LSTM-AVOA model, with significantly superior prediction accuracy and stability in multi-well production forecasting scenarios, highlighting the technical potential of this hybrid model in dynamic oil and gas indicator prediction.

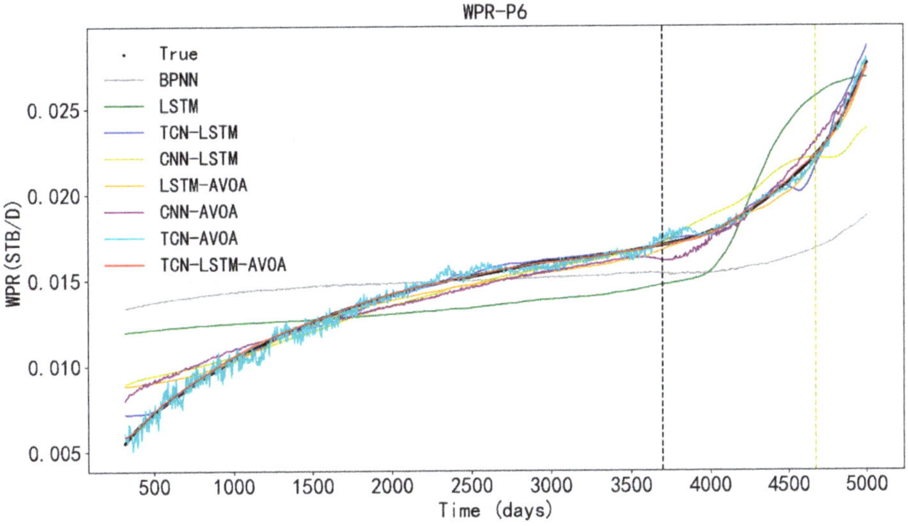

Figure 16. Matching results of water production rate for well P6.

The performance of the eight models on well BHP-I1 on the validation set is illustrated in Figure 17. As shown in Figure 17, the horizontal and vertical axes represent the actual and predicted values of BHP-I1, respectively. The data points of the TCN-LSTM, TCN-LSTM-AVOA, and TCN-AVOA models are closely clustered around the line $y = x$, indicating that these three models demonstrate superior fitting performance for BHP-I1. In contrast, the scatter plots of other models exhibit more dispersed distributions with poorer performance.

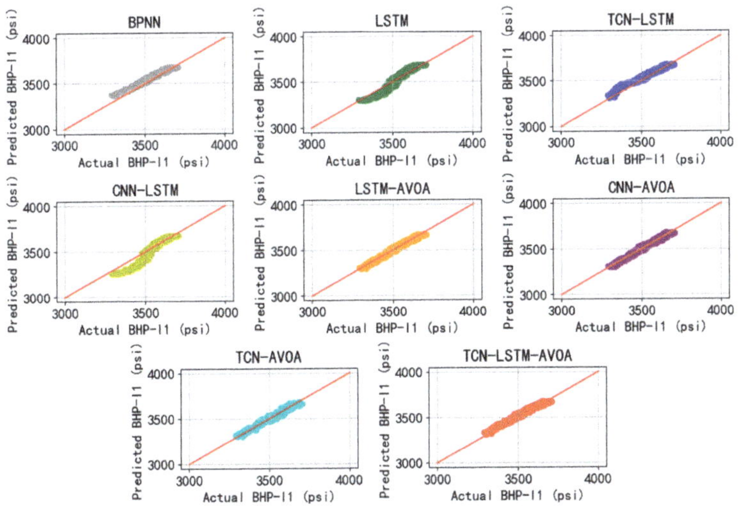

Figure 17. Cross plot of actual and predicted bottom-hole flowing pressure for well I1 on the validation set across 8 models.

 https://doi.org/10.3390/pr14101558

3.4. Evaluation Metric Analysis

The evaluation metrics of different models across three datasets, namely the total testing set, BHP-I1, and GPR-P5, are compared in Tables 8–10.

Table 8. Comparison of evaluation metrics on total testing set.

Model	RMSE	MAE	MAPE	R^2
BPNN	137.8566	66.7908	19.86%	0.5497
LSTM	84.7137	44.3560	32.21%	0.7217
TCN-LSTM	23.4386	8.0179	3.22%	0.9712
CNN-LSTM	35.6232	14.4933	23.19%	0.9163
LSTM-AVOA	38.2645	16.8729	21.96%	0.9449
CNN-AVOA	34.6366	21.1203	27.11%	0.9566
TCN-AVOA	17.8476	11.0184	5.96%	0.9904
TCN-LSTM-AVOA	7.0806	3.4780	1.81%	0.9975

Table 9. Comparative evaluation metrics of multiple models for BHP-I1 on the testing set.

Model	RMSE	MAE	MAPE	R^2
BPNN	58.2887	51.4562	1.40%	0.9419
LSTM	115.5605	67.8102	2.03%	0.7717
TCN-LSTM	35.9883	23.0696	0.67%	0.9779
CNN-LSTM	64.1424	33.8349	1.02%	0.9297
LSTM-AVOA	32.6771	22.5183	0.65%	0.9817
CNN-AVOA	47.6868	39.3035	1.09%	0.9611
TCN-AVOA	30.6685	25.1515	0.69%	0.9839
TCN-LSTM-AVOA	20.9949	17.8203	0.49%	0.9925

Table 10. Comparative evaluation metrics of multiple models for GPR-P5 on the testing set.

Model	RMSE	MAE	MAPE	R^2
BPNN	107.3144	64.3951	34.66%	0.3758
LSTM	98.5411	65.2667	39.91%	0.4737
TCN-LSTM	27.4883	9.3823	3.57%	0.9590
CNN-LSTM	27.5142	9.5254	3.67%	0.9590
LSTM-AVOA	49.4312	27.2590	15.61%	0.8676
CNN-AVOA	31.7034	22.4378	15.88%	0.9455
TCN-AVOA	11.3903	8.9487	6.95%	0.9930
TCN-LSTM-AVOA	7.3345	3.3350	1.51%	0.9971

In Table 8 (total testing set), the TCN-LSTM-AVOA model achieves the best performance with the lowest RMSE (7.0806), MAE (3.4780), MAPE (1.81%), and the highest R^2 (0.9975), significantly outperforming all other models. TCN-AVOA also achieves strong results, ranking second only to TCN-LSTM-AVOA. CNN-LSTM, while improving over the baseline LSTM and BPNN, yielding a substantially higher error (RMSE = 35.6) compared to the TCN-based models. In contrast, traditional models exhibit poor performance, like BPNN and LSTM, with the highest RMSE and MAE values being shown by BPNN, indicating its limitations in handling complex data patterns.

Moving to single-well performance (Table 9, BHP-I1 on the testing set), TCN-LSTM-AVOA again dominates with the lowest RMSE (20.9949), MAE (17.8203), MAPE (1.51%), and the highest R^2 (0.9925), confirming its superior ability to match the BHP-I1 dataset. TCN-LSTM and TCN-AVOA also perform well, though not as effectively as TCN-LSTM-AVOA. Traditional models like LSTM and BPNN exhibit significantly larger errors, particularly LSTM, which shows instability, with an exceptionally high RMSE of 115.5605 on

 https://doi.org/10.3390/pr14101558

this dataset. For the GPR-P5 dataset (Table 10), TCN-LSTM-AVOA maintains its leading position with an RMSE of 7.3345, an MAE of 3.3350, and an R^2 of 0.9971. TCN-AVOA follows closely, while LSTM and BPNN continue to underperform with substantial errors. Although slight improvements in certain metrics are shown by LSTM-AVOA and CNN-AVOA compared to their base versions, inferiority to TCN-based models like TCN-AVOA remains for them.

Overall, the best performance across all three datasets is consistently achieved by the TCN-LSTM-AVOA model, with particular excellence being shown in the total and GPR-P5 dataset. Its RMSE, MAE, and MAPE are significantly lower than those of other models, and its R^2 values approach one, demonstrating exceptional predictive accuracy and robustness. This success likely stems from the synergistic integration of TCN's ability to extract temporal features from production data, LSTM's strength in processing sequential dependencies, and the AVOA optimization algorithm's effectiveness in refining hyper-parameters. Furthermore, the application of AVOA enhances model performance, as evidenced by improvements in LSTM-AVOA and CNN-AVOA compared to their original versions. The combination of AVOA with TCN yields the most substantial performance gains, highlighting the importance of architectural design and optimization in complex predictive tasks.

That the TCN-LSTM-AVOA hybrid model achieves the highest accuracy and stability in multi-well production prediction for heterogeneous reservoirs compared to seven benchmark models is demonstrated by the experimental results, with effective integration among TCN, LSTM, and AVOA components being indicated. However, the forecasting models in this study rely solely on historical production data, lacking integration with key geological constraints like permeability and porosity. To build more physically realistic and generalizable models, future research should prioritize developing hybrid modeling frameworks that explicitly incorporate geological parameters. This could involve treating parameters such as permeability and porosity as additional input features to condition the deep learning models or exploring physics-informed architectures that encode geological priors. Such advancements are essential for reliably extending this data-driven approach to diverse and complex reservoir types.

4. Conclusions

Gaussian noise with a mean of 0 and a standard deviation of 0.05 was added to the heterogeneous reservoir dataset, and experiments were conducted using this modified dataset. This study integrates optimization algorithms with neural networks to propose the TCN-LSTM-AVOA model for reservoir production forecasting. Superior performance is demonstrated by the TCN-LSTM-AVOA model, which is evaluated on heterogeneous reservoir datasets, compared to conventional models (BPNN, LSTM) and hybrid models (CNN-LSTM, TCN-LSTM, etc.). The following conclusions can be drawn:

(1) The TCN-LSTM-AVOA model is developed by incorporating the Artificial Vulture Optimization Algorithm into the existing TCN-LSTM architecture. Experimental results prove that the proposed model achieves superior performance on the testing set, with RMSE = 7.0806, MAE = 3.4780, MAPE = 1.81%, and R^2 = 0.9975, exhibiting significantly smaller errors compared with seven benchmark models;

(2) The challenges of temporal processing and global optimization in reservoir production data are addressed by the TCN-LSTM-AVOA model through the synergistic combination of the temporal feature extraction capability of TCN, the time-dependent relationship modeling strength of LSTM, and the global search strategy of the AVOA algorithm. This integration optimizes the hyperparameters (91 convolutional kernels, 90 LSTM units, learning rate 0.0018, batch size 139), substantially reducing manual

https://doi.org/10.3390/pr14101558

hyperparameter tuning efforts and accelerating model convergence. Experimental validation confirms that the model achieves optimal performance in predicting the bottom-hole flowing pressure, gas production rate, and water production rate under these parameter configurations;

(3) A novel research approach and technical method for dynamic production forecasting in a reservoir are offered by the TCN-LSTM-AVOA model. It improves the prediction accuracy and, at the same time, has strong robustness in reservoir prediction. And it presents a more efficient solution for future forecasting and management of dynamic reservoir production data.

Author Contributions: X.X.: Writing—original draft, Methodology, Investigation, Funding acquisition. Z.X.: Writing—review and editing, Investigation, Formal analysis. S.L.: Writing—review and editing, Formal analysis. G.Y.: Software, Conceptualization. J.C.: Validation, Conceptualization. All authors have read and agreed to the published version of the manuscript.

Funding: This work was supported by the National Oil & Gas Major Project (Grant No. 2025ZD1401502).

Data Availability Statement: The dataset used and/or analyzed during the current study is available from the corresponding author on reasonable request.

Conflicts of Interest: The authors declare no competing interests.

Abbreviations

The following abbreviations are used in this manuscript:

TCN	Temporal Convolutional Network
LSTM	Long Short-Term Memory network
AVOA	African Vulture Optimization Algorithm
ML	Machine Learning
SEPD	Stretched Exponential Production Decline
MTS	Multivariate Time Series
VAR	Vector Autoregression
RF	Random Forest
DT	Decision Tree
GBR	Gradient Boosting Regression
KNN	K-Nearest Neighbors
SVM	Support Vector Machine
PSO	Particle Swarm Optimization
SSA	Sparrow Search Algorithm
PCA	Principal Component Analysis
BPNN	Backpropagation Neural Network
CNN	Convolutional Neural Network
GRU	Gated Recurrent Unit
RMSE	Root Mean Square Error
MAE	Mean Absolute Error
MAPE	Mean Absolute Percentage Error
R^2	Coefficient of Determination
BHP	Bottom-Hole Flowing Pressure
GPR	Gas Production Rate
WPR	Water Production Rate
P2, P3, P4, P5, P6	Five Production Wells
I1	One Injection Well

 https://doi.org/10.3390/pr14101558

References

1. Zhu, D.Y.; Bai, B.J.; Hou, J.R. Polymer gel systems for water management in high-temperature petroleum reservoirs: A chemical review. *Energy Fuels* **2017**, *31*, 13063–13087. [CrossRef]
2. Zhu, D.Y.; Zhang, J.; Li, G.H.; Gao, Y.Q.; Tan, H.G.; Li, H.Y.; Cheng, H.B.; Lu, C.Y. Development of oil sludge preformed particle gels (OS-PPG) for petroleum reservoir conformance control. *Energy Fuels* **2026**, *40*, 2027–2037. [CrossRef]
3. Zhu, D.Y. New Advances in oil, gas and geothermal reservoirs. *Energies* **2025**, *18*, 4789. [CrossRef]
4. Khalili, Y.; Ahmadi, M. Reservoir modeling & simulation: Advancements, challenges, and future perspectives. *J. Chem. Pet. Eng.* **2023**, *57*, 343–364. [CrossRef]
5. Kukharova, T.; Martirosyan, A.; Asadulagi, M.-A.; Ilyushin, Y. Development of the separation column's temperature field monitoring system. *Energies* **2024**, *17*, 5175. [CrossRef]
6. Yehia, T.; Abdelhafiz, M.; Hegazy, G.; Elnekhaily, S.A.; Mahmoud, O. A comprehensive review of deterministic decline curve analysis for oil and gas reservoirs. *Geoenergy Sci. Eng.* **2023**, *226*, 211775. [CrossRef]
7. Pratama, M.A.; Al Qoroni, O.; Rahmatullah, I.K.; Jameel, M.F.; Weijermars, R. Probabilistic production forecasting and reserves estimation: Benchmarking Gaussian decline curve analysis against the traditional Arps method (Wolfcamp shale case study). *Geoenergy Sci. Eng.* **2024**, *232*, 212373. [CrossRef]
8. Zhu, D.Y. New advances in low-energy processes for geo-energy development. *Energies* **2025**, *18*, 2357. [CrossRef]
9. Tripathi, B.K.; Kumar, I.; Kumar, S.; Singh, A. Deep learning–based production forecasting and data assimilation in unconventional reservoir. *SPE J.* **2024**, *29*, 5189–5206. [CrossRef]
10. Zhen, Y.; Fang, J.Y.; Zhao, X.M.; Ge, J.W.; Xiao, Y.F. Temporal convolution network based on attention mechanism for well production prediction. *J. Pet. Sci. Eng.* **2022**, *218*, 111043. [CrossRef]
11. Dou, H.E.; Zhang, H.J.; Shen, S.B. Correct understanding and application of waterflooding characteristic curve. *Pet. Explor. Dev.* **2019**, *46*, 755–762. [CrossRef]
12. Liu, X.T.; Yang, J.; Cheng, Z.P.; Wang, S.Y. A new method of predicting oilfield output and its application. *Pet. Explor. Dev.* **2002**, *29*, 74–76. [CrossRef]
13. Zhang, N.; Huangfu, H.Y. Application and evaluation of production rate predict of an oilfield. *J. Chengdu Univ. Technol.* **2002**, *29*, 193–195. [CrossRef]
14. Wang, N.T.; Du, L.Y.; He, H.B.; Lin, H.Y.; Zhu, M.Q. New method for analysis of oil and gas well production decline. *Nat. Gas Geosci.* **2020**, *31*, 335–339. [CrossRef]
15. Gao, W.J.; Yin, R.; Yang, J. Establishment and theoretical basis of the new water-flooding characteristic curve. *Acta Petrol. Sin.* **2020**, *41*, 342–347. [CrossRef]
16. Ren, F.L.; Ren, S.D.; Li, J.J. Prediction of oil production based on linear regression method and time series method. *Henan Sci.* **2018**, *36*, 817–822. [CrossRef]
17. Bing, S.X. Quantitative forecasting model for recoverable reserves of single well. *Pet. Geol. Recov. Eff.* **2013**, *20*, 85–88. [CrossRef]
18. Zhang, R.; Hu, J. Production performance forecasting method based on multivariate time series and vector autoregressive machine learning model for waterflooding reservoirs. *Pet. Explor. Dev.* **2021**, *48*, 201–211. [CrossRef]
19. Fang, M.J.; Shi, H.Y.; Li, H.; Liu, T.J. Application of machine learning for productivity prediction in tight gas reservoirs. *Energies* **2024**, *17*, 1916. [CrossRef]
20. Du, Z.Y.; Xu, L.L.; Yin, S.X.; Dong, S.N.; Dai, Z.X.; Ma, Y.; Thanh, H.V.; Soltanian, R. Enhanced prediction and uncertainty analysis for hydrogen production rate in depleted oil and gas reservoirs using advanced machine learning techniques. *Geoenergy Sci. Eng.* **2025**, *249*, 213795. [CrossRef]
21. Yahya, W.; Yang, B.L.; AlRassas, A.; Wang, Y.T.; Al-Khafaji, H.; Al Dawood, R. Developing robust machine learning techniques to predict oil recovery: A comprehensive field and experimental study. *Geoenergy Sci. Eng.* **2025**, *250*, 213853. [CrossRef]
22. Han, B.; Bian, X.Q. A hybrid PSO-SVM-based model for determination of oil recovery factor in the low-permeability reservoir. *Petrol.* **2018**, *4*, 43–49. [CrossRef]
23. Guo, Z.X.; Ma, B.; Zhang, S.; Zhang, S.; Deng, H.; Chen, D.; Chen, Y.Y.; Zhou, S.K. Research status and prospects of deep learning in oil and gas production prediction. *Nat. Gas Ind.* **2024**, *44*, 88–98. [CrossRef]
24. Fan, L.; Zhao, M.M.; Yin, C.; Peng, X.Y. Analysis method of oilfield production performance based on BP neural network. *Fault-Block Oil Gas Field* **2013**, *20*, 204–206. [CrossRef]
25. Liu, H.H.; Li, W.; Zhang, S.L. *Oil Production Predicting with Modified BP Neural Network Method*; Atlantis Press: Dordrecht, The Netherlands, 2014; pp. 146–148.
26. Zhou, J.J.; He, Y.; Zhu, K.J. Research on oilfield production forecasting based on principal component BP neural network. *Stat. Decis.* **2008**, 35–37. [CrossRef]
27. Gers, F.A.; Schmidhuber, J.; Cummins, F. Learning to forget: Continual prediction with LSTM. *Neural Comput.* **2000**, *12*, 2451–2471. [CrossRef] [PubMed]
28. Schmidhuber, J.; Hochreiter, S. Long short-term memory. *Neural Comput.* **1997**, *9*, 1735–1780. [CrossRef]

https://doi.org/10.3390/pr14101558

29. Gao, Y.J.; Tang, L.H.; Wang, Z.P.; Xie, X.Q.; Xu, H.T. A method for oilfield production prediction based on recurrent neural network and data differential processing. *China Offshore Oil Gas* **2023**, *35*, 126–136. [CrossRef]
30. Gu, J.W.; Zhou, M.; Li, Z.T.; Jia, X.J.; Liang, Y. Oil well production forecast with long-short term memory network model based on data mining. *Spec. Oil Gas Reserv.* **2019**, *26*, 77–81+131. [CrossRef]
31. Hou, C.H. New well oil production forecast method based on long-short term memory neural network. *Pet. Geol. Recov. Eff.* **2019**, *26*, 105–110. [CrossRef]
32. Zhao, H.T.; Li, J.Z.; Yang, Y.; Zhao, H.X.; Fang, X.L.; Yu, W.Q. Production and water cut performance predicting method based on long-short term memory neural network model. *Pet. Geol. Eng.* **2022**, *36*, 75–80. [CrossRef]
33. Fragalla, M.A.M.; Yan, W.; Deng, J.; Xue, L.; Hegair, F.; Zhang, W.; Li, G.C. Time series production forecasting of natural gas based on transformer neural networks. *Geoenergy Sci. Eng.* **2025**, *248*, 213749. [CrossRef]
34. Zha, W.S.; Liu, Y.P.; Wan, Y.J.; Luo, R.L.; Li, D.L.; Yang, S.; Xu, Y.M. Forecasting monthly gas field production based on the CNN-LSTM model. *Energy* **2022**, *260*, 124889. [CrossRef]
35. Li, X.C.; Ma, X.F.; Xiao, F.C.; Xiao, C.; Wang, F.; Zhang, S.C. Multistep ahead multiphase production prediction of fractured wells using bidirectional gated recurrent unit and multitask learning. *SPE J.* **2023**, *28*, 381–400. [CrossRef]
36. Fang, J.Y.; Yan, Z.; Lu, X.Y.; Xiao, Y.F.; Zhao, Z. An oil production prediction approach based on variational mode decomposition and ensemble learning model. *Comput. Geosci.* **2024**, *193*, 105734. [CrossRef]
37. Chen, G.X.; Tian, H.L.; Xiao, T.; Xu, T.F.; Lei, H.W. Time series forecasting of oil production in enhanced oil recovery system based on a novel CNN-GRU neural network. *Geoenergy Sci. Eng.* **2024**, *233*, 212528. [CrossRef]
38. Li, W.R.; Wang, L.J.; Dong, Z.Z.; Wang, R.F.; Qu, B.C. Reservoir production prediction with optimized artificial neural network and time series approaches. *J. Pet. Sci. Eng.* **2022**, *215*, 110586. [CrossRef]
39. Huo, F.C.; Chen, Y.; Ren, W.J.; Dong, H.L.; Yu, T.; Zhang, J.F. Prediction of reservoir key parameters in 'sweet spot'on the basis of particle swarm optimization to TCN-LSTM network. *J. Pet. Sci. Eng.* **2022**, *214*, 110544. [CrossRef]
40. Zhang, L.; Dou, H.E.; Wang, T.Z.; Wang, H.L.; Peng, Y.; Zhang, J.F.; Liu, Z.S.; Mi, L.; Jiang, L.W. A production prediction method of single well in water flooding oilfield based on integrated temporal convolutional network model. *Pet. Explor. Dev.* **2022**, *49*, 1150–1160. [CrossRef]
41. Abdollahzadeh, B.; Gharehchopogh, F.S.; Mirjalili, S. African vultures optimization algorithm: A new nature-inspired metaheuristic algorithm for global optimization problems. *Comput. Ind. Eng.* **2021**, *158*, 107408. [CrossRef]
42. Lu, X.Y.; Wang, Z.M.; Zhao, M.H.; Peng, S.Z.; Geng, S.; Ghorbani, H. Data-driven insights into climate change effects on groundwater levels using machine learning. *Water Resour. Manag.* **2025**, *39*, 3521–3536. [CrossRef]
43. Bednarski, B.P.; Singh, A.D.; Zhang, W.H.; Jones, W.M.; Naeim, A.; Ramezani, R. Temporal convolutional networks and data rebalancing for clinical length of stay and mortality prediction. *Sci. Rep.* **2022**, *12*, 21247. [CrossRef] [PubMed]
44. Ghorbani, H.; Ghorbani, P.; Ghorbani, S.; Khlghatyan, N.; Babayan, B.G.; Varkonyi-Koczy, A.V. A robust approach for estimation of the bone age. In *Proceedings of the 2022 IEEE 20th Jubilee International Symposium on Intelligent Systems and Informatics (SISY), Subotica, Serbia, 15–17 September 2022*; IEEE: New York, NY, USA, 2022; pp. 385–390.
45. Dao, F.; Zeng, Y.; Zou, Y.D.; Qian, J. Wear fault diagnosis in hydro-turbine via the incorporation of the IWSO algorithm optimized CNN-LSTM neural network. *Sci. Rep.* **2024**, *14*, 25278. [CrossRef]
46. Chen, W.H.; Zhou, H.T.; Cheng, L.S.; Xia, M. Wind turbine blade icing diagnosis using convolutional LSTM-GRU with improved African vultures optimization. *IEEE Open J. Instrum. Meas.* **2022**, *1*, 3500809. [CrossRef]
47. Shaddeli, A.; Soleimanian Gharehchopogh, F.; Masdari, M.; Solouk, V. An improved African vulture optimization algorithm for feature selection problems and its application of sentiment analysis on movie reviews. *Big Data Cogn. Comput.* **2022**, *6*, 104. [CrossRef]
48. Li, H.I.; Li, S.Z.; Ghorbani, H. Data-driven novel deep learning applications for the prediction of rainfall using meteorological data. *Front. Environ. Sci.* **2024**, *12*, 1445967. [CrossRef]
49. Wang, D.W.; Zhao, Y.B.; Shang, L.Y.; Wang, F.G.; Zhang, Z. Application of Pearson correlation coefficient algorithm in selecting chemical sand control measure wells in B oilfield. *Fine Spec. Chem.* **2022**, *30*, 26–28. [CrossRef]
50. Macêdo, D.; Zanchettin, C.; Ludermir, T. Sigmoidal learning rate optimizer for deep neural network training using a two-phase adaptation approach. *Appl. Soft Comput.* **2024**, *167*, 112264. [CrossRef]

 https://doi.org/10.3390/pr14101558

processes

MDPI

Article

Development and Application of Software for Calculating the Crack Arrest Toughness of Impurity-Containing Carbon Dioxide Pipelines Based on the BTCM

Xinze Li [1,2,*], Dezhong Wang [1], Xingyu Jiang [1], Yuetian Yu [1] and Xiaokai Xing [1,2]

1 School of Engineering, China University of Petroleum (Beijing) at Karamay, Karamay 834000, China; wangdezhong@st.cupk.edu.cn (D.W.); 2024020043@st.cupk.edu.cn (X.J.); 2025217048@st.cupk.edu.cn (Y.Y.); xingxiaokai@vip.sina.com (X.X.)

2 Xinjiang Key Laboratory of Multi-Medium Pipeline Safety Transportation, Urumqi 830011, China

* Correspondence: lixinze@cupk.edu.cn

Abstract

To ensure the safety of supercritical CO_2 pipelines and address the limitations of full-scale fracture tests, such as high risk and substantial investment, software for evaluating the crack arrest toughness of CO_2 pipelines containing impurities was developed based on the Battelle Two-Curve Model (BTCM) in this study. The software is programmed in Python (v.3.12.4), with a graphical user interface (GUI) built using PyQt6 (v.6.10.0) and a three-tier architecture design. It integrates the resistance curve model and the decompression wave model. To determine the thermodynamic state of the fluid, a large property database covering pure components and various mixtures is embedded, incorporating state equations such as PR, HEOS, and GERG-2008. The software can generate pressure drop curves, decompression curves, and resistance curves. The pressure plateau can be quickly identified by examining the pressure drop curve. Whether the pipeline can achieve self-crack arrest can be rapidly judged by comparing the positional relationships between the decompression curve and the resistance curve. To verify the accuracy of the software's calculation results, comparisons were conducted with previous decompression wave experimental data, full-scale burst test data of a CO_2 pipeline, and the international HLP model. The calculation error of the software is within 10%. The development and application of this software provide a convenient, efficient, and accurate practical tool for the calculation of crack arrest toughness and crack arrest evaluation of supercritical CO_2 pipelines.

Keywords: carbon dioxide pipeline; pipeline crack arrest; decompression wave; BTCM; software development

1. Introduction

Since the signing of the Paris Agreement, unprecedented attention has been paid globally to climate change, and a fundamental consensus has been formed on the long-term goal of controlling carbon emissions and achieving carbon neutrality [1–4]. Carbon Capture, Utilization and Storage (CCUS) is recognized as a key pathway to achieve carbon neutrality, within which CO_2 transportation serves as a critical intermediate link connecting various carbon handling processes. Among the methods for CO_2 transport, pipeline transportation is widely acknowledged as the most economical and efficient option for long-distance, large-scale scenarios, with supercritical CO_2 pipeline transportation being considered the most efficient method currently available [5–8].

Under supercritical CO_2 transport conditions, the pipeline interior is maintained at high pressure. Compared to natural gas, CO_2 exhibits a more pronounced Joule-Thomson effect. In the event of a pipeline leak caused by damage, defects, corrosion, or seismic events, the leakage leads to a rapid local temperature drop in the pipe material, a sharp decrease in toughness, and the initiation of cracks. The phase transition during supercritical CO_2 decompression results in a saturation pressure, and the inability of the crack-tip pressure to release instantaneously can lead to long-running ductile fracture propagation [9,10]. Consequently, engineering requirements for the fracture arrest toughness of CO_2 pipelines are significantly more stringent than those for natural gas pipelines.

Ductile fracture propagation in pipelines is highly hazardous, as it may cause damage to the pipeline system, harm the environment, and pose risks to personnel. Specialized research and analysis are required to determine the minimum fracture arrest toughness requirement, ensuring that the pipeline material can arrest fracture propagation autonomously upon crack initiation. The core of pipeline fracture control lies in specific design optimization and material selection to ensure the pipeline possesses sufficient fracture resistance to prevent crack propagation, thereby safeguarding structural integrity and operational safety. Among these, autonomous fracture arrest—achieved solely through the material's inherent toughness without relying on external devices such as arrestors—represents a highly valuable approach to fracture control [11–13].

Currently, the Battelle Two-Curve Method (BTCM) is an effective criterion for verifying whether a pipeline can achieve autonomous crack arrest through its toughness. This method is a semi-empirical formula derived from full-scale fracture tests conducted on low-grade steel pipelines for natural gas transportation. It assesses the pipeline's arrest toughness by comparing the crack propagation speed curve with the decompression wave speed curve. Clearly, materials with higher mechanical properties and toughness are more capable of resisting ductile fracture propagation than those with inferior properties. Research on CO_2 decompression behavior and toughness-based arrest is a key and challenging focus in the safety design, risk assessment, and integrity management of CO_2 pipelines.

Regarding decompression wave propagation characteristics, international experimental studies have been conducted to clarify the mechanisms and influencing factors of CO_2 decompression waves. Europe and the United States have carried out early experimental work on CO_2, with representative studies by Botros [14,15] and Cosham [16,17]. Full-scale fracture tests performed by Cosham et al. indicated that the decompression wave "pressure plateau" is caused by the formation of gas–liquid saturated phases due to pressure instability inside the pipe, preventing timely pressure reduction. The shock tube experimental setup established by Botros et al. was used to study the influence of temperature, pressure, and purity on the decompression wave speed in CO_2 pipelines. It was found that the initial decompression wave speed is positively correlated with the initial pressure of CO_2, while the initial temperature and concentration of CO_2 affect the magnitude of the decompression wave "pressure plateau".

Existing models for decompression waves in CO_2 pipelines are primarily categorized into two types: the Homogeneous Equilibrium Model (HEM) and the Two-Fluid Model (TFM). Among the relatively well-established decompression wave calculation models are GASDECOM, GASMISC, and REFPORP, among others [18]. Based on a one-dimensional Homogeneous Equilibrium Model, the influences of pipeline diameter and wall thickness on decompression wave propagation were investigated by Picard et al. [19]. Aursand et al. [20] examined the relationship between decompression wave curves and the content/type of impurities in CO_2 mixtures using different equations of state (EOS). An HEM was developed, the results of which showed good agreement with experimental data, and the decompression theory was optimized by considering factors such as friction and heat transfer. A decompression wave prediction model for CO_2 containing impurities was established by Jie et al. [21]. By comparing with experimental data published by Cosham et al. [16,17], the effects of different EOS on the decompression wave speed in CO_2 mixtures were analyzed, taking pipe wall roughness into account. The results indicated that the plateau pressure in the gaseous phase decompression wave was overestimated, and the assumptions of the homogeneous equilibrium model were no longer valid when wall friction was considered. Through studying the decompression process of CO_2 mixtures with various impurities, Botros et al. [22] found that calculation results using the GERG-2008 EOS showed better agreement with experimental data compared to those obtained with the PR EOS. It was found by Guo et al. [23] that the variation trend of decompression wave propagation speed at different locations along the pipeline was almost identical. Furthermore, pipelines initially containing supercritical CO_2 were more susceptible to long-running fracture propagation compared to those containing dense-phase CO_2. Research by Li et al. [24,25] revealed that the introduction of non-polar impurities (impurity components with coinciding positive and negative charge centers, lacking a dipole moment, and exhibiting relatively stable chemical properties) led to an increase in the plateau pressure of the decompression wave curve for supercritical CO_2 pipelines. This increased the likelihood of the curve intersecting with the pipeline crack propagation speed curve, thereby heightening the risk of pipeline fracture. Conversely, the presence of polar impurities was found to slightly reduce the plateau pressure of the decompression wave curve in supercritical CO_2 pipelines, consequently reducing the risk of long-running fracture propagation.

Furthermore, full-scale fracture testing serves as a critical method for investigating pipeline crack propagation and arrest. For natural gas pipelines, extensive research and testing have been conducted on fracture control in both conventional and high-strength steel pipelines transporting lean and rich gases. Over the past 40 years, hundreds of full-scale burst tests have been performed to calibrate empirical and semi-empirical models for predicting the toughness values required for fracture arrest [12,26]. In contrast, only a limited number of full-scale burst tests have been conducted worldwide for CO_2 transportation pipelines, as listed in Table 1. Key differences among several major full-scale test programs (e.g., COOLTRANS, SARCO$_2$, CO$_2$SAFE-ARREST) are reflected in their test parameters, crack arrest performance, and conclusions regarding modifications to the Battelle Two-Curve Method (BTCM). The COOLTRANS program involved three burst tests on X65 grade pipelines of different specifications (Φ 914 mm $\times$ 25.4 mm, Φ 609.6 mm $\times$ 19 mm). BTCM predictions were found to be non-conservative in all cases, requiring the application of a fixed correction factor ranging from 1.2 to 1.8. The SARCO$_2$ program conducted two tests using X65 grade, Φ 609.6 mm pipelines with wall thicknesses of 12.5–13.7 mm. The crack arrest locations differed between the tests: in the first test, cracks arrested after short-distance propagation on both sides of the pipe body; in the second test, the crack on the north side arrested at a crack arrester, while the crack on the south side arrested in the second pipe segment following the initiation pipe. This program concluded that applying a fixed correction factor to the BTCM was difficult, suggesting instead a correction factor range between 1.6 and 2.2. The CO$_2$SAFE-ARREST program utilized X65 grade, Φ 610 mm pipelines with wall thicknesses of 13.5–15.0 mm, performing two tests with different backfill conditions. The first test employed clay backfill at a burial depth of 1 m, resulting in crack arrest in the third pipe segment on either side of the initiation pipe. The second test used asymmetric backfill, leading to crack arrest in the fourth pipe segment. Even with a correction factor of 1.7, the BTCM prediction remained non-conservative.

Table 1. Full-scale burst test data of CO_2 pipelines.

Test Name	Steel Grade	Diameter/mm	Wall Thickness/mm	Pressure/MPa	Temperature/°C	CO_2 Concentration	Fracture Propagation Velocity/($m\cdot s^{-1}$)	Type of Crack Arrest
Battelle 1 [27]	X70	510	11.1	14	42.3	100%	-	Crack arrestor/Without crack arrestor
Battelle 2 [27]	X52	510	7.6	9	34.9	100%	-	Crack arrestors at both ends
COOLTRANS 1 [28]	X65	914	25.4	14.8	16.8	100%	90–140	Ductile fracture arrest
COOLTRANS 2 [29]	X65	914	25.4	15.1	8.2	100%	80–120	Ductile fracture arrest
COOLTRANS 3 [17]	X65	610	19.70	15	15.2	87.5%	100–130	Ductile fracture arrest
$SARCO_2$ 1 [17]	X65	610	12.2/13.7	13	13	96.2%	110–130	Circumferential weld crack arrest
$SARCO_2$ 2 [30]	X65	610	12.5–13.7	12.7	17.8	94%	105–137	Crack arrestor/Ductile fracture arrest
CO_2SAFE-ARREST 1 [31]	X65	610	13.7	15	11.6	91.1%	75–130	Ductile fracture arrest
CO_2SAFE-ARREST 2 [32]	X65	610	14.25–15	14.9	12.7	89.8%	50–270	Ductile fracture arrest
PipeChina [33]	X65	610	7–7.6	11.8	13.7	95%	100–250	Ductile fracture arrest

Numbers 1–3 denote Experiment 1, Experiment 2 and Experiment 3, respectively.

Industrial-scale full-scale burst testing for CO_2 pipelines is associated with high costs, significant challenges in data acquisition, lengthy preparation periods, and the need for high-precision data acquisition equipment [34]. Currently, full-scale pipeline fracture control testing facilities are only available in the UK, Italy, Russia, China, and Norway. The primary differences between CO_2 and natural gas pipelines concerning ductile fracture propagation control are: (a) CO_2 pipelines operate at higher pressures to maintain the CO_2 in a dense-phase/liquid state. (b) The decompression behavior of CO_2 mixtures differs due to the liquid-to-gas phase change. (c) Impurities have a more pronounced effect on the decompression behavior of CO_2 mixtures compared to natural gas mixtures. Because the decompression process following a leak in supercritical CO_2 maintains a high-pressure plateau due to phase transition, the BTCM tends to yield unsafe (non-conservative) predictions for fracture arrest in CO_2 pipelines. Consequently, corrections based on full-scale burst tests are necessary, particularly for high-grade and high-toughness materials, requiring adjustments to the predicted toughness values. Standards such as ISO 27913 recommend that when considering carbon dioxide mixtures, the minimum correction factor is 1.2, but it shall be verified based on applicable test data [35].

In the context of the high costs and significant safety risks associated with current full-scale pipeline experiments, the development of specialized software dedicated to calculating the fracture arrest toughness of CO_2 pipelines is considered to possess considerable practical significance and engineering value. Traditional physical testing methods not only require substantial investments of funding and time but are also accompanied by high operational risks and uncertainties in outcomes. Particularly in high-pressure application scenarios such as supercritical CO_2 transport, the experimental challenges and safety risks are further exacerbated. By establishing a scientifically reliable numerical simulation tool, a portion of the high-cost physical tests can be effectively substituted, thereby providing a low-cost, high-efficiency, and repeatedly verifiable computational solution for material selection, structural design, and safety assessment of CO_2 pipelines. The successful development of such software is expected to vigorously promote the safety development of key long-distance pipeline infrastructure within the Carbon Capture, Utilization and Storage (CCUS) technological framework. Consequently, it will provide crucial technical support for the implementation of green energy strategies and the achievement of carbon neutrality goals.

2. Pipeline Fracture Arrest Assessment Model

Since the concept of fracture mechanics was introduced by Griffith, the theoretical framework in this field has been continuously supplemented and refined by scholars for over a century. Among these contributions, Irwin's extension of Griffith's theoretical system laid the foundation for the establishment of ductile fracture theory. In 1972, the Battelle Two-Curve Method (BTCM) was proposed by Maxey [36], which remains a fundamental approach widely applied in the field of pipeline ductile fracture control. Over the following decades, extensive research has been conducted by scholars on pipeline decompression wave propagation behavior and crack propagation mechanisms, leading to significant progress both theoretically and practically.

It was proposed by Maxey that a correspondence exists between the driving force for ductile fracture propagation and the gas pressure within the crack tip plane. Under steady-state conditions, the pressure value at the crack tip is related to the corresponding propagation speed of the crack itself. Based on this hypothesis, the BTCM assesses whether the pipeline's arrest toughness can prevent fracture propagation by comparing the crack propagation speed curve with the decompression wave speed curve [37]. When a pipeline undergoes rapid decompression due to leakage or rupture, a decompression wave is generated at the leak point and propagates towards both ends of the pipeline. If the decompression wave speed exceeds the crack propagation speed, arrest will occur; conversely, if the decompression wave speed is lower, the crack will continue to propagate indefinitely until an obstruction is encountered.

Whether a crack continues to propagate is determined by the relative positions of the crack propagation velocity curve and the gas decompression curve. Ductile crack arrest is achieved when the two curves do not intersect, as the decompression wave speed exceeds the crack speed at any pressure. If the curves intersect, indefinite propagation occurs at the pressure where the speeds are equal, until an obstruction is met. The minimum required arrest toughness value for the pipeline can be determined when the two curves are tangent, as illustrated in Figure 1.

The decompression curve is expressed as a functional relationship between the pipeline internal pressure or hoop pressure and the decompression wave velocity, and it is a key factor promoting crack propagation. Driving factors for decompression wave velocity include the temperature of the transported medium, the local speed of sound, and the outflow velocity of the fluid in the pipeline. The crack propagation velocity curve is defined as a functional relationship between the pipeline internal pressure or hoop pressure and the crack propagation velocity, characterizing the fracture velocity at different toughness levels. Driving factors for crack propagation velocity include the stress in the pipeline, the pipeline geometry, the crack arrest toughness of the pipe material, and the pipeline backfill constraints (e.g., soil, water, air). Obviously, materials with higher mechanical properties and toughness are more resistant to ductile fracture propagation than those with inferior properties.

The BTCM was initially developed based on specific assumptions, including pure methane flow within the pipeline (with only minimal impurities) and the gas being in a state of adiabatic isentropic flow. Under this premise, the relationship between the decompression wave speed and pressure can be derived, as shown in Equation (1).

$$P_d = P_0\left[\frac{2}{\gamma+1} + \left(\frac{\gamma-1}{\gamma+1}\right)\frac{V}{V_a}\right]^{\frac{2\gamma}{\gamma-1}} \tag{1}$$

In the formula, P_d is the gas decompression pressure in the pipeline, MPa; P_0 is the initial pressure of the pipeline before fracture, MPa; V is the decompression wave velocity at the pressure P_d, m/s, V_a is the speed of sound in the gas under initial pressure and temperature, m/s; γ is the initial specific heat ratio of the gas, J/(mol·K).

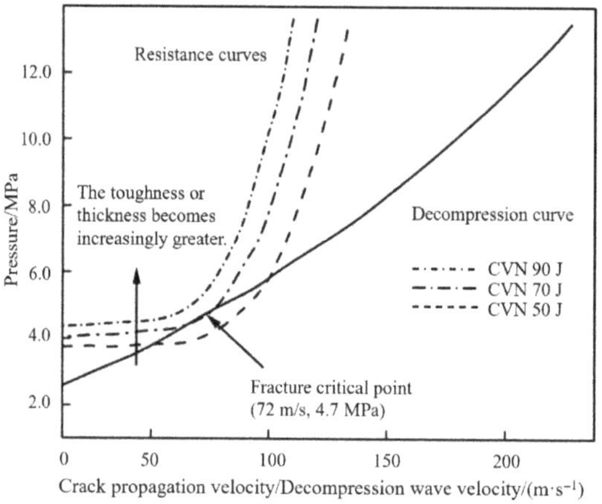

Figure 1. Schematic of the fracture arrest assessment using BTCM.

2.1. Resistance Curve Model

Currently, the BTCM based on Charpy impact toughness remains the most standardized and widely applied method for fracture arrest design in industry. The specific approach is based on the premise that crack velocity depends on flow stress and fracture resistance. By analyzing existing full-scale experimental data, the relationship between crack velocity and pressure is determined, as shown in Equation (2). The expressions for the arrest hoop stress and arrest pressure are provided in Equations (3) and (4), respectively. Equation (4) gives the intercept of the crack propagation velocity curve on the pressure axis [26].

$$V_f = C\frac{\overline{\sigma}}{\sqrt{CVN/A}}\left(\frac{\sigma_{\mathrm{d}}}{\sigma_a} - 1\right)^{\frac{1}{6}} \tag{2}$$

In the equation, V_f is the crack propagation speed, m/s; C is the backfill constraint coefficient, taken as 3.79 for air backfill, 2.75 for soil backfill, and 2.34 for water backfill; CVN is the Charpy impact energy (full-size Charpy V-notch specimen), J; A is the cross-sectional area of the Charpy V-notch specimen, mm^2; $\overline{\sigma}$ is the flow stress (=specified minimum yield strength of the pipe material + 68.95 MPa), MPa; σ_{d} is the arrest stress, MPa; σ_a is the arrest hoop stress, MPa.

$$\sigma_{\mathrm{a}} = \frac{2\overline{\sigma}}{3.33\pi}\cdot arc\cos\left(e^{-\frac{10CVN\cdot E\cdot\pi}{24A\cdot\overline{\sigma}^2\cdot\sqrt{R\cdot t}}}\right) \tag{3}$$

$$P_a = \frac{\sigma_a t}{R} \tag{4}$$

In the equation, E denotes the elastic modulus, MPa; R represents the average radius of the pipeline, mm; t indicates the wall thickness of the pipeline, mm; P_a is the crack tip arrest pressure, MPa.

The Battelle Two-Curve Model (BTCM) method has been validated on low-to-medium grade pipeline steels with a Charpy impact energy of 100 J. However, the toughness value of standard pipe materials currently produced can be guaranteed to exceed 250 J, indicating significant differences between the materials used for model validation in the 1970s and modern steel materials. In recent years, the results of several full-scale burst tests conducted on high-grade and high-toughness natural gas pipelines have been inconsistent with the predictions of BTCM to a certain extent. For the application of the BTCM method in modern steel materials, especially high-grade and high-toughness materials, modifications to the predicted toughness values are required, which are generally referred to as BTCM correction factors.

The latest versions of standards issued by the International Organization for Standardization (ISO) and Det Norske Veritas (DNV) have proposed a new method for evaluating the resistance of pipeline ductile fracture propagation events during the transportation of carbon dioxide mixtures. For carbon dioxide mixtures, the plateau period in their decompression curves is extremely long, and the tangent point of the resistance curve often occurs within this plateau region. The crack arrest pressure can be easily calculated using the BTCM equation, where the saturation pressure is the starting point of the plateau on the right side of the decompression curve. The calculation methods recommended in ISO 27913 are shown in Equations (5) and (6).

$$1000 \cdot \frac{CVN \cdot E}{A_c \cdot \overline{\sigma}^2 \cdot \sqrt{R \cdot t}} = \frac{24}{\pi} \cdot \ln\left(\sec\left(\frac{\pi}{2} \cdot \frac{c_{cf} \cdot 3.33\sigma_a}{\overline{\sigma}}\right)\right) \quad (5)$$

$$\sigma_a = \frac{p_s \cdot OD}{2 \cdot t} \quad (6)$$

In the equations, CVN denotes the Charpy V-notch impact energy (full-size specimen), J; E represents the elastic modulus, MPa; A_C is the shear area of the impact specimen (80 mm^2 for full-size specimens), mm^2; $\overline{\sigma}$ refers to the flow stress ($\overline{\sigma}$ = yield stress + 68.95 MPa), MPa; R is the average radius of the steel pipe, mm; T denotes the wall thickness of the steel pipe, mm; OD is the outer diameter of the steel pipe, mm; C_{cf} is the correction factor ($C_{cf} \geq 1.2$); and p_s represents the maximum saturation pressure, MPa.

As mentioned earlier, standards such as ISO 27913 recommend that when considering carbon dioxide mixtures, the minimum correction factor is 1.2, but it shall be verified based on applicable test data [35]. By applying the BTCM prediction to 5 selected tests from the published full-scale tests of carbon dioxide mixture pipelines (see Table 2), it is found that there is a 60% error in the BTCM prediction, and all errors are related to the predicted crack propagation. Through trial calculations of different correction factors, it is determined that the correction factor ranges between 1.6 and 2.2, which is much higher than the minimum value of 1.2 recommended by ISO standards. The evaluation results of the correction factor are shown in Figure 2. This range is bounded by an upper limit and a lower limit, where the upper limit is a conservative threshold above which all crack propagation can be avoided. It should be noted that due to the limited number of existing tests on CO_2 pipelines and the large dispersion of existing test results, more research and tests are required to determine a reliable correction factor.

Table 2. Five full-scale burst tests of CO_2 gas mixtures listed in published literature and correction factors.

	Test 1	Test 2	Test 3	Test 4	Test 5
Test name	SARCO$_2$ 1	SARCO$_2$ 2	NG 1	NG 2	NG 3
Outer diameter/mm	610	610	914	914	610
Wall thickness/mm	12.7–13.7	12.5–13.7	25.4	25.4	19.1
Test pressure/MPa	12.95	12.77	14.83	15.06	15.02
Test temperature/°C	13	17	13.1	10.5	15.0

Table 2. *Cont.*

	Test 1	Test 2	Test 3	Test 4	Test 5
Gas composition	CO_2: 96.2% N_2: 3.8%	CO_2: 94.0% N_2: 6.0%	CO_2: 90.9% H_2: 1.0% N_2: 4.0% O_2: 1.8% CH_4: 2.2%	CO_2: 93.7% H_2: 1.1% N_2: 3.4% O_2: 0.0% CH_4: 1.8%	CO_2: 90.3% H_2: 1.1% N_2: 6.6% O_2: 0.0% CH_4: 2.0%
Saturation pressure/MPa	6.1	7.2	8.06	7.38	7.38
Correction factor of arrest (min)	1.68	1.62	1.65	2.81	3.35

Numbers 1–3 denote Experiment 1, Experiment 2 and Experiment 3, respectively.

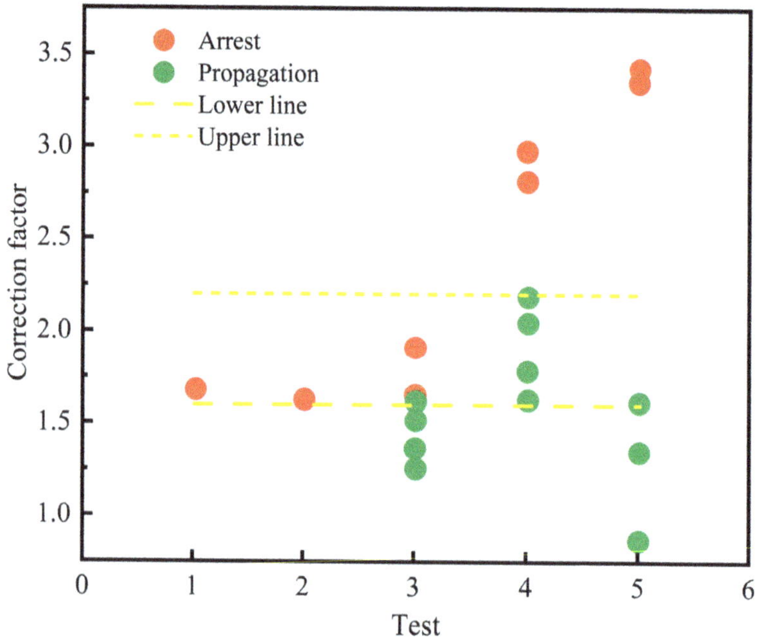

Figure 2. Evaluation of the BTCM correction coefficient in the transportation of CO_2 mixtures.

2.2. Decompression Wave Model

A decompression wave refers to a pressure wave generated by a sudden pressure drop in a pipeline due to reasons such as pipeline leakage, pipeline fracture, or blockage inside the pipeline. The propagation speed and attenuation characteristics of the pressure wave are crucial for the safe operation of the pipeline.

The characteristics of decompression waves, including their propagation speed and pressure variation, are influenced by multiple factors, such as the composition, phase state, pressure, and temperature of the fluid, as well as the pipe material and pipeline dimensions [38,39]. The composition of the fluid directly affects the fluid's physical property parameters, leading to changes in the characteristics of decompression waves [20,40]. The higher the initial pressure or flow velocity of the fluid, the greater the energy stored in the fluid; thus, the larger the energy released when the pipeline ruptures, resulting in a higher propagation speed of the decompression wave [18]. When the temperature of the fluid changes, it affects the density and viscosity of the fluid, altering the propagation characteristics of the decompression wave. In gas–liquid two-phase flow, the interaction between the two phases makes the propagation of decompression waves more complex, and the propagation characteristics of decompression waves in supercritical fluids also have their uniqueness [41,42]. In small-diameter pipelines, due to the significant non-isentropic effect, isentropic and non-isentropic models need to be established to calculate the speed of decompression waves [43].

Since full-scale pipeline experiments require higher costs than numerical simulations and are difficult to repeat, scholars have mostly adopted numerical simulation methods to study the propagation characteristics of decompression waves. Existing CO_2 pipeline decompression wave models are mainly divided into two categories: one is the Homogeneous Equilibrium Model (HEM), and the other is the Two-Fluid Model (TFM). These models simulate the influence of different physical conditions on the propagation of decompression waves, such as pipeline diameter, wall thickness, impurity content and types, providing important references for theory and practical operations.

The propagation speed of a decompression wave depends on the outflow velocity at the leak orifice and the local sound speed of the fluid. The local sound speed of the fluid depends on the thermodynamic state of the fluid. Due to the extremely strong Joule-Thomson effect of CO_2, once a CO_2 pipeline ruptures, the sudden drop in pressure and temperature causes the CO_2 inside the pipe to undergo a phase transition into the gas–liquid two-phase region. The mixing of gas and liquid phases greatly changes the flow structure of the medium, resulting in its compressibility being much greater than that of single-phase gas or liquid, causing a sudden drop in the decompression wave. In this case, the establishment of a decompression wave model for CO_2 pipelines containing impurities needs to be based on the homogeneous equilibrium model framework, combined with the gas state equation and sound speed equation. The model makes the following assumptions: (a) The CO_2 inside the pipe is one-dimensional isentropic flow, ignoring the effects of heat conduction and convection; (b) The pipeline is a horizontal pipe, without considering the effects of height difference and pipe diameter; (c) The initial state of CO_2 inside the pipe is a thermodynamic equilibrium state; (d) There is no slip between the gas and liquid phases.

The decompression wave reflects the relationship between the medium pressure inside the pipe and the decompression wave speed. The calculation method for the decompression wave speed (W, m/s) is shown in Equation (7). After pipeline leakage, CO_2 inside the pipeline flows from the leak orifice to the atmosphere. At the moment of pipeline leakage, the outflow velocity of the fluid at the leak orifice is 0. As the leakage process proceeds, according to the method of characteristics in computational fluid dynamics, the outflow velocity (U, m/s) of the fluid inside the pipe is given in the form of Riemann invariants, and the calculation method is shown in Equation (8). For large-diameter pipelines, the decompression process can be regarded as one-dimensional isentropic and adiabatic, and the final simplified form is shown in Equation (9). The calculation method for the local sound speed (a, m/s) is shown in Equation (10). The local sound speed of the fluid depends on the thermodynamic state of the fluid. During calculation, the initial temperature at the time of leakage is taken as the reference, and the pressure is trial-calculated with ΔT as the temperature reduction step. According to the assumed isentropic process conditions, when the entropy values corresponding to the two temperatures in the calculation are equal, the pressure value is output, and then the sound speed, medium outflow velocity, and decompression wave speed are calculated. When the calculated decompression wave speed or pressure is zero, the program terminates. The program block diagram is shown in Figure 3.

$$W = a - U \tag{7}$$

$$U = -\left[\int_{p_\eta}^{p} \frac{a\,\mathrm{d}\rho}{\rho_l}\right]_s \quad (8)$$

$$\Delta U = \frac{\Delta p}{a_\eta \rho_{l\eta}} \quad (9)$$

$$a = \left[\left(\frac{\partial p}{\partial \rho_l}\right)_s\right]^{1/2} \quad (10)$$

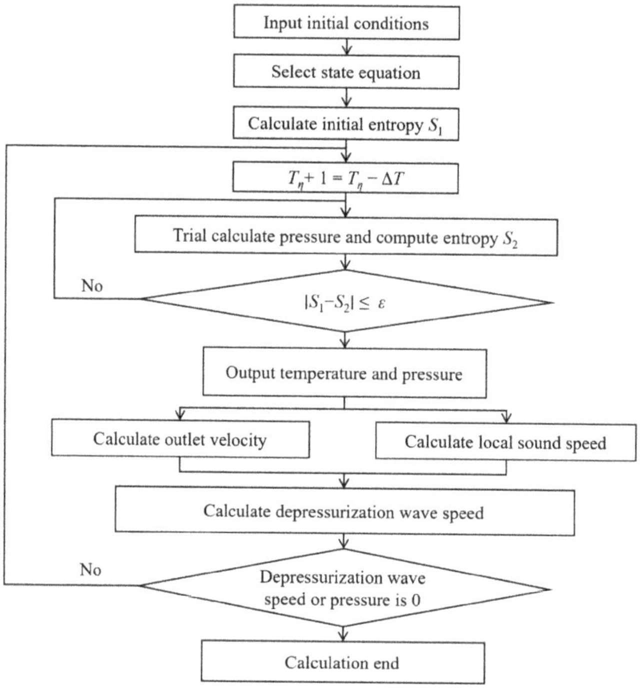

Figure 3. Model calculation flowchart.

In the equation, W denotes the propagation speed of the decompression wave, m/s; a represents the local sound speed of the medium inside the pipe, m/s; U is the outflow velocity of the fluid inside the pipe, m/s. p is the fluid pressure, kPa; ρ_l is the average density of the fluid, kg/m^3. The subscript η indicates physical parameters under different calculation steps; the subscript s denotes the isentropic process.

2.2.1. Calculation of Local Sound Speed Based on Isochoric Thermodynamics

For the calculation of decompression wave speed, scholars have focused on the calculation of local sound speed. The influence of discontinuous sound speed on the propagation speed of decompression waves was discussed by Lund et al. [44]. A calculation model for local sound speed was proposed by Flatten et al. [45] to calculate the decompression wave speed. However, when calculating supercritical CO_2, especially for CO_2 fluid near the critical point (P_c = 7.38 MPa, T_c = 31.1 °C), the model often fails to converge.

The sound speed calculation in this model is based on the parameter tracking method of isochoric thermodynamics, which can avoid non-convergence problems. From the perspective of isochoric thermodynamics, Bell et al. [46] calculated the thermodynamic parameters of mixtures. By taking the Helmholtz energy density as the basic thermodynamic potential and converting the mole fraction into molar density (or concentration), the problem of difficult differential calculation was effectively avoided. This method takes molar density as an independent variable, which does not affect other variables, and treats the concentration difference in each component as an independent variable, simplifying the mathematical expression and calculation process, and it is no longer necessary to consider whether a phase transition occurs during the flow process.

The calculation process of local sound speed is shown in Figure 4. Different algorithms are used to solve the nonlinear equations and differential equations in the Helmholtz energy density expression. It is found that far from the critical region, the solution of the differential equations converges rapidly; near the critical region, the solution of the differential equations is numerically unstable, and at this time, the solution effect of the nonlinear equations is better. By substituting the calculation results into the explicit form of the GERG-2008 equation of state, the physical parameters such as medium temperature, pressure, density, adiabatic coefficient, and compression factor can be solved, and thus the local sound speed can be obtained.

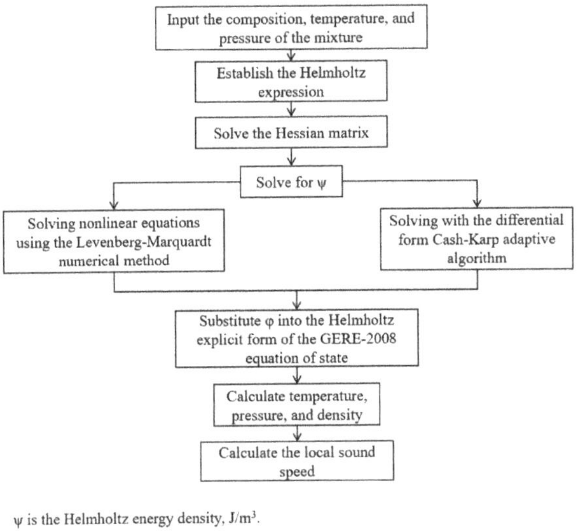

Figure 4. Flowchart for local sound speed calculation based on isochoric thermodynamics.

2.2.2. Equation of State

To determine the thermodynamic state of the fluid, three equations of state, namely the PR equation, HEOS equation, and GERG-2008 equation, are selected as thermodynamic models for calculation when studying the characteristics of decompression waves.

(1) PR Equation

The PR equation of state was proposed by Peng and Robinson in 1976 [47] and is used to describe the behavior of common gases or liquids, and can be applied to simulate a variety of complex fluid systems. Its equations are shown in Equations (11)–(14).

$$p = \frac{RT}{V-b} - \frac{a(T)}{V(V+b)+b(V-b)} \tag{11}$$

$$a(T) = a_c \alpha(T) \tag{12}$$

$$\alpha(T) = \left[1 + k(1 - T_r^{0.5})\right]^2 \tag{13}$$

$$k = 0.37464 + 1.54226\omega - 0.26992\omega^2 \tag{14}$$

In the equations, p is the system pressure, MPa; R is the gas constant with a value of 8.314 J/(mol·K); T is the thermodynamic temperature, K; V is the gas volume, m^3/mol (note: corrected from m^2/mol for physical consistency); b is the characteristic parameter of the substance; $\alpha(T)$ is a function of temperature; T_r is the reduced temperature, K; a_c is the acentric factor; and k is the coefficient of $\alpha(T)$.

Since the isentropic partial derivative has the relationship shown in Equation (15) with the isothermal partial derivative and temperature derivative, substituting Equation (15) into Equation (10) yields Equation (16).

$$\left(\frac{\partial p}{\partial \rho}\right)_S = \left(\frac{\partial p}{\partial \rho}\right)_T + \frac{T}{c_V}\left(\frac{\partial p}{\partial T}\right)_\rho^2 \tag{15}$$

$$c = \left\{-\frac{V^2}{M}\left[-\frac{RT}{(V-b)^2} + a_c\alpha\frac{2V+2b}{(V^2+2bV-b^2)^2}\right] + \frac{T}{c_V}\left[\frac{R}{V-b} - \frac{a}{V^2+2bV-b^2}\cdot\frac{d\alpha}{dT}\right]^2\right\}^{1/2} \tag{16}$$

$$\frac{d\alpha}{dT} = -\frac{m}{T_c\sqrt{T/T_c}}\left(1 + m\left(1 - \sqrt{\frac{T}{T_c}}\right)\right) \tag{17}$$

(2) HEOS Equation

The Helmholtz Energy Equation of State (HEOS) is used to describe the thermodynamic properties of substances under different states. Based on the concept of Helmholtz free energy, it can effectively characterize the phase behavior of gases, liquids, and solids. In the initial stage of supercritical CO_2 pipeline leakage, the rapid drop in pressure at the leak orifice leads to a significant temperature change, and the sharp pressure reduction causes a phase transition of supercritical CO_2. According to the principle of isochoric thermodynamics, the initial stage of pipeline segment leakage is approximated as an isochoric process. The change in internal energy is mainly caused by temperature variation, so the temperature change under specific heat exchange and internal energy change can be calculated, thereby determining the phase transition behavior during the leakage process. The central thermodynamic potential of isochoric thermodynamics is the Helmholtz energy density, and its expression is shown in Equation (18) [48].

$$\Psi \equiv \frac{A}{V} \tag{18}$$

In the equation, Ψ denotes the energy density, J/m^3; A represents the total Helmholtz free energy, J; V is the total volume of the mixture, m^3. Among these parameters, A can be expressed as a function of two independent variables, namely density ρ and temperature T, and its expression is shown in Equation (19).

$$A(\rho, T) = A^{o}(\rho, T) + A^{r}(\rho, T) \tag{19}$$

In the equation, ρ denotes the molar density of the mixture, mol/m^3; the superscript o indicates the ideal state; and r represents the part related to the compressibility of the fluid.

The Helmholtz free energy equation is a fundamental thermodynamic equation, from which other thermodynamic parameters can be derived. Among these, the expression for pressure is shown in Equation (20). By dividing Ψ into an ideal part and a real part (see Equation (21), the pressure can be further expressed as Equation (22) [49].

$$p = -\Psi + \sum_{i=1}^{N} \rho_i \mu_i \tag{20}$$

$$\Psi = \Psi^{o} + \Psi^{r} \tag{21}$$

$$p = \rho RT - \Psi^{r} + \sum_{i=1}^{N} \rho_i (\frac{\partial \Psi^{r}}{\partial \rho_i})_{T,\rho_{j \neq i}} \tag{22}$$

In the equation, p denotes the medium pressure, kPa; Ψ^r represents the real Helmholtz energy density, which itself is a function of T and ρ, with the unit of J/m^3. The subscript i denotes different components.

A numerical method must be adopted to solve for Ψ. Taking two components as an example, to ensure numerical stability in the numerical calculation and solution process, a Hessian matrix is introduced. The expression of the Hessian matrix for the local curvature of $\Psi(T,\rho)$ is shown in Equation (23) [46].

$$\mathbf{H}_{\Psi}(T, \vec{\rho}) = \begin{bmatrix} \frac{RT}{\rho_1} & 0 \\ 0 & \frac{RT}{\rho_2} \end{bmatrix} + \begin{bmatrix} \left(\frac{\partial^2 \Psi^{r}}{\partial \rho_1^2}\right)_T & \left(\frac{\partial^2 \Psi^{r}}{\partial \rho_1 \partial \rho_2}\right)_T \\ \left(\frac{\partial^2 \Psi^{r}}{\partial \rho_2 \partial \rho_1}\right)_T & \left(\frac{\partial^2 \Psi^{r}}{\partial \rho_2^2}\right)_T \end{bmatrix} \tag{23}$$

Combined with the isochoric phase equilibrium characteristics, different algorithms can be adopted to solve the nonlinear equations and differential equations. Equations (24)–(26) form a system of nonlinear equations, which can be solved using the Levenberg–Marquardt numerical method.

$$\left(\frac{\partial \Psi^{r}}{\partial \rho_1}\right)'_{T,\rho_2} + RT \ln \rho_1' = \left(\frac{\partial \Psi^{r}}{\partial \rho_1}\right)'_{T,\rho_2} + RT \ln \rho_1'' \tag{24}$$

$$\left(\frac{\partial \Psi^{r}}{\partial \rho_2}\right)'_{T,\rho_1} + RT \ln \rho_2' = \left(\frac{\partial \Psi^{r}}{\partial \rho_2}\right)''_{T,\rho_1} + RT \ln \rho_2'' \tag{25}$$

$$p' = p'' \tag{26}$$

In the equations, ρ denotes the liquid-phase molar density, mol/m^3; ρ'' represents the vapor-phase molar density, mol/m^3; p' is the liquid-phase pressure, kPa; p'' is the vapor-phase pressure, kPa. The subscript 1 indicates component 1, and 2 indicates component 2.

In addition, the differential expression for the isochoric phase equilibrium process is shown in Equation (27).

$$\begin{bmatrix} \mathbf{H}'_{\Psi,1} \cdot \vec{\rho}'' & \mathbf{H}'_{\Psi,2} \cdot \vec{\rho}'' \\ \mathbf{H}'_{\Psi,1} \cdot \vec{\rho}' & \mathbf{H}'_{\Psi,2} \cdot \vec{\rho}' \end{bmatrix} (\frac{\mathrm{d}\vec{\rho}}{\mathrm{d}p})'_{T,\sigma} = \begin{bmatrix} 1 \\ 1 \end{bmatrix} \tag{27}$$

In the equation, $\mathbf{H}_{\Psi,i}$ denotes the i-th row of the Hessian matrix; σ represents the derivative taken along the phase envelope. The first-order derivative of the concentration in the vapor phase can be obtained through the linear system of equations (Equation (28).

$$\mathbf{H}''_{\Psi} \left(\frac{\mathrm{d}\vec{\rho}}{\mathrm{d}p}\right)''_{T,\sigma} = \mathbf{H}'_{\Psi} \left(\frac{\mathrm{d}\vec{\rho}}{\mathrm{d}p}\right)'_{T,\sigma} \tag{28}$$

In the equation, $\left(\frac{\mathrm{d}\vec{\rho}}{\mathrm{d}p}\right)'_{T,\sigma}$ has been obtained through Equation System (27), and the remaining parts can be solved by combining the differential equation system (Equation (27) using the adaptive Cash-Karp algorithm.

Both of the above solution methods can be used to calculate the Helmholtz energy density. For the solution of nonlinear equation systems, high requirements are imposed on the initial conditions, and convergence cannot generally be guaranteed. If convergence is achieved, the accuracy of the solution is independent of the initial solution. For the solution of differential equation systems, accurate initial values are required, and the accuracy of the solution depends on the quality of the solution from the previous state. During the simulation and calculation process in this study, the above different calculation methods were selected according to the magnitude of the calculation result error. It was found that far from the critical region, the solution using the differential form converges rapidly; near the critical region, the solution of the differential equation system is numerically unstable, and at this time, the solution effect using the nonlinear equation system is better.

(3) GERG-2008 Equation

For the calculation of the thermodynamic properties of mixtures, the GERG-2008 equation of state is adopted in the model. Developed by the Groupe Européen de Recherches Gazières (GERG), the GERG-2008 equation of state is a multi-parameter Helmholtz equation. It was designed for various natural gas types, including those with high contents of nitrogen, carbon dioxide, ethane, and other components, and exhibits high accuracy. Covering a wide range of temperature and pressure conditions, it is applicable to gaseous, liquid, supercritical, and gas–liquid two-phase states, and has been recognized by the International Organization for Standardization (ISO) as a standard equation for calculating the physical properties of natural gas. All thermodynamic parameters can be derived by substituting the calculation results of the HEOS equation into the explicit Helmholtz form of the GERG-2008 equation of state.

Based on the HEOS equation, the GERG-2008 equation of state incorporates the interactions between multi-component substances, and its expression is shown in Equation (29). The specific expression for the Helmholtz free energy of the ideal mixture gas is presented in Equation (30), while the expression for the residual Helmholtz free energy of the mixture gas is given in Equation (31) [49].

$$\varphi(\delta,\tau,\overline{x}) = \varphi^{o}(\rho,T,\overline{x}) + \varphi^{r}(\delta,\tau,\overline{x}) \tag{29}$$

In the equation, φ^{o} denotes the Helmholtz free energy of the ideal fluid in the mixture; φ^{r} represents the Helmholtz free energy of the residual fluid in the mixture; ρ and T are the density and temperature of the mixture, respectively; $\overline{x}$ is the molar ratio of each substance.

$$\varphi^{o}(\rho,T,\overline{x}) = \sum_{i=1}^{N} x_i[\varphi_i^{o}(\rho,T) + \ln x_i] \tag{30}$$

In the equation, $\varphi_i^{o}(\rho,T)$ is the dimensionless form of the Helmholtz free energy of the ideal gas for component i; N is the number of components in the mixture; x_i is the mole fraction of component i.

$$\varphi^{r}(\delta,\tau,\overline{x}) = \sum_{i=1}^{N} x_i\varphi_i^{r}(\delta,\tau) + \sum_{i=1}^{N-1}\sum_{j=i+1}^{N} x_i x_j F_{ij}\varphi_{ij}^{r}(\delta,\tau) \tag{31}$$

In Equation (31), the first term on the right-hand side is the linear summation of the product of the residual properties of pure gases and their mole fractions; the second term represents the influence of interactions between all binary mixtures on the Helmholtz free energy.

3. Development of Pipeline Fracture Arrest Assessment Software

To address the lack of effective quantitative calculation and analysis methods in the field of pipeline resistance to ductile fracture propagation, this study established a computational analysis technique for assessing the resistance of CO_2 pipelines to ductile fracture extension. This was achieved by comparing gas decompression models with experimental data and defining the applicable range of correction factors. The developed technique is not only applicable to CO_2 pipelines but can also be extended to long-distance natural gas pipelines, demonstrating its cross-media practicality.

Based on this methodology, a computational software for determining the fracture arrest toughness of CO_2 pipelines containing impurities was developed using computer programming, grounded in the Battelle two-curve model. This software provides theoretical support for CO_2 pipeline design by enabling: (1) rapid calculation of the minimum impact toughness required to resist ductile fracture propagation, verifying whether the pipeline possesses sufficient toughness to arrest fractures autonomously, and (2) delivering precise data for the design phase, thereby enhancing the safety of pipeline construction.

3.1. Development Environment and Technology

The software uses Python (v.3.12.4) as the primary development language. Known for its concise and readable syntax, Python offers robust scientific computing and data processing libraries such as NumPy (v.1.26.4), SciPy (v.1.13.0), and Pandas (v.2.2.2), which can efficiently support numerical simulations and data analysis tasks in this study. Thus, it is well-suited as the foundational language for the software development. The graphical user interface (GUI) was developed using the PyQt6 (v.6.10.0) framework and designed based on a modular approach and a three-tier architecture (data layer, logic layer, and presentation layer). The overall framework structure of the software is detailed in Figure 5.

Within the three-tier architecture, the bottom layer is the Data Layer, which is primarily responsible for reading various critical data from external sources, including pipeline information data, thermophysical property databases for fluids, and equations of state databases. The middle layer is the Logic Layer; it handles the core tasks of data analysis and computation. This layer can first perform macro-definitions for gas mixtures and select appropriate equations of state, then proceed to calculate gas physical properties, decompression wave curves, and resistance curves. The top layer is the Presentation Layer, encompassing the interactive interface and built-in functions directly accessible to users. At this level, users can set pipeline-related parameters and obtain the final calculation results. The pseudocode for each layer within the three-tier structure, along with the pseudocode for their interaction workflow, is shown in Figure 6.

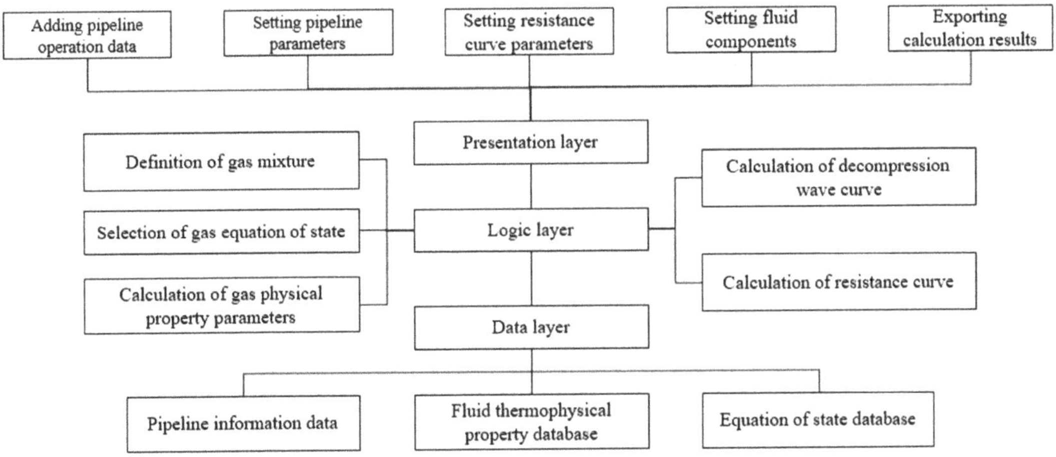

Figure 5. Overall software framework structure.

```
class DataLayerWaveSpeed:
  method initialize(inputParameters):  # Receive standardized parameters from the logic layer
    initialize fluid Equation of State (EOS), initial temperature-pressure, and mole fractions
    verify the validity of thermodynamic parameters

  method calculateWaveSpeed(entropy, pressure):  # Core calculation logic
    call the CoolProp library to solve for physical properties (density, entropy, etc.)
    calculate wave speed based on pressure-density variations

  method getCalculationResults():  # Return standardized data to the logic layer
    organize results (pressure, temperature, wave speed, etc.) into a DataFrame
```

(**a**) Data layer

```
class LogicLayerCalculation:
  method receiveGuiInput():  # Obtain raw input from the presentation layer
    read temperature, pressure, fluid composition, and structural parameters
    validate input legality (sum of mole fractions = 100%, non-negative parameters)

  method formatParameters():  # Convert to format recognizable by the data layer
    pressure unit conversion (MPa → Pa), temperature unit conversion (°C → K)
    construct fluid composition string (e.g., "CO2[99.85]&N2[0.15]")

  method invokeDataLayerCalculation():
    instantiate data layer object and pass formatted parameters
    execute sound velocity and thermodynamic parameter calculations

  method processCalculationResults():  # Prepare display data for the presentation layer
    convert units (Pa → MPa, K → °C)
    calculate derived parameters such as fracture velocity and Bernoulli velocity

  method passResultsToGui():
    return table data and chart data to the presentation layer
```

(**b**) Logic layer

```
class PresentationLayerGUI:
  method initializeInterface():  # Construct user interaction controls
    create parameter input fields, fluid composition table, calculation button, and
chart canvas
    bind button click events (e.g., "Run Calculation" → call business logic layer)

  method receiveUserInput():
    collect parameter values from controls (temperature, pressure, composition,
structural parameters)

  method invokeBusinessLogicLayer():
    pass raw input data to the business logic layer

  method displayResults():  # Receive processed data from the business logic layer
    display thermodynamic parameters, sound velocity, and other results in the
table
    plot "sound velocity vs. pressure" and "temperature vs. pressure" charts

  method provideExportFunctionality():
    support exporting result table as CSV and charts as PNG
```

(**c**) Presentation layer

```
# Core Interaction Flow
if user clicks "Run Calculation":
  Presentation Layer → Collect input parameters → Pass to Logic Layer
  Logic Layer → Validate + Format parameters → Invoke Data Layer calculation
  Data Layer → Execute physical property calculation → Return raw results to Logic Layer
  Logic Layer → Process results (unit conversion, derived calculation) → Pass to Presentation Layer
  Presentation Layer → Display table + charts → Support user export
```

(**d**) Interaction workflow

Figure 6. Pseudocode for individual layers and their interaction workflow in the three-tier architecture.

The GUI is divided into three main functional areas. The upper-left section is the input parameter area, containing thermodynamic parameters (e.g., temperature, pressure, pressure step size), EOS equation of state selection, a fluid composition table supporting custom input or quick selection of "pre-/post-combustion components" (the table automatically validates that the mole fractions sum to 100%), and input fields for structural parameters like wall thickness and elastic modulus. All parameters are directly editable. The lower-left section is the results display area, featuring a text box for calculation status, a results data table showing six key parameters (including pressure, temperature, speed of sound, and density), along with "Save Results" (export to CSV) and "Save Image" (export chart) buttons. The right section is the chart display area, with the BTCM curve (including the crack propagation velocity comparison line) displayed above and the pressure drop curve below. The charts support real-time refreshing and high-resolution export. The key operational workflow is as follows: First, select the fluid composition (either custom or by quick-selecting preset components). Next, input the thermodynamic and structural parameters. Then, choose the equation of state (HEOS/REFPROP; REFPROP offers GERG/PR sub-options). Click "Run Calculation" to initiate the computation. The results table and charts will then be displayed, and users can click the export buttons to save data or images as needed. The software GUI is shown in Figure 7.

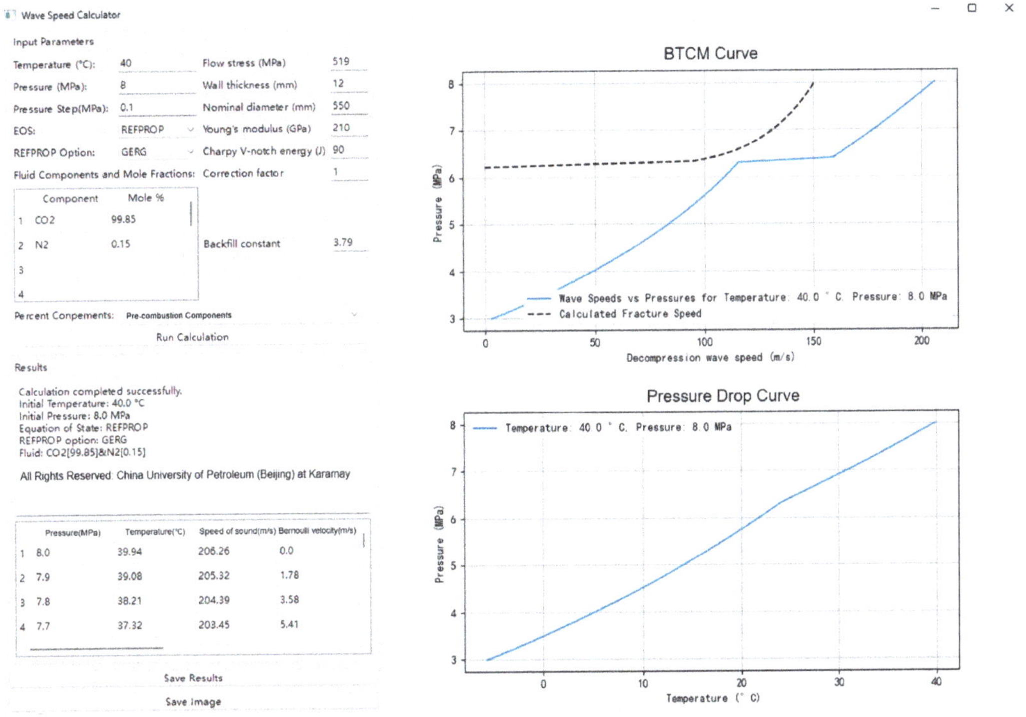

Figure 7. Layout of the software GUI.

3.2. *Software Usage*

3.2.1. Input Parameter Settings

The parameter input interface is shown in Figure 8. Users only need to set the initial pressure and temperature of the pipeline transport medium, the calculation step size, select an appropriate equation of state, and define the gas components and their mole fractions. The software will then output the decompression wave curve and the pressure drop curve.

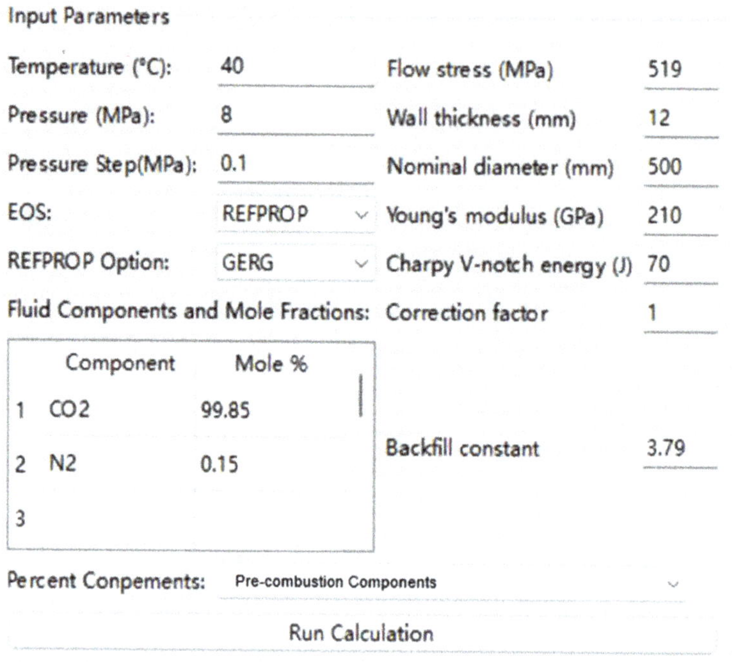

Figure 8. Input parameter panel of the software.

For supercritical CO_2 pipelines, the operating temperature is typically maintained between 40 °C and 50 °C, and the operating pressure between 9 MPa and 15 MPa. A smaller step size can improve calculation accuracy; a value between 0.1 MPa and 0.5 MPa is recommended. When simulating non-hydrocarbon pure components, the HEOS equation of state is recommended. When simulating mixtures or hydrocarbon components, the REFPROP database is recommended, allowing selection of either the GERG-2008 or Peng-Robinson (PR) equation of state based on specific requirements. The gas composition and mole fractions can be adjusted according to the actual gas components. For convenience, the software includes typical gas compositions for three common capture scenarios: pre-combustion capture, post-combustion capture, and oxy-fuel combustion capture.

To calculate the pipeline fracture arrest toughness, users need to input parameters such as flow stress, pipe diameter, wall thickness, elastic modulus, Charpy impact energy, correction factors, and backfill coefficients (accounting for environments such as soil, water, or air).

3.2.2. Output Results Visualization

Upon completion of the simulation, the software outputs parameters such as the decompression wave propagation velocity and local speed of sound as pressure and temperature decrease. The results output interface is shown in Figure 9. Additionally, both the BTCM curve and the pressure drop curve are simultaneously generated, as illustrated in Figure 10.

When the fracture arrest curve lies above the decompression wave curve, the pipeline can arrest a crack relying on its inherent fracture arrest toughness. Conversely, if the fracture arrest curve falls below the decompression wave curve, the pipeline will be unable to arrest the crack, leading to crack propagation. The presence of an inflection point on the pressure drop curve indicates that the internal gas has entered the gas–liquid two-phase region.

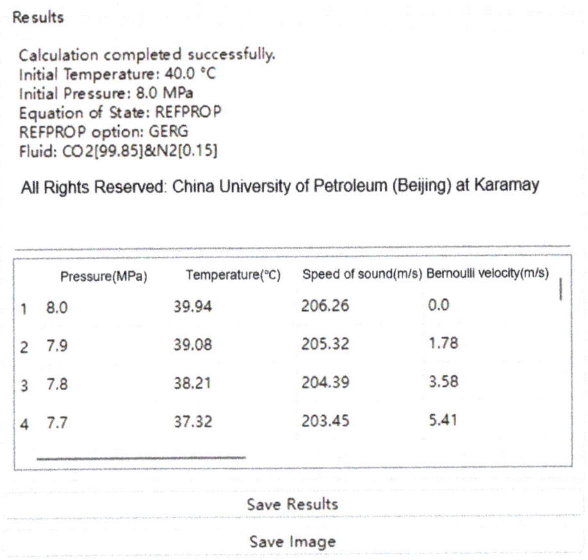

Figure 9. Results display panel of the software.

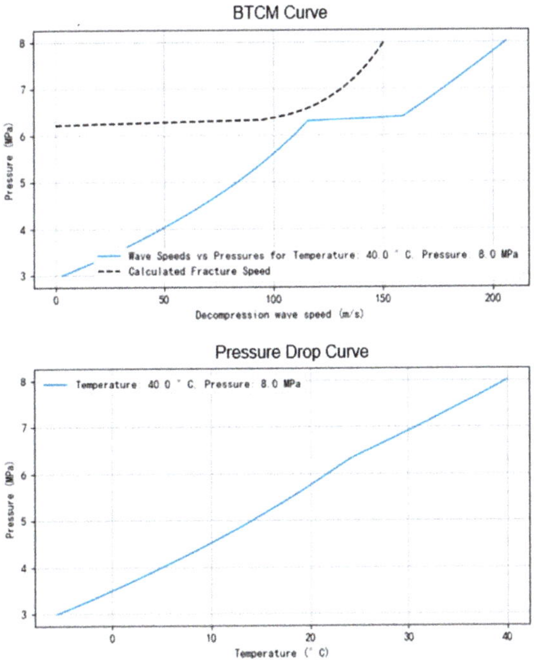

Figure 10. BTCM curve and pressure drop curve.

3.3. *Software Advantages and Potential Applications*

The core advantage of this software lies in its two precise models: an accurate thermodynamic model and an accurate fracture arrest toughness evaluation model.

Together, these form the foundation supporting the software's core functionality and application value.

The accurate thermodynamic model provides the essential basis for physical property calculations required for decompression wave analysis and pipeline fracture arrest assessment. The software incorporates an extensive built-in database of physical properties covering pure components and various mixtures, integrated with mainstream equations of state such as HEOS (Helmholtz Energy Equation of State). Utilizing these resources, the software can accurately calculate a range of key physical parameters—including fluid density, speed of sound, enthalpy, entropy, and phase equilibrium boundaries—across a broad range of temperatures and pressures. This ensures the scientific reliability and engineering value of the subsequent fracture arrest toughness predictions.

The accurate fracture arrest toughness evaluation model operates on the core principle of comparing the relative positions of the decompression wave curve and the pipeline's fracture arrest toughness curve to determine the crack propagation tendency. Based on this model, the software delivers core functionalities such as decompression wave velocity calculation and toughness assessment, providing direct support for pipeline fracture arrest analysis.

Leveraging these advantages, the software is primarily applicable to the safety design and assessment of ductile fracture control in high-pressure gas pipelines within the energy, chemical engineering, and CCUS sectors. It can precisely characterize decompression wave propagation behavior and evaluate whether a pipeline meets safe operating conditions. Specific application scenarios include the safety design of new pipelines, integrity assessment of existing pipelines, and evaluation of pipeline repurposing for different transport media.

4. Verification of Software Calculation Accuracy

4.1. Experimental Validation of the Decompression Wave Prediction Model

To verify the accuracy and reliability of the built-in decompression wave model, the software's calculation results were compared and analyzed against experimental data. This study selected relevant experimental data from sources such as Botros for reference. Comparative analyses were conducted on the predicted decompression wave curves for two typical scenarios: pure CO_2 and impurity-containing CO_2 (with a CO_2 volume fraction of 96.67% and an O_2 volume fraction of 3.33%) pipeline leaks. The initial conditions were 36.5 °C and 34.0 MPa, and 35.1 °C and 14.6 MPa, respectively. The specific comparison results are shown in Figure 11.

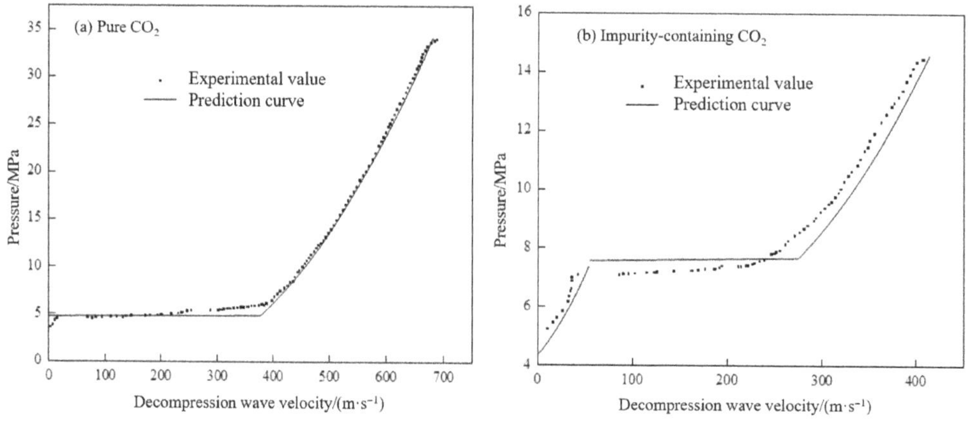

Figure 11. Comparison of predicted decompression wave curves with experimental data.

The decompression wave curves calculated by the model show a consistent trend with the experimental values. A pressure plateau appears in both cases upon entering the gas–liquid two-phase region. For the prediction of the decompression wave curve in pure CO_2 pipeline leaks, the model's predictions fit the experimental values well across various pressure zones. For the impurity-containing CO_2 pipeline leak scenario, the model's predicted pressure plateau is slightly higher than the experimental value. A higher predicted plateau pressure corresponds to a higher saturation pressure, which in turn leads to a greater required fracture arrest toughness for the pipeline. From a pipeline design perspective, this result is more conservative.

4.2. Engineering Verification of the Pipeline Fracture Arrest Assessment Model

4.2.1. Project Overview

On 15 May 2023, China conducted its first full-scale burst test of a CO_2 pipeline [33]. The test section consisted of pipe segments made of X65 grade steel with an outer diameter of 323.9 mm (1 initiation pipe and 6 test pipes). The wall thickness of the initiation pipe was 7.0 mm. Symmetrically arranged on both sides of the initiation pipe were pipes with increasing toughness, having wall thicknesses of 7.0 mm, 7.2 mm, and 7.6 mm, respectively. The specific arrangement and parameters are shown in Figure 12 and Table 3. The test gas composition was 95% CO_2 + 4% N_2 + 1% H_2. The initial pressure of the gas inside the steel pipe was 11.85 MPa, and the initial temperature was 12.6 °C.

Figure 12. Schematic diagram of the distribution of steel pipes in the burst test.

Table 3. Material properties of the burst test pipeline.

Pipe No.	Length/m	Wall Thickness/m	Average CVN/J	Yield Strength/MPa
T1E3	11.93	7.6	156	550
T1E2	11.68	7.2	140	502
T1E1	11.91	7.0	86	548
T1I	11.92	7.0	70	583
T1W1	11.90	7.0	78	586
T1W2	11.69	7.2	156	526
T1W3	11.64	7.6	148	536

The test involved the installation of 8 high-frequency pressure sensors and 6 temperature sensors. Pressure sensors were installed on the east and west sides of T1W3, and on the west sides of T1W2, T1W1, T1E1, T1E2, and T1E3. One insertion temperature sensor (at the midpoint of T1I) was inserted into the pipe interior, and five patch-type temperature sensors (installed at T1W1, T1W2, T1E1, T1E2, and T1E3) were attached to the external surface of the pipe to measure temperature changes inside and on the surface of the pipeline.

After initiation, the crack propagated from the center of the initiation pipe towards both sides, as shown in Figure 13. On the west side of the initiation point, after penetrating the initiation pipe, the crack continued to extend along test pipe T1W1. The average crack propagation speed in this section was 124.7 m/s, and arrest was ultimately achieved at the girth weld between test pipe T1W1 and the second test pipe, T1W2. On the east side of the initiation point, the crack first propagated continuously within test pipe T1E1 at an average speed of 132.5 m/s, then entered test pipe T1E2 and extended a further 0.75 m before finally being arrested by the inherent toughness of the base pipe material.

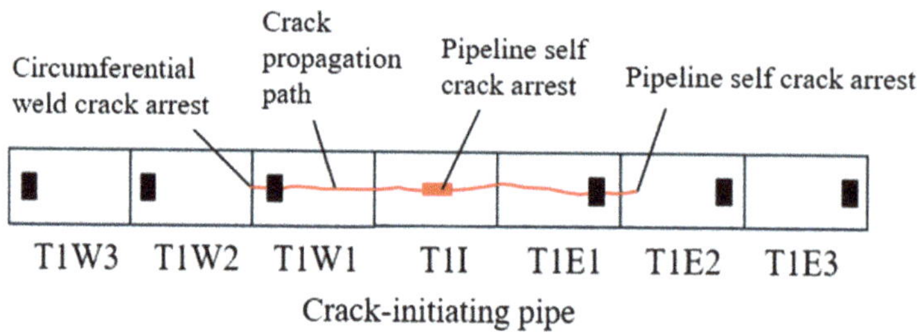

Figure 13. Schematic diagram of pipeline fracture.

4.2.2. Comparison of Software Calculation Results

The decompression wave curve from the on-site burst test was plotted based on pressure change data captured by dynamic pressure sensors. The software calculated the predicted decompression wave curve using the composition, initial pressure, and temperature of the test gas. As shown in Figure 14, the predicted decompression wave curve shows a high degree of overlap with the experimental curve. The calculated saturation pressure plateau value is close to the experimental value, both around 6.9 MPa, with an error within 5%, demonstrating good predictive accuracy. A detailed error analysis between the calculated and experimental values of the decompression wave curve during the pressure plateau period is listed in Table 4. A noticeable discrepancy exists between the experimental data and the software's predictions when the decompression wave velocity exceeds 300 m/s. This discrepancy is primarily attributed to the limited number of pressure monitoring points in the field test. The sparse data points in this high-velocity range struggle to adequately capture the continuous relationship between pressure and wave velocity, consequently amplifying experimental errors.

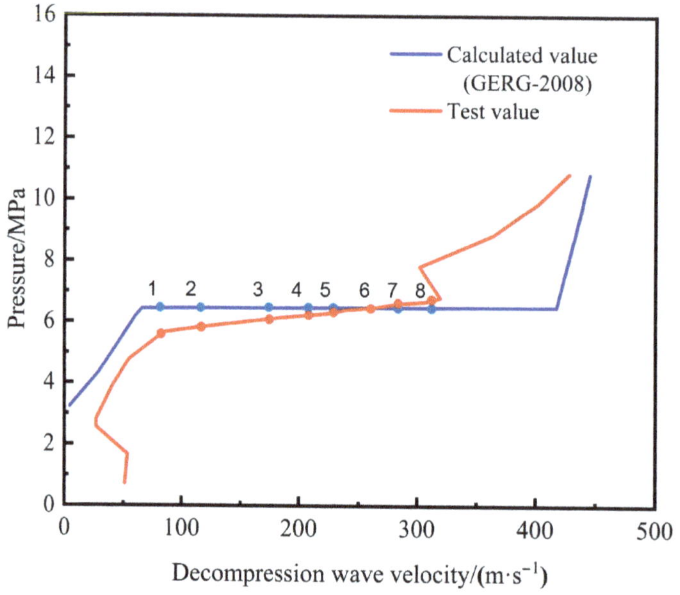

Figure 14. Comparison of decompression wave curves: software prediction vs. experimental results.

Table 4. Error analysis between calculated and experimental values during the pressure plateau period.

No.	Decompression Wave Velocity/(m·s^{-1})	Calculated Pressure/MPa	Experimental Pressure/MPa	Error
1	84.51	6.45	5.66	13.80%
2	116.90	6.45	5.84	10.38%
3	184.63	6.45	6.15	4.84%
4	225.48	6.45	6.29	2.50%
5	238.67	6.45	6.41	0.62%
6	264.56	6.45	6.49	0.65%
7	285.43	6.45	6.62	2.64%
8	316.20	6.45	6.66	3.27%
Average error	-	-	-	4.87%

Furthermore, by inputting the specific parameters of pipe segments T1I, T1W1, and T1E1 into the software, it generated their respective resistance curves, as shown in Figure 15. In the figure, the resistance curves of segments T1I, T1W1, and T1E1 are arranged from right to left, demonstrating sequentially increasing toughness, which is consistent with field observations. Additionally, all three curves intersect the decompression curve, confirming that crack propagation occurred in the pipe segments, thereby validating the accuracy of the software's predictions.

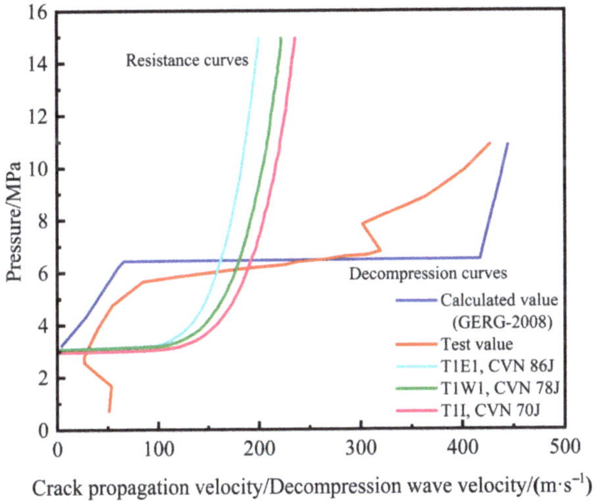

Figure 15. Comparison of BTCM curves predicted by software and obtained from experiments.

4.3. Comparison with International HLP Model Calculation Results

The self-developed software was used to analyze and calculate the fracture arrest toughness for two CO_2 pipeline projects (Project A and Project B). The results were compared with those obtained from the High-strength Line Pipe (HLP) model developed by the JFE Holdings, Inc. (Tokyo, Japan).

For project A, the pipeline material is X52 grade steel with an outer diameter of 406.4 mm, a wall thickness of 7.1 mm, a design pressure of 4.5 MPa, and a design temperature range of 5–50 °C. The gas composition is listed in Table 5. A comparison of the fracture arrest toughness calculation results between this software and the HLP model is shown in Figure 16. The HLP model calculated a value of 14 J, while this software calculated 14.6 J, resulting in a relative error of 4.29%. Both the graphical representation and the numerical results show good agreement between the two methods.

Table 5. Gas composition of project A.

Mixture Units	Molar Percentage/%
Ar	0.0055
CO	0.589
CO_2	91.4619
C_1	0.0121
H_2	0.2697
H_2S	0.0001
N_2	7.6617

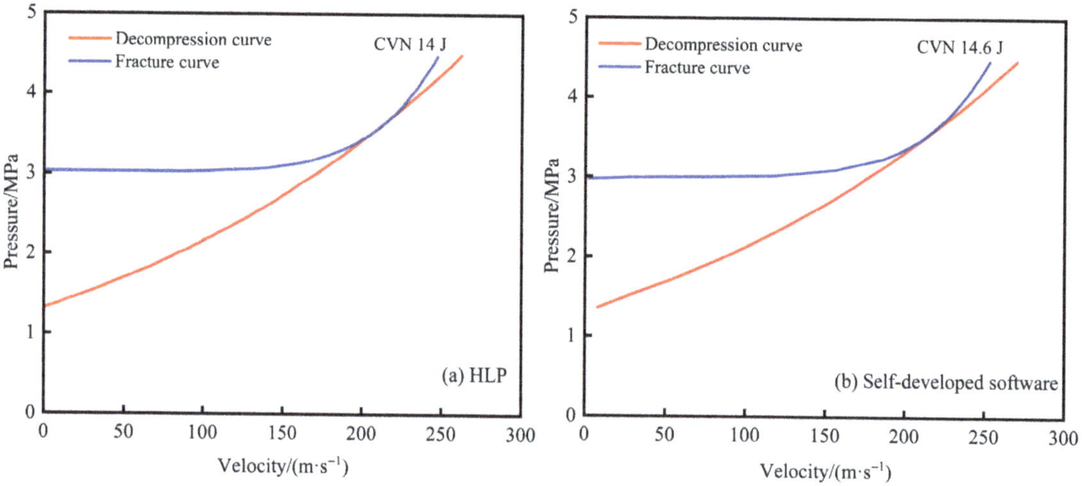

Figure 16. Comparison of fracture arrest toughness calculation results for project A.

For project B, the pipeline utilizes X52 grade steel with an outer diameter of 273.1 mm, a wall thickness of 11.1 mm, a design pressure of 12.0 MPa, and a design temperature range of 13.1–50 °C. The source gas composition is detailed in Table 6. A comparison of the fracture arrest toughness results calculated by this software and the HLP model is presented in Figure 17. The HLP model yields a result of 18 J, whereas this software calculates 16.3 J, resulting in a relative error of 9.44%. The graphical representations from both methods show good agreement.

Table 6. Gas composition of project B.

Mixture Units	Molar Percentage/%
CO_2	99.5
N_2	0.5

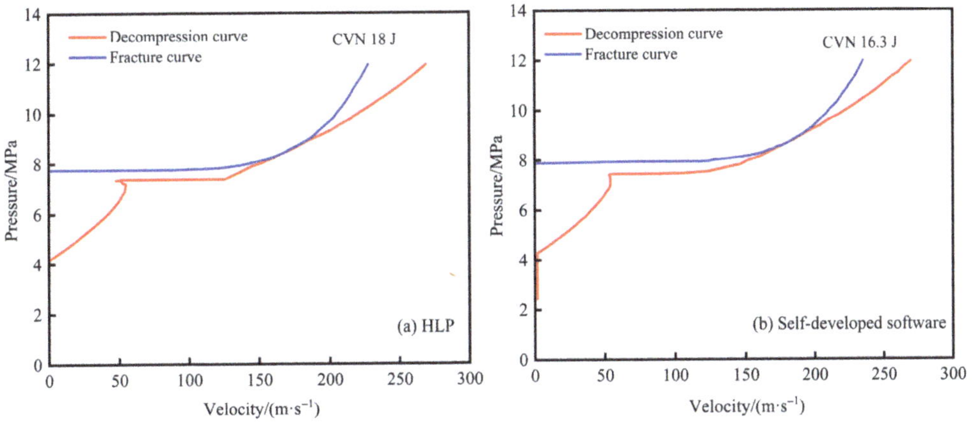

Figure 17. Comparison of fracture arrest toughness calculation results for project B.

5. Parameter Sensitivity Analysis

5.1. Decompression Wave Propagation Characteristics

5.1.1. Gas Composition

Gas composition varies depending on the carbon capture technology employed. Post-combustion capture typically yields the highest concentration of carbon dioxide, while pre-combustion capture introduces the highest level of impurities. The composition and content of gases from three different carbon capture processes are detailed in Table 7.

Table 7. Gas composition and concentrations under different carbon capture methods.

Carbon Capture Method	Component Concentration (Vol. Fraction)/%						
	CO_2	N_2	O_2	Ar	H_2S	CO	H_2
Pre-Combustion Capture	96.00	0.32	0	0	1.84	0.22	1.62
Post-Combustion Capture	99.85	0.15	0	0	0	0	0
Oxy-Fuel Combustion Capture	91.00	4.20	1.8	3.00	0	0	0

The decompression characteristics of CO_2 released from pipelines vary significantly depending on the carbon capture method. Using an initial temperature of 40 °C and a medium pressure of 12 MPa as the starting conditions, the decompression wave propagation characteristics for different capture methods are shown in Figure 18.

For post-combustion captured CO_2, the decompression curve follows the bubble point line during depressurization. In contrast, for pre-combustion captured CO_2, the decompression curve follows the dew point line. For oxy-fuel combustion captured CO_2, the initial decompression wave velocity decreases from approximately 345 m/s (as in pre- or post-combustion capture) to 256 m/s, representing a reduction of about 26%. Meanwhile, the plateau pressure of the decompression wave significantly increases from around 7.00 MPa (pre- or post- combustion capture) to 8.53 MPa, a rise of 22%, and the plateau range is narrowed. During the depressurization process, the pipeline medium no longer follows the gas–liquid equilibrium line but enters the two-phase region.

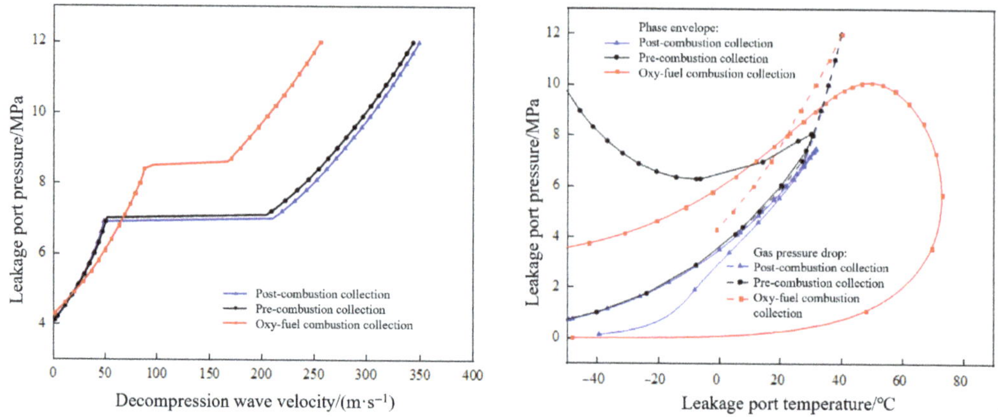

(**a**) Decompression wave curve (**b**) Phase envelope and pressure drop curve

Figure 18. Propagation characteristics of pipeline decompression waves under different carbon capture methods.

5.1.2. Initial Phase State

Using the gas composition from post-combustion capture as the simulation subject, the decompression wave propagation characteristics following a pipeline fracture were calculated for both supercritical phase and dense-phase CO_2. The initial conditions for supercritical CO_2 were set at 40 °C and 12 MPa, and for dense-phase CO_2 at 20 °C and 9 MPa. The calculation results are presented in Figure 19.

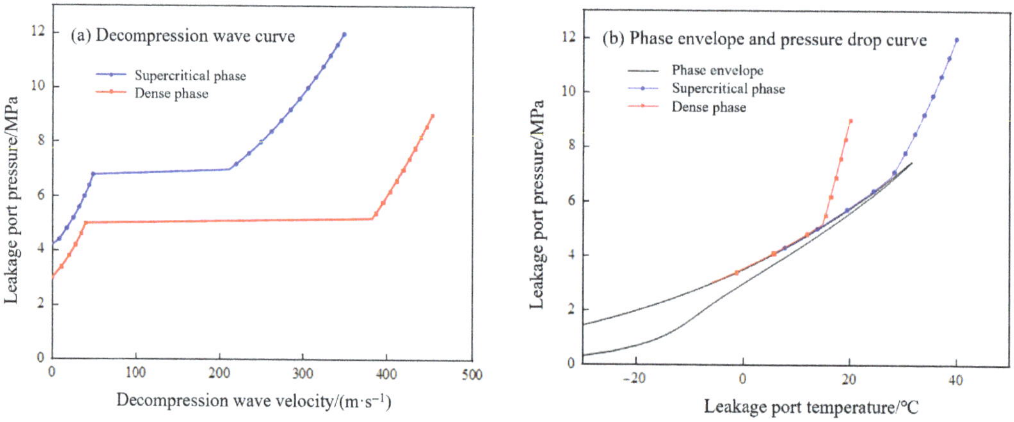

Figure 19. Propagation characteristics of pipeline decompression waves under different initial phases.

As shown in Figure 19a, the initial decompression wave velocity for a dense-phase CO_2 pipeline leak (approximately 452 m/s) is higher than that for a supercritical CO_2 pipeline leak (approximately 350 m/s). This is because dense-phase CO_2 possesses a higher density and initial speed of sound, allowing the decompression wave to maintain a higher velocity even when it enters the gas–liquid two-phase region. Furthermore, the pressure plateau for the dense-phase CO_2 pipeline leak is longer than that for the supercritical CO_2 leak and occurs at a lower pressure.

From Figure 19b, it can be observed that the saturation pressure corresponding to the intersection point of the supercritical CO_2 pipeline pressure drop curve and the CO_2

bubble point curve is higher (6.89 MPa), approximately 26% higher than that of the dense-phase CO_2 pipeline. A higher saturation pressure at this intersection point leads to a higher pressure plateau in the decompression wave curve. This indicates that during the operation of a supercritical CO_2 pipeline, compared to dense-phase transport, a leak event would require a higher fracture arrest toughness value from the pipeline, result in a lower likelihood of autonomous crack arrest, and pose a greater risk of ductile crack propagation.

5.1.3. Pipeline Initial Pressure

The decompression wave propagation characteristics were calculated and analyzed for a supercritical CO_2 pipeline at different initial pressures of 9 MPa, 12 MPa, and 15 MPa. The results are shown in Figure 20.

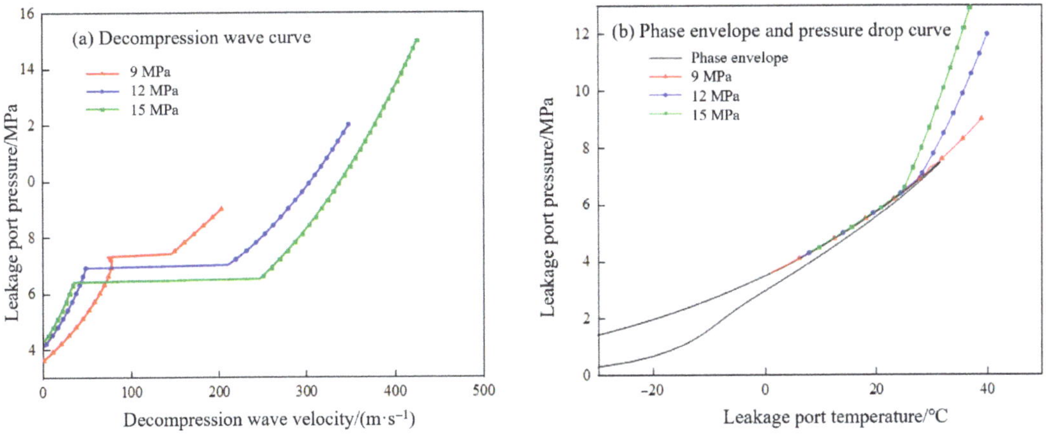

Figure 20. Propagation characteristics of pipeline decompression waves under different initial leakage pressures.

The corresponding initial decompression wave velocities were approximately 203 m/s, 349 m/s, and 424 m/s, respectively, while the pressure plateau levels were 7.4 MPa, 7.0 MPa, and 6.4 MPa, respectively. This indicates that a higher initial leak pressure results in a greater initial decompression wave velocity, a lower pressure plateau, a reduced required fracture arrest toughness for the pipeline, and consequently, a lower risk of fracture. Appropriately increasing the transportation pressure within the allowable range of operational conditions can effectively lower the pressure plateau and enhance the safety factor of pipeline operation.

5.1.4. Pipeline Initial Temperature

The decompression wave propagation characteristics of a supercritical CO_2 pipeline were calculated and analyzed at different initial temperatures of 35 °C, 40 °C, and 45 °C. The results are shown in Figure 21.

The corresponding initial decompression wave velocities were approximately 391 m/s, 349 m/s, and 308 m/s, respectively, while the plateau pressures were approximately 6.5 MPa, 6.9 MPa, and 6.9 MPa, respectively. Therefore, for a supercritical CO_2 pipeline, a lower operating temperature corresponds to a smaller required fracture arrest toughness in the event of a pipeline leak. Under actual pipeline operating conditions, moderately reducing the operating temperature can effectively lower the pressure plateau and reduce the required fracture arrest toughness of the pipeline. This helps prevent ductile crack propagation and enhances operational safety.

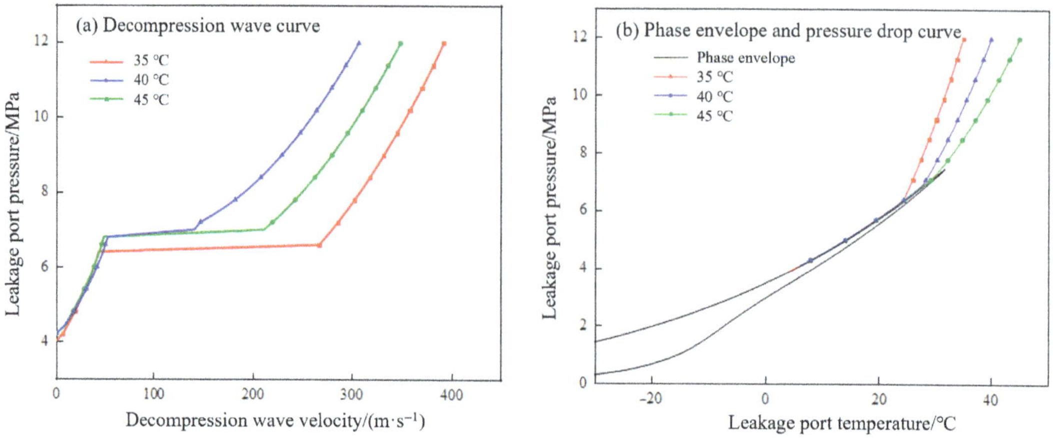

Figure 21. Propagation characteristics of pipeline decompression waves under different initial leakage temperatures.

5.2. Pipe Material Toughness Analysis

A fracture arrest analysis was conducted for pipelines of three different steel grades: X65, X60, and X56. The outer diameter of the pipelines was set at 355.6 mm, and the design pressure was set at 13 MPa. Calculations were performed to determine the arrest pressure corresponding to different Charpy impact energies for each given steel grade and various wall thicknesses. The Charpy impact energy was generally within the range of 80 J to 200 J, and the arrest pressure values were determined based on the Charpy impact energies within this range.

The critical pressure of the carbon dioxide pipeline medium is 7.38 MPa. Considering the minor effects caused by fluctuations in the medium composition, an arrest pressure of no less than 8.0 MPa is adopted for fracture arrest control.

The calculated results for the arrest pressure of the 355.6 mm outer diameter pipeline with different steel grades and wall thicknesses are listed in Table 8. The relationship between CVN and arrest pressure for the three wall thicknesses under the X65, X60, and X56 steel grades is shown in Figure 22. Therefore, by increasing the wall thickness, the minimum wall thicknesses required for the straight pipe with a 355.6 mm outer diameter to meet the fracture arrest control criteria for the X65, X60, and X56 steel grades are 12.5 mm, 12.5 mm, and 14.2 mm, respectively.

Table 8. Table of arrest pressure calculations for 355.6 mm outer diameter pipelines with different steel grades and wall thicknesses.

Steel Grade	Outer Diameter/mm	Wall Thickness/mm	Arrest Pressure/MPa (80–200 J)	Meets Arrest Requirement? (Arrest Pressure > 8 MPa)
X65	355.6	8.0	5.6–5.8	No
X65	355.6	11.0	7.5–8.0	No
X65	355.6	12.5	8.4–9.1	Yes
X60	355.6	8.0	5.3–5.4	No
X60	355.6	11.0	7.1–7.5	No
X60	355.6	12.5	8.0–8.5	Yes
X56	355.6	8.8	5.5–5.7	No
X56	355.6	12.5	7.7–8.1	No
X56	355.6	14.2	8.7–9.2	Yes

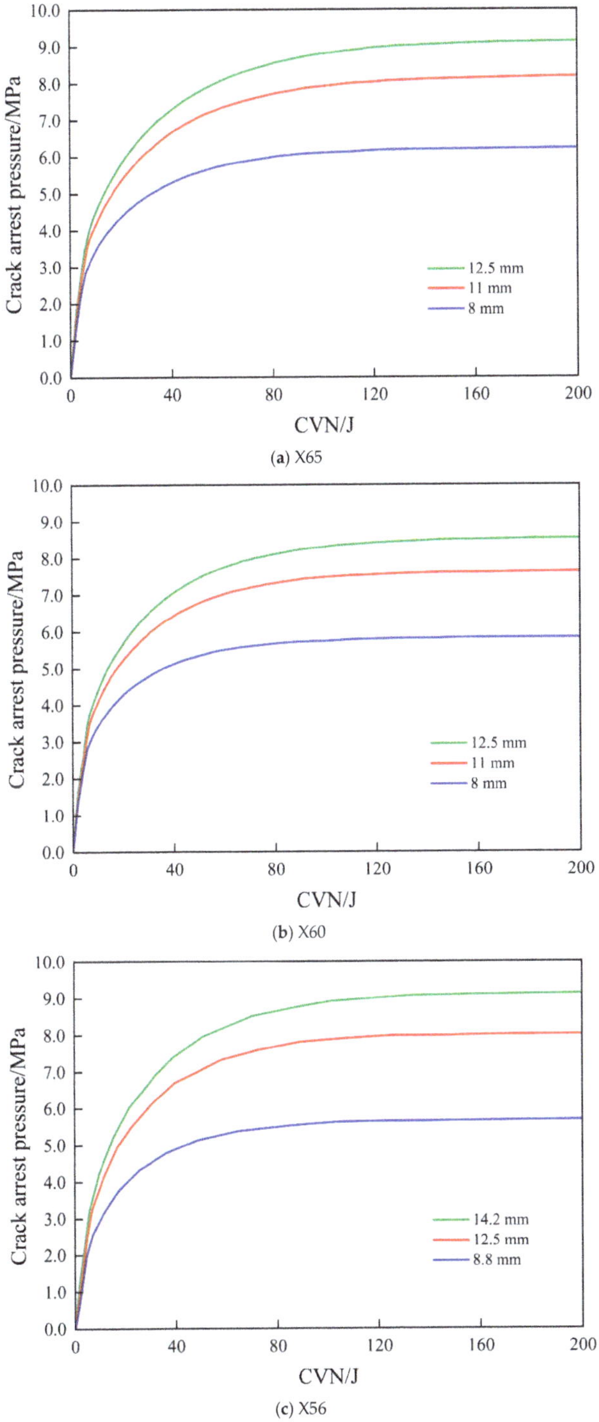

Figure 22. Relationship between charpy impact energy and arrest pressure for D355.6 pipe with different steel grades.

6. Conclusions and Prospects

(1) Computational software for determining the fracture arrest toughness of CO_2 pipelines containing impurities has been developed. Centered on the Battelle Two-Curve Method (BTCM) and incorporating various thermodynamic models such as the Peng-Robinson (PR) and Helmholtz Energy Equation of State (HEOS) equations, this software successfully couples the effects of impurities on both the fluid decompression behavior of CO_2 and the material toughness of the pipe. It can accurately determine the fracture arrest toughness boundary for a specific pipe material under given operating conditions. The calculated pressure plateau shows an error within 5% compared to decompression wave test data and full-scale burst test data. The calculated minimum toughness value exhibits an error within 10% compared to results from the international HLP model.

(2) For post-combustion captured CO_2 and pre-combustion captured CO_2, the pressure drop curves follow the bubble point line and dew point line, respectively, during decompression. In contrast, oxy-fuel combustion captured CO_2 shows approximately 26% lower initial decompression wave velocity and about 22% higher pressure plateau. As decompression progresses, the fluid inside the pipe no longer follows the vapor-liquid equilibrium line but directly enters the two-phase region. Compared to the dense phase, supercritical CO_2 exhibits a saturation pressure at the intersection of the pressure drop curve and the CO_2 bubble point line that is about 26% higher, resulting in a higher pressure plateau. Under different initial pressures of 9 MPa, 12 MPa, and 15 MPa, the pressure plateaus are 7.4 MPa, 7.0 MPa, and 6.4 MPa, respectively. Under different initial temperatures of 35 °C, 40 °C, and 45 °C, the pressure plateaus are approximately 6.5 MPa, 6.9 MPa, and 6.9 MPa, respectively. Lower operating temperatures and higher operating pressures can effectively reduce the pressure plateau.

(3) This software can be used as an auxiliary tool for calculating and evaluating the crack arrest toughness of supercritical CO_2 pipelines, providing data support for pipe material selection and operating condition optimization. During application, the recommended value range for the minimum correction factor is 1.6–2.2. Since the core model of the developed software is built based on existing test data and theoretical derivations, it cannot fully replicate the complex operating environment of pipelines in actual engineering. Without verification by professional engineers in combination with specific scenarios, it shall not replace full-scale burst tests as the basis for final decision-making.

(4) Future work plans include introducing more advanced dynamic fracture mechanics models to describe the microscopic mechanisms of crack tip propagation more precisely. We intend to integrate more complex Equations of State (EOS) and a broader impurity database into the software to enhance its predictive capability under extreme compositions or operating conditions. Concurrently, we will study the feasibility of deploying the software at pipeline control centers and develop interfaces with SCADA systems. This upgrade would shift the software from offline manual input to real-time data interaction. By utilizing sensors to monitor gas composition, operational pressure, and temperature at different pipeline sections in real-time, and employing intelligent pigs for periodic wall thickness inspection, the software could enable real-time, continuous fracture arrest assessment for pipelines. Setting alarm thresholds and triggering warnings would provide a means for predicting pipeline safety risks.

(5) In the next step, the software will be developed to include a solver for calculating the minimum toughness required to arrest pipeline crack propagation. The tangent point

between the fracture velocity curve and the decompression curve of supercritical CO_2 pipelines occurs in the pressure plateau region. Therefore, it is not necessary to determine the complete decompression wave velocity curve when measuring the crack arrest toughness of supercritical CO_2 pipelines; it only needs to be ensured that the crack arrest pressure (the intercept of the crack propagation velocity curve on the pressure axis) is higher than the saturation pressure (consistent with the pressure plateau region on the gas decompression curve). Generally, the pipeline's operating pressure, temperature, medium composition, and other parameters are design input conditions that cannot be easily modified, yet they are key factors for determining the decompression curve. To meet the pipeline crack arrest toughness requirements under the established decompression curve and determine the minimum toughness required to prevent pipeline crack propagation, optimization can be achieved by adjusting pipeline material properties (especially fracture toughness and minimum yield strength) and wall thickness. Of course, in engineering practice, the optimal pipe specification should also be ultimately determined by considering factors such as the price of different steel grades and the quality assurance of welding processes.

Author Contributions: Conceptualization, X.L. and D.W.; methodology, D.W. and X.X.; software, X.L., D.W. and X.J.; validation, D.W., X.J. and Y.Y.; formal analysis, D.W.; investigation, X.J. and Y.Y.; resources, X.L., D.W. and X.X.; data curation, X.L., D.W. and X.J.; writing—original draft preparation, X.L., D.W. and Y.Y.; writing—review and editing, X.L., D.W. and Y.Y.; visualization, X.L., D.W. and Y.Y.; supervision, D.W.; project administration, X.L.; funding acquisition, X.L. All authors have read and agreed to the published version of the manuscript.

Funding: Supported by the National Natural Science Foundation of China for Regional Fund (52562047); Natural Science Foundation of Xinjiang Uygur Autonomous Region (2023D01A19); Xinjiang Uygur Autonomous Region "Tianchi talents" Introduction Plan Project (TCYC12); Xinjiang Tianshan Innovation Team for Research and Application of High-Efficiency Oil and Gas Pipeline Transportation Technology (2022TSYCTD0002); Xinjiang Uygur Autonomous Region "One Case, One Policy" Strategic Talent Introduction Project (XQZX20240054).

Data Availability Statement: The original contributions presented in this study are included in the article. Further inquiries can be directed to the corresponding author.

Conflicts of Interest: The authors declare no conflicts of interest.

References

1. Van Vuuren, D.P.; Stehfest, E.; Gernaat, D.E.; Van Den Berg, M.; Bijl, D.L.; De Boer, H.S.; Daioglou, V.; Doelman, J.C.; Edelenbosch, O.Y.; Harmsen, M.; et al. Alternative pathways to the 1.5 C target reduce the need for negative emission technologies. *Nat. Clim. Change* **2018**, *8*, 391–397. [CrossRef]
2. Peters, G.P.; Andrew, R.M.; Canadell, J.G.; Friedlingstein, P.; Jackson, R.B.; Korsbakken, J.I.; Le Quéré, C.; Peregon, A. Carbon dioxide emissions continue to grow amidst slowly emerging climate policies. *Nat. Clim. Change* **2019**, *10*, 3–6. [CrossRef]
3. Höhne, N.; Gidden, M.J.; Elzen, M.D.; Hans, F.; Fyson, C.; Geiges, A.; Jeffery, M.L.; Gonzales-Zuñiga, S.; Mooldijk, S.; Hare, W.; et al. Wave of net zero emission targets opens window to meeting the Paris Agreement. *Nat. Clim. Change* **2021**, *11*, 820–822. [CrossRef]
4. Grubler, A.; Wilson, C.; Bento, N.; Boza-Kiss, B.; Krey, V.; McCollum, D.L.; Rao, N.D.; Riahi, K.; Rogelj, J.; De Stercke, S.; et al. A low energy demand scenario for meeting the 1.5 °C target and sustainable development goals without negative emission technologies. *Nat. Energy* **2018**, *3*, 515–527. [CrossRef]
5. Chaudhry, A.U.; Abdullah, F.T.; Khan, M.S.; Rabbani, H.S.; Lal, B. Feasibility of Long-Distance Transportation of Captured CO_2 using Transmission Pipelines: An Overview of Techno-economic Models. In *Gas Hydrate in Carbon Capture, Transportation and Storage*; CRC Press: Boca Raton, FL, USA, 2024; pp. 33–43.
6. Joarder, M.S.A.; Islam, M.S.; Hasan, M.H.; Kutub, A.; Kabir, M.F.; Rashid, F.; Joarder, T.A. A comprehensive review of carbon dioxide capture, transportation, utilization, and storage: A source of future energy. *Environ. Sci. Pollut. Res.* **2025**, *32*, 9299–9332. [CrossRef]

7. Akinmoladun, A.; Tomomewo, O.S. Advances and future perspectives in post-combustion carbon capture technology using chemical absorption process: A review. *Carbon Capture Sci. Technol.* **2025**, *16*, 100461. [CrossRef]
8. Ighomuaye, E.; Asimatey, P.D.; Annankra, J.A. Strategies for Industrial Integration of Carbon Capture, Utilization, and Storage Technologies: A Critical Review. *J. Eng. Comput. Sci.* **2025**, *4*, 284–293.
9. Chen, L.; Hu, Y.; Liu, Z.; Yan, X.; Yu, S.; Ding, J.; Liu, P.; Yu, J.; Chen, S. Experimental research on the fracture and arrest process of supercritical CO_2 pipelines. *Int. J. Press. Vessel. Pip.* **2024**, *212*, 105314. [CrossRef]
10. Yu, S.; Yan, X.; He, Y.; Yu, J.; Chen, S. Study on the leakage morphology and temperature variations in the soil zone during large-scale buried CO_2 pipeline leakage. *Energy* **2023**, *288*, 129674. [CrossRef]
11. Xi, D.; Lu, H.; Xu, Z.-D.; Jiang, X.; Peng, H.; Fang, H. Prevention of natural gas pipeline cracking. In *Advances in Natural Gas: Formation, Processing, and Applications*; Volume 6: Natural Gas Transportation and Storage; Elsevier: Amsterdam, The Netherlands, 2024; pp. 293–313.
12. Gu, X.T.; Guo, Y. Crack Arrest Toughness of High Grade Gas Pipeline. *Mater. Sci. Forum* **2017**, *898*, 758–765. [CrossRef]
13. Zhen, Y.; Jiao, Z.; Cao, Y.; Niu, R. Unified correlation of constraints with crack arrest toughness for high-grade pipeline steel. *Int. J. Press. Vessel. Pip.* **2021**, *193*, 104454. [CrossRef]
14. Botros, K.K.; Geerligs, J.; Rothwell, B.; Robinson, T. Measurements of decompression wave speed in pure carbon dioxide and comparison with predictions by equation of state. *J. Press. Vessel. Technol.* **2015**, *138*, 031302. [CrossRef]
15. Botros, K.K.; Igi, S.; Kondo, J. Measurements of decompression wave speed in natural gas containing 2–8%(mole) hydrogen by a specialized shock tube. In Proceedings of the 2016 11th International Pipeline Conference, Calgary, AB, Canada, 26–30 September 2016; American Society of Mechanical Engineers: New York, NY, USA, 2016; Volume 50275, p. V003T05A001.
16. Cosham, A.; Jones, D.G.; Armstrong, K.; Allason, D.; Barnett, J. The decompression behaviour of carbon dioxide in the dense phase. In Proceedings of the 2012 9th International Pipeline Conference, Calgary, AB, Canada, 24–28 September 2012; American Society of Mechanical Engineers: New York, NY, USA, 2012; Volume 45141, pp. 447–464.
17. Cosham, A.; Jones, D.G.; Armstrong, K.; Allason, D.; Barnett, J. Analysis of a dense phase carbon dioxide full-scale fracture propagation test in 24 inch diameter pipe. In Proceedings of the 2016 11th International Pipeline Conference, Calgary, AB, Canada, 26–30 September 2016; American Society of Mechanical Engineers: New York, NY, USA, 2016; Volume 50275, p. V003T05A012.
18. Zhang, Z.; Lu, Z.; Wang, J.; Liu, D.; Xiao, C.; Fan, D. Effects of initial flow velocity on decompression behaviours of GLE CO_2 upstream and downstream the pipeline. *Int. J. Greenh. Gas Control* **2022**, *118*, 103690. [CrossRef]
19. Picard, D.J.; Bishnoi, P.R. The importance of real-fluid behavior in predicting release rates resulting from high-pressure sour-gas pipeline ruptures. *Can. J. Chem. Eng.* **1989**, *67*, 3–9. [CrossRef]
20. Aursand, E.; Aursand, P.; Hammer, M.; Lund, H. The influence of CO_2 mixture composition and equations of state on simulations of transient pipeline decompression. *Int. J. Greenh. Gas Control* **2016**, *54*, 599–609. [CrossRef]
21. Jie, H.E.; Xu, B.P.; Wen, J.X.; Cooper, R.; Barnett, J. Predicting the decompression characteristics of carbon dioxide using computational fluid dynamics. In Proceedings of the 2012 9th International Pipeline Conference, Calgary, AB, Canada, 24–28 September 2012; American Society of Mechanical Engineers: New York, NY, USA, 2012; Volume 45141, pp. 585–595.
22. Botros, K.K.; Geerligs, J.; Rothwell, B.; Robinson, T. Effects of Argon as the primary impurity in anthropogenic carbon dioxide mixtures on the decompression wave speed. *Can. J. Chem. Eng.* **2016**, *95*, 440–448. [CrossRef]
23. Guo, X.; Xu, S.; Chen, G.; Yan, X.; Cao, Q. Fracture criterion and control plan on CO_2 pipelines: Theory analysis and full-bore rupture (FBR) experimental study. *J. Loss Prev. Process. Ind.* **2021**, *69*, 104394. [CrossRef]
24. Gu, S.; Li, Y.; Teng, L.; Wang, C.; Hu, Q.; Zhang, D.; Ye, X.; Wang, J.; Iglauer, S. An experimental study on the flow characteristics during the leakage of high pressure CO_2 pipelines. *Process. Saf. Environ. Prot.* **2019**, *125*, 92–101. [CrossRef]
25. Teng, L.; Zhang, D.; Li, Y.; Wang, W.; Wang, L.; Hu, Q.; Ye, X.; Bian, J.; Teng, W. Multiphase mixture model to predict temperature drop in highly choked conditions in CO_2 enhanced oil recovery. *Appl. Therm. Eng.* **2016**, *108*, 670–679. [CrossRef]
26. Dakhel, A.Y.; Lukács, J. Full-scale tests of transporting pipeline sections—A review and consequences to our investigations. *Des. Mach. Struct.* **2023**, *13*, 24–44. [CrossRef]
27. Maxey, W.A. Long shear fractures in CO_2 lines controlled by regulating saturation, arrest pressures. *Oil Gas J.* **1986**, *84*, 31.
28. Cosham, A.; Jones, D.G.; Armstrong, K.; Allason, D.; Barnett, J. Ruptures in gas pipelines, liquid pipelines and dense phase carbon dioxide pipelines. In Proceedings of the 2012 9th International Pipeline Conference, Calgary, AB, Canada, 24–28 September 2012; American Society of Mechanical Engineers: New York, NY, USA, 2012; Volume 45141, pp. 465–482.
29. Cosham, A.; Jones, D.G.; Armstrong, K.; Allason, D.; Barnett, J. Analysis of two dense phase carbon dioxide full-scale fracture propagation tests. In Proceedings of the 2014 10th International Pipeline Conference, Calgary, AB, Canada, 29 September–3 October 2014; American Society of Mechanical Engineers: New York, NY, USA, 2014; Volume 46124, p. V003T07A003.
30. Biagio, M.D.; Lucci, A.; Mecozzi, E.; Spinelli, C.M. Fracture propagation prevention on CO_2 pipelines: Full scale experimental testing and verification approach. In Proceedings of the Pipeline Technology Conference, Berlin, Germany, 1–5 May 2017.

31. Linton, V.; Leinum, B.H.; Newton, R.; Fyrileiv, O. CO2SAFE-ARREST: A full-scale burst test research program for carbon dioxide pipelines—Part 1: Project overview and outcomes of test 1. In Proceedings of the 2018 12th International Pipeline Conference, Calgary, AB, Canada, 24–28 September 2018; American Society of Mechanical Engineers: New York, NY, USA, 2018; Volume 51883, p. V003T05A008.
32. Michal, G.; Davis, B.; Østby, E.; Lu, C.; Røneid, S. CO2SAFE-ARREST: A Full-Scale Burst Test Research Program for Carbon Dioxide Pipelines—Part 2: Is the BTCM Out of Touch with Dense-Phase CO_2. In Proceedings of the 2018 12th International Pipeline Conference, Calgary, AB, Canada, 24–28 September 2018; American Society of Mechanical Engineers: New York, NY, USA, 2018; Volume 51883, p. V003T05A009.
33. Zhang, D.; Wang, R.; Ouyang, X.; Fu, Y.; Wang, H.; Yan, F.; Liu, X. Experimental study of the full-scale burst failure behavior of carbon dioxide steel pipeline. *J. Pipeline Sci. Eng.* **2025**, *5*, 100221. [CrossRef]
34. Chen, L.; Hu, Y.; Yang, K.; Yan, X.; Yu, S.; Yu, J.; Chen, S. Fracture process characteristic study during fracture propagation of a CO_2 transport network distribution pipeline. *Energy* **2023**, *283*, 129060. [CrossRef]
35. *ISO 27913:2024*; Carbon Dioxide Capture, Transportation And geological Storage—Pipeline Transportation Systems. International Organization for Standardization (ISO): Geneva, Switzerland, 2024.
36. Maxey, W.; Kiefner, J.; Eiber, R.; Duffy, A. Ductile fracture initiation, propagation, and arrest in cylindrical vessels. In *Fracture Toughness: Part II*; ASTM International: West Conshohocken, PA, USA, 1972; pp. 70–81.
37. Rothwell, A.B. The application of the Battelle "Short Formula" to the determination of ductile fracture arrest toughness in gas pipelines. In Proceedings of the 2000 3rd International Pipeline Conference, Calgary, AB, Canada, 1–5 October 2000; American Society of Mechanical Engineers: New York, NY, USA, 2000; Volume 40245, p. V001T02A019.
38. Chen, L.; Hu, Y.; Yang, K.; Yan, X.; Yu, S.; Yu, J.; Chen, S. Experimental study of decompression wave characteristics and steel pipe applicability of supercritical CO_2 pipeline. *Gas Sci. Eng.* **2023**, *121*, 205203. [CrossRef]
39. Gu, S.; Li, Y.; Teng, L.; Hu, Q.; Zhang, D.; Ye, X.; Wang, C.; Wang, J.; Iglauer, S. A new model for predicting the decompression behavior of CO_2 mixtures in various phases. *Process. Saf. Environ. Prot.* **2018**, *120*, 237–247. [CrossRef]
40. Mahgerefteh, H.; Brown, S.; Martynov, S. A study of the effects of friction, heat transfer, and stream impurities on the decompression behavior in CO_2 pipelines. *Greenh. Gases Sci. Technol.* **2012**, *2*, 369–379. [CrossRef]
41. Terenzi, A. Influence of real-fluid properties in modeling decompression wave interacting with ductile fracture propagation. *Oil Gas Sci. Technol.* **2005**, *60*, 711–719. [CrossRef]
42. Teng, L.; Li, Y.; Zhao, Q.; Wang, W.; Hu, Q.; Ye, X.; Zhang, D. Decompression characteristics of CO_2 pipelines following rupture. *J. Nat. Gas Sci. Eng.* **2016**, *36*, 213–223. [CrossRef]
43. Botros, K.K.; Geerligs, J.; Fletcher, L.; Rothwell, B.; Venton, P.; Carlson, L. Effects of pipe internal surface roughness on decompression wave speed in natural gas mixtures. In Proceedings of the 8th International Pipeline Conference, Calgary, AB, Canada, 27 September–1 October 2010; Volume 44212, pp. 907–922.
44. Lund, H. Relaxation Models for Two-Phase Flow with Applications to CO_2 Transport. Ph.D. Thesis, NTNU—Institutt for Energi-og Prosessteknikk (EPT), Trondheim, Norway, 2013.
45. Flåtten, T.; Lund, H. Relaxation two-phase flow models and the subcharacteristic condition. *Math. Model. Methods Appl. Sci.* **2011**, *21*, 2379–2407. [CrossRef]
46. Bell, I.H.; Deiters, U.K. On the construction of binary mixture p-x and T-x diagrams from isochoric thermodynamics. *AIChE J.* **2018**, *64*, 2745–2757. [CrossRef]
47. Peng, D.Y.; Robinson, D.B. A new two-constant equation of state. *Ind. Eng. Chem. Fundam.* **1976**, *15*, 59–64. [CrossRef]
48. Simão, J.S.D.; Emmanuel, L.; João, A.A.; Manuel, E.J.L.; Nzinga, E.J.; Cangue, F.R.; Barros, A.A.C. Analysis of the thermodynamic behavior of gaseous mixtures using equations of state. *S. Afr. J. Chem. Eng.* **2024**, *49*, 339–347. [CrossRef]
49. Kunz, O.; Wagner, W. The GERG-2008 wide-range equation of state for natural gases and other mixtures: An expansion of GERG-2004. *J. Chem. Eng. Data* **2012**, *57*, 3032–3091. [CrossRef]

processes

MDPI

Article

Staged Effective Medium Modeling and Experimental Validation for Rock Thermal Conductivity

Yanming Chen, Michael T. Myers *, Lori Hathon, Gabriel C. Unomah and David Myers

Rock Mechanics, Petrophysics and Sample Characterization Laboratory, Department of Petroleum Engineering, University of Houston, Houston, TX 77204, USA; lahathon@central.uh.edu (L.H.); gcunomah@central.uh.edu (G.C.U.); dmyers6@uh.edu (D.M.)
* Correspondence: mtmyers@central.uh.edu

Abstract

The thermal conductivity (λ) of porous rocks as a function of total porosity, grain size, and fluid saturation is measured and modeled by combining high-precision experiments with a Staged Differential Effective Medium (SDEM) modeling framework. A 1-D divided-bar apparatus with computer-controlled guard heaters with an integrated ultrasonic pulse-transmission system was developed to measure the thermal conductivity and P and S-wave velocities simultaneously. Measurements were made on Fontainebleau sandstone cores and quartz sand packs of varying grain size and effective stresses up to 2000 psi. The sample properties were measured in both dry and water-saturated states. The SDEM model performs significantly better at predicting the saturated thermal conductivities in the sand packs. For the sand packs, the thermal conductivity and compressional velocity are the highest and most stress-sensitive for the fine-grained material. In contrast, the shear velocity is largest in the coarse-grained material. The SDEM model is adapted from previous acoustic models for use in understanding thermal conductivity. These joint models accurately reproduce the evolution of both thermal conductivity and bulk modulus during increasing compaction and varying saturation. A single parameter fits both the dry and saturated data, which allows Gassmann-style fluid substitution for the thermal conductivity. This model improves the prediction of in situ thermal conductivity from sonic well logs.

Keywords: acoustic velocity; thermal conductivity; effective medium

1. Introduction

Motivation

The thermal conductivity of rocks is defined as the ability of the material to conduct heat, or the rate of heat transfer through a unit area under a unit temperature gradient ($W \cdot m^{-1} \cdot K^{-1}$), which is a key parameter in geoscience and engineering that governs heat transfer [1]. Thermal conductivity controls geothermal gradients, energy transport in reservoirs and repositories, and heat dissipation in tunnels, foundations, and buried infrastructure [2,3]. Geothermal energy relies on extracting subsurface heat; the thermal conductivity of the host rocks directly determines the efficiency of heat flow toward production wells and thus the sustainability and power potential of geothermal systems [4–6]. Typical thermal conductivities of sedimentary formations span approximately 1–6 $W \cdot m^{-1} \cdot K^{-1}$, reflecting substantial lithological variability: from high-conductivity quartz-rich rocks

(~7.7 $W \cdot m^{-1} \cdot K^{-1}$) to low-conductivity clay-rich units (~2 $W \cdot m^{-1} \cdot K^{-1}$) [7,8]. The commonly used thermal conductivity of natural quartz is ~7.7 $W \cdot m^{-1} \cdot K^{-1}$ [2,8]. However, when field and lab data are extrapolated to zero porosity in quartz-rich sandstones, it typically falls below this value due to the influence of lower thermal conductivity minerals, thermal contact resistance at grain boundaries, the assumption of perfect grain packing at zero porosity, and the exclusion of the influence of micropores and cements. Such variability introduces significant uncertainty into geothermal and heat flow modeling, as even modest deviations in thermal conductivity can markedly influence subsurface temperature predictions. Consequently, recent investigations underscore the need to accurately constrain thermal conductivity distributions to improve the reliability of modeled geothermal gradients and subsurface thermal regimes [9,10].

Borehole-Based Measurements

In situ determinations can be performed using downhole thermal conductivity logging tools [4,11–13]. These measurements are typically spatially discontinuous, technically demanding, and economically restrictive. In shallow subsurface investigations, the thermal response test (TRT) has become a widely adopted technique for assessing the effective thermal properties of borehole environments [14]. For deeper boreholes, optical frequency-domain reflectometry (OFDR) enables direct, high-resolution profiling of thermal conductivity along the borehole wall [15]. Measurements in unconsolidated sediments require controlled variation in water content and the application of incremental compaction pressures to account for the strong dependence of heat transfer on fluid saturation and packing density [16]. These methods have led to wide variations in the values of thermal conductivity measurements, which inhibit model calibration and increase prediction errors in modeling subsurface heat flow [17].

Laboratory Measurements

Laboratory thermal conductivity measurements are broadly classified into steady-state and transient-state methods, depending on whether a constant or time-dependent temperature gradient is imposed across the sample [1,17]. Steady-state techniques, such as the divided-bar and guarded-hot-plate methods, determine thermal conductivity once thermal equilibrium is achieved, yielding high accuracy but requiring longer equilibration times and precise control of boundary conditions [18]. In contrast, transient methods, including the line-source (needle probe) and laser-flash techniques, infer conductivity from the dynamic temperature response to a heat pulse or step input, allowing for faster measurements and application to a broader range of materials [17,19]. Steady-state methods are limited by sample geometry constraints and susceptibility to heat losses, while transient methods are affected by uncertainties in contact resistance, heat capacity estimation, and assumptions of homogeneity [20,21]. Often, datasets are obtained using diverse experimental setups and methodological approaches [16]. Datasets are reported inconsistently, as some studies report only median values, while others provide full ranges or extreme bounds (minimum and maximum) with an uncertainty range of approximately 5–50% [16]. In addition, this variability in thermal conductivity values reflects the influence of the microstructural and saturation properties of the rock [22]. These differences in data reporting and measurement techniques preclude rigorous statistical analysis.

Correlations

In attempts to understand this variability in the measurements, a substantial body of research has employed empirical mixing laws to estimate the effective thermal conductivity of rocks from petrophysical parameters. In two-phase systems comprising a solid matrix and pore space, the simplest theoretical limits are provided by the Wiener

https://doi.org/10.3390/pr14091437

bounds [23], which represent end-member limits in which heat flow is either parallel or perpendicular to bedding. Numerous studies have used indirect methods for estimating thermal conductivity from petrophysical parameters such as bulk density, porosity, and mineralogical composition [24–26]. These regression-based empirical formulations are inherently limited, as their validity is generally restricted to the specific lithologies and environmental conditions for which they were calibrated. Comprehensive summaries are provided in [24].

Empirical Models

These predictive uncertainties in the bulk thermal conductivity of a rock are governed by its mineralogical composition, porosity, pore fluid characteristics, and the geometric configuration of its pore network. Sandstones, shales, and carbonates with comparable bulk compositions may nonetheless exhibit distinctly different thermal conductivities due to variations in pore structure and connectivity [27]. In particular, the presence of anisotropic pore systems and microcrack networks can enhance or inhibit heat transfer depending on their orientation relative to the thermal gradient [20,24]. Furthermore, fluid saturation exerts a strong influence, as the substitution of air by water or brine substantially increases the effective thermal conductivity through improved heat transfer across pore spaces [4,8,28]. The application of induced stress closes microcracks and increases grain-to-grain contact areas, enhancing heat transfer efficiency and improving thermal conductivity [29]. Temperature exerts an opposite effect on thermal conductivity: higher temperatures enhance phonon scattering, intensifying lattice vibrations and disrupting the orderly transfer of thermal energy through the crystal lattice, thereby diminishing effective heat transport [30,31]. In porous, fluid-saturated rocks, temperature-induced changes in fluid properties (e.g., density) may partially counteract this reduction by improving conductive pathways within the pore network [4,20]. Consequently, accurate modeling requires applying temperature-dependent corrections to measured thermal conductivity to obtain precise estimates of subsurface heat flow. These interrelated thermal effects highlight the importance of integrating mineralogical, textural, geomechanical, and petrophysical parameters to predict the thermal properties of geological formations in geothermal systems and basin-scale heat flow models [30–32].

Bounds

Attempts to reduce this uncertainty have been improved by establishing bounds on the possible values of the thermal conductivity. For isotropic media, the Hashin–Shtrikman bounds provide the most restrictive theoretical limits [33]. Between the Wiener and Hashin–Shtrikman bounds, various rock physics models have been applied to estimate an "effective" thermal conductivity of rocks. The Maxwell–Eucken models provide two complementary formulations for heat transfer in rocks, distinguishing whether the solid matrix or the pore fluid forms the continuous phase containing isolated spherical inclusions [34–36]. Both models rely on dilute-sphere assumptions in which inclusions are non-interacting and randomly distributed, making them most valid at low porosity where phase interactions remain negligible. Although simple, they offer only first-order estimates of effective thermal conductivity, but they underpin more advanced effective medium theories commonly used in geothermal and basin-scale heat flow modeling [8,24]

Self-Consistent Models

An example of a self-consistent formulation is Bruggeman's widely used symmetrical effective medium model that predicts composite thermal conductivity by treating both the solid and fluid phases as statistically equivalent constituents embedded within an infinite surrounding medium. This approach incorporates higher inclusion concentrations and

 https://doi.org/10.3390/pr14091437

phase-interaction effects, yielding more realistic bulk thermal conductivity estimates in heterogeneous or isotropic materials [24,31]. Differential Effective Medium models composite behavior by incrementally introducing inclusions into a uniform host and updating the effective properties at each step. This enables modeling the thermal conductivity of mixtures more realistically than static mixing laws [37,38]. Although DEM models often yield improved predictions for rocks with moderate porosity or anisotropy, they may diverge from observed thermal conductivities when microstructural heterogeneity or fabric orientation strongly control heat transport [24,39]. The geometric mean model [40] estimates bulk thermal conductivity as the geometric average of the matrix and pore conductivities, weighted by their volume fractions, assuming an isotropic, randomly distributed two-phase system. Although it lacks explicit microstructural parameters, it often matches the performance of more complex EMT formulations for consolidated rocks with moderate porosity. This typically predicts sandstone thermal conductivity within about ten to twenty percent [8]. Other studies show that weighted geometric means often provide the best empirical fits to laboratory data [41]. More sophisticated models yield only marginal improvements [3,42]. Zimmerman modified the Hashin–Shtrikman upper bound by incorporating a textural parameter, namely, the pore aspect ratio, which directly influences the distribution of heat flow paths [43].

Pore Structure-Based Models

Pimienta et al. developed effective medium theory-based models to predict thermal conductivity in porous and microcracked rocks by explicitly incorporating crack density and aspect ratio into the upscaling scheme [44]. Model 2 embeds compliant cracks into a high-pressure host medium, assuming all cracks are closed, reducing assumptions about mineralogy and pore shape and enabling more reliable inversion of crack parameters as pressure decreases. In contrast, Model 3 assumes a monomineralic matrix with spherical, stress-insensitive pores and uses the Mori–Tanaka homogenization scheme, introducing additional assumptions and generating larger uncertainties in predicted crack geometry and thermal conductivity. The authors also show that although thermal conductivity can be reasonably inferred from elastic wave velocity and porosity, the predictions remain limited by assuming a simplified pore geometry, idealized microstructural assumptions, and the inherent non-uniqueness of the inverse crack-parameter problem. These models offer useful theoretical constraints on thermal conductivity, but none adequately capture the complexity of natural rocks [3,42]. This level of uncertainty implies that improved links to other readily measured parameters are needed.

Joint Modeling of Modulus and Thermal Conductivity

Elastic-wave velocity in rocks exhibits similar sensitivity to mineral composition, porosity, fluid saturation, and fabric anisotropy as thermal conduction, and seismic velocity measurements are often employed as indirect proxies for estimating thermal conductivity [45]. Empirical relationships between thermal conductivity and P-wave velocity (Vp) have been established for specific lithologies, indicating that for dry sandstones and volcanites, thermal conductivity can be estimated from Vp with an accuracy of approximately $\pm 0.5\ W \cdot m^{-1} \cdot K^{-1}$ [28]. They also showed that increasing water saturation leads to simultaneous increases in both thermal conductivity and P-wave velocity, with porosity exerting a dominant influence on the strength of this relationship. However, Mielke et al. reported weak correlations between thermal conductivity and P-wave velocity in clastic sediments, as evidenced by low coefficients of determination ($R^2 = 0.04$–0.48) across sandstone types [28]. This indicates that acoustic velocity alone provides only limited predictive capability for thermal conductivity, reflecting the complex influence of porosity, mineralogy, and microstructural heterogeneity. Subsequent investigations have proposed linear and

https://doi.org/10.3390/pr14091437

non-linear empirical correlations between these properties. Such models remain largely statistical in nature and lack explicit physical grounding [46]. Several studies have explored the interdependence of thermal conductivity and elastic-wave velocity through their mutual sensitivity to porosity, mineralogy, fluid saturation, and confining stress [47,48]. These approaches rely primarily on direct regression between velocity and thermal conductivity.

Staged Differential Effective Medium Models (SDEM models)

As described above, the classical mixing laws and conventional effectivemedium formulations show systematic biases and fail to capture detailed mineralogical and fabric effects. The SDEM technique allows interpolation between Voight and Reuss bounds and also allows varying minerals and porosity changes to be included through multiple integration stages using a Differential Effective Medium approximation. This modeling technique has been extensively validated for acoustic moduli and resistivity [49–52]. An important contribution of this type of modeling is the development and validation of a Staged Differential Effective Medium (SDEM) model that accurately predicts effective thermal conductivity (ETC). This model allows fluid substitution similar to a Gassman substitution, which provides an important validation of the modeling and the measurement capabilities.

Two systems were investigated, a lithified sandstone, Rupelian Fontainebleau (Young's modulus ~10^6), with porosity varying between 9% and 12%, and low modulus (~50×10^3) sand packs with varying porosity (between 17% and 37%) due to different levels of compaction. Both rock types are quartz-dominated but with different fabric geometries.

2. Materials and Methods

A custom-designed, calibrated, divided-bar 1-D system for high-precision thermal conductivity measurements of small cylindrical rock discs, equipped with multiple computer-controlled guard heaters, was developed. The samples are confined between two heated reference samples in a 1-D axial setup, which minimizes any buoyancy-driven circulation. The thermal conductivity apparatus and pulse-transmission ultrasonic setup were then used to measure thermal conductivity and acoustic velocities under increasing isostatic stress (200–2000 psi) for Fontainebleau sandstones and laboratory-constructed sand packs (of varying grain size and mineralogy) in both dry and saturated states.

2.1. *Sample Description and Preparation*

Two different sample sets, Fontainebleau sandstone and Brazos River sand packs, were chosen to reflect a common mineralogy (quartz dominated) but different porosities and levels of cementation. The sand packs allowed a range in grain sizes and porosities to be generated.

2.1.1. Fontainebleau Sandstone

The Rupelian Fontainebleau sandstone, deposited in a marine shoreface subaerial–eolian environment, is quartz-rich (quartz > 99%) and known for its high degree of cementation and uniform grain fabric [53]. Three cylindrical cores (2.54 cm in diameter and ~6 cm in length) were drilled from large rectangular blocks using a water-cooled diamond-tipped coring bit. The samples were precision-ground and polished at both ends (right cylindrical geometry) to ensure good thermal contact for thermal conductivity measurements and reliable coupling of acoustic transducers to the samples. The core plugs were vacuum-oven-dried at 60 °C. The dry weight, bulk volume, and dry bulk density were measured using calipers and a precision balance. Grain density and porosity were determined using a nitrogen expansion (Boyle's Law) porosimeter. The samples were saturated by evacuating for one hour, followed by the injection of degassed water. The sample then remained at 800 psi pressure for 12 h to ensure full saturation.

https://doi.org/10.3390/pr14091437

2.1.2. Sand Packs

The sand-pack specimens were prepared from Brazos River sand, which is quartz-rich and characterized by subangular grains.

As shown in Table 1, three grain-size fractions (45–75 µm, 180–212 µm, and 355–425 µm) were prepared by sieving to ensure consistent particle-size distributions. Samples were sleeved in heat-shrink Viton (2.5 cm inner diameter and ~3 cm length). The initial porosity was almost independent of grain size. This allows direct grain size comparisons.

Table 1. Grain Size-Dependent Sand Pack Properties.

Sample	Mineralogy	Porosity (%)	Grain Size (µm)
1	Brazos River	36.97	45–75
2	Sand over 95%	37.10	180–212
3	quartz	36.23	355–425

A series of isostatic compaction experiments were performed on the 355–425 µm sand to produce a wide range of porosities for testing. Both thermal conductivity and ultrasonic wave velocities were measured in both dry and saturated states. After compaction, the sample porosity is given by:

$$\phi_{final} = 1 - \frac{(1 - \phi_{ini}) l_{ini}}{l_{final}} \quad (1)$$

where ϕ_{ini} is the initial porosity (fraction), ϕ_{final} is the porosity after compaction, l_{ini} is the initial sample length, and l_{final} is the final sample length (after compaction). The calculation assumes that the grain volume is constant.

2.2. Thermal Conductivity (λ) Measurement and Calibration

2.2.1. Measurement

A steady-state divided-bar apparatus, integrated with a servo-controlled pressure vessel, was used to measure thermal conductivity. The apparatus combines two subsystems: (1) a divided-bar thermal conductivity assembly with actively tuned guard heaters across a 1-inch diameter sample column and (2) a triaxial press employing a programmable isostatic pressure (+1% error) to the sample stack (Figure 1). The core sample was placed between two high-thermal-conductivity aluminum discs (Figure 1). A constant heat flux was applied to the top disc by an electric heater, while the bottom disc was held at a lower reference temperature (Figure 1). The temperature drop across the sample was measured by precision RTD sensors embedded in the aluminum discs, and a one-dimensional Fourier's law (steady state) was used to compute the sample's thermal conductivity (Figure 2). This divided-bar stack consists of (in order): a heat source (temperature-controlled bath or heater), an upper heat flow meter (aluminum disc), the rock sample, a lower heat flow meter, and a heat sink [54]. The stack was surrounded by thermal insulation to minimize convective heat losses. Guard heaters were applied to eliminate radial heat flow.

https://doi.org/10.3390/pr14091437

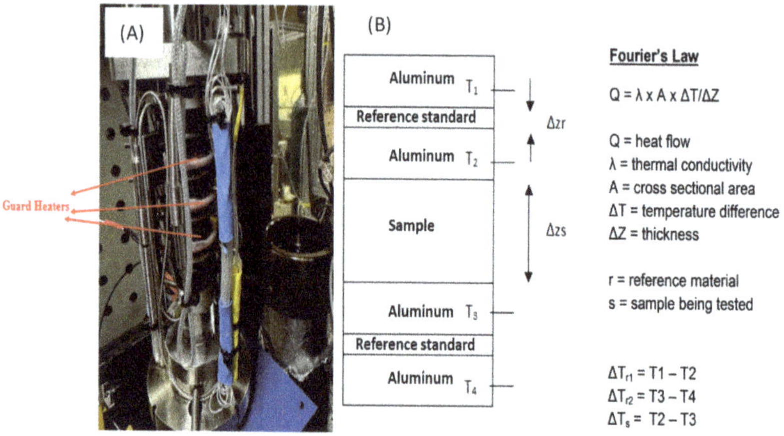

Figure 1. Viton sleeved high-pressure thermal conductivity measurement stack with embedded reference discs, thermocouples, and heater wiring (**A**). Sketch of the experimental sample and reference stack (**B**).

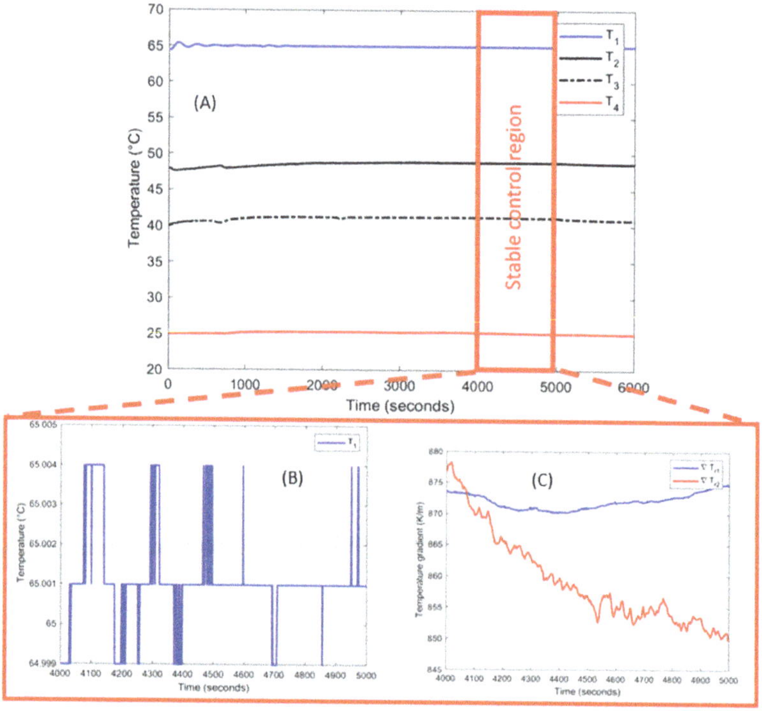

Figure 2. Temperature control performance illustrating the PID regulation (**A**). Zoomed view of a representative time interval of the equilibrated measurement, T_1 showing the high thermal stability (**B**). The gradient evolution for top and bottom reference zones during PID control (**C**).

The improved divided-bar apparatus uses three independently PID (Proportionality–Integral–Derivative)-controlled guard heaters (two for the reference sections and one for

https://doi.org/10.3390/pr14091437

the sample) to minimize radial heat losses. The error signal for the steady-state control condition is determined by the temperature gradients across the standards:

$$e(t) = \nabla G_{top} - \nabla G_{bottom} \quad (2)$$

where $e(t)$ is the input error signal, and ∇G_{top} and the ∇G_{bottom} terms are the measured temperature gradients across the top and bottom standards. The error signal measures the imbalance in heat flow. The time-discretized PID for the top and bottom sections is based on the PID control, u(t), and is adjusted to supply heat flux as:

$$u(t) = K_p e[K] + K_I T_S \sum_{I=0}^{k} e[i] + K_d \frac{e[k] - e[k-1]}{T_s} \quad (3)$$

where $e[K]$ is the current error at time step K, i.e., the difference between the target gradient (setpoint) and the measured gradient, $K_p e[K]$ is the proportional error term, which applies a correction proportional to the errors (in response to deviations), $K_I T_S \sum_{I=0}^{k} e[i]$ is the integral term, which accumulates past errors to eliminate steady-state offsets, T_s is the sampling time (time interval between control updates), and $K_d \frac{e[k]-e[k-1]}{T_s}$ is the derivative term, which anticipates future error trends by estimating the rate of change in error, which helps to dampen oscillations. The proportional, integral, and derivative gains (K_p, K_I, K_d) tune the responsiveness, stability, and smoothness of temperature control. The middle (sample) section's heater power is modeled and balanced in the divided-bar thermal conductivity apparatus using Newton's law of cooling to account for radial heat loss across the sample:

$$\in = \alpha(T_c - T_{env}) = \alpha \Delta T \quad (4)$$

where $\in$ is the radial heat flux lost from the sample to the environment, not the main conductive heat passing through the stacked setup, α is the heat transfer coefficient, T_c is the surface temperature of the middle section, and T_{env} is the ambient temperature. The term $\alpha\Delta T$ quantifies radial heat losses for each section. The proportionality expresses how radial heat losses from each section of the divided-bar apparatus relate to their local temperature differences (ΔT) and are assumed equal. The equilibrium in the mid-section power (P_{mid}) is assumed to be the average of the top and bottom sections.

$$P_{mid} = \in_{mid} = \left(\in_{top} \frac{\Delta T_{mid}}{\Delta T_{top}} + \in_{bottom} \frac{\Delta T_{mid}}{\Delta T_{bot}} \right) / 2. \quad (5)$$

This corrects for heat losses in the mid-section by maintaining constant axial heat flux through the top and bottom standards.

The measurement stack is illustrated in Figure 1. The standard gradients are calculated between T_1 and T_2 (upper gradient) and T_3 and T_4 (lower gradient), and between T_2 and T_3 (the sample gradient). A fused quartz standard sample was used for the PID-controlled heater calibration. Figure 2 shows an example of the time-dependent evolution of the temperature for the four thermocouples monitored over a 6000 s interval, where T_1 (blue) corresponded to the top aluminum section maintained at approximately 65 °C, T_2 (solid black) to the upper reference standard near 48 °C, T_3 (black dashed) to the sample region around 41 °C, and T_4 (red) to the bottom aluminum section stabilized near 25 °C. All temperatures reach a steady state between 4000 and 5000 s after an initial transient, indicating excellent thermal stability (Figure 2A). The small fluctuations (<0.01 °C) suggest that the PID controllers are maintaining a nearly constant temperature with minimal oscillation, which is a key condition for accurate steady-state thermal conductivity measurement. The zoomed-in view of T_1 between 4000 and 5000 s shows variations within ±0.0025 °C, or

https://doi.org/10.3390/pr14091437

a total range of $\Delta T_1 \leq 0.005$ °C. These small spikes shown in Figure 2B correspond to micro-adjustments by the PID heater to maintain the boundary conditions. The nearly flat profile confirms that thermal equilibrium has been reached, and sensor noise dominates over real thermal drift. The temperature gradient across each section is held nearly constant, with the upper gradient (∇T_1) staying around 875 K/m and the lower gradient (∇T_2) showing a slight downward drift but remaining within 845–880 K/m, which represents a change of 0.243 °C between the top and bottom reference sections, for approximately a 2% error (Figure 2C). The measured error of 2.0% indicates only a minor mismatch between the top and bottom temperature gradients. These results confirm that the PID-controlled divided-bar system achieved thermal equilibrium, stable gradients, and minimal thermal gradient asymmetry.

2.2.2. Calibration

Prior to conducting sample measurements, the apparatus was systematically calibrated using reference materials of known thermal conductivity, fused silica and Teflon, to evaluate stress- and temperature-sensitivity. The sample–meter-disc interfaces were subjected to a controlled axial preload of approximately 200 psi isostatic stress to establish consistent and reproducible initial thermal contact resistance. Figure 3 illustrates the relationship between the top temperature and heater voltage (control power) for the top and bottom guard heaters during guard heater calibration; fused quartz is shown Figure 3A and Teflon in Figure 3B. Both materials exhibit a linear increase in voltage with temperature, confirming predictable, stable PID control behavior. The slopes (~0.017–0.020 V/°C) indicate that fused quartz requires slightly higher guard power than Teflon to maintain equivalent gradients, consistent with its higher thermal conductivity (~1.4 W/m·K) compared to Teflon ~0.25 W/(m·K). The $R^2 > 0.97$ for all regressions validates the proportional control tuning between heating zones. The minor offset between the top and bottom heater voltages confirms balanced but distinct zone responses, as required for thermal symmetry.

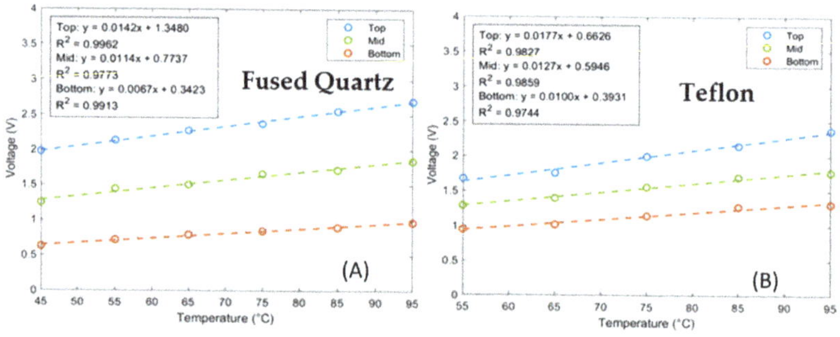

Figure 3. Calibration measurements confirming heater voltages for two standards. Guard heater voltage versus top temperature (3-Zone Control) for fused quartz (**A**) and Teflon (**B**).

The stress-dependent measurements were performed at controlled confining pressures (0–1500 psi) using argon as the confining gas. During each test, a small differential (deviatoric) axial stress of ~50 psi was applied to ensure mechanical stability and the accuracy of the axial LVDTs. Data acquisition was conducted at multiple pressure increments after confirming steady-state heat flow. Figure 4 shows the small, gradual increase in measured thermal conductivity with temperature, indicating the calibration stability. It also shows a moderate increase in thermal conductivity up to 1500 psi, consistent with reduced interfacial resistance at high pressure. These trends confirm that fused quartz is an excellent calibration reference, owing to its low temperature–pressure sensitivity and reproducible

 https://doi.org/10.3390/pr14091437

response. Figure 5 shows results for Teflon, indicating a minor increase in thermal conductivity with temperature (from ~0.247 W/m·K to ~0.275 W/m·K), though with slightly larger deviations, reflecting greater sensitivity to temperature fluctuations and a lower thermal diffusivity than quartz. It also shows that Teflon's measured thermal conductivity increases only slightly with increasing confining pressure, likely due to improved thermal contact between the stack elements. The combined results verify that the divided-bar system maintains a consistent heat-flux balance and predictable thermal performance.

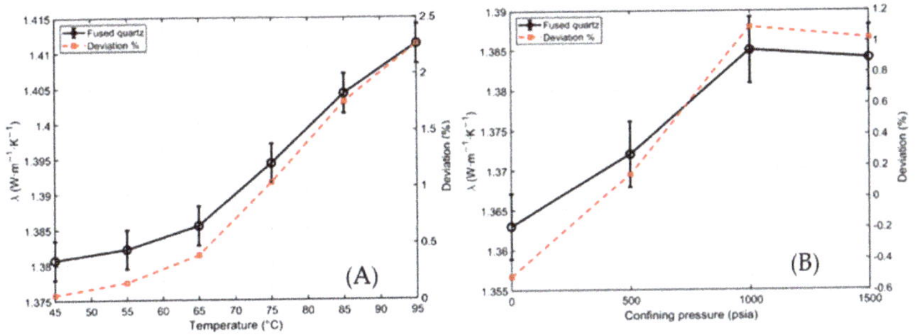

Figure 4. Fused quartz measurement with respect to temperature (**A**) and confining pressure (**B**). The small variation in thermal conductivity with temperature and pressure validates the equipment stability.

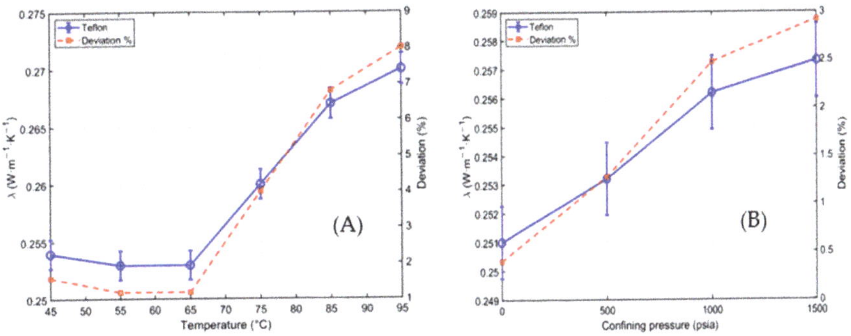

Figure 5. Results of the measurement of Teflon with respect to temperature (**A**) and confining pressure (**B**). The higher thermal conductivity is consistent with the higher voltages required in Figure 3.

2.2.3. Measurements

Each sample of Fontainebleau sandstone and the sand packs precisely machined to a length of 1.27 cm and a diameter of 2.5 cm, and then mounted in the divided-bar assembly to obtain stress-dependent thermal conductivity measurements under controlled isostatic loading conditions. Ultrasonic P-wave (compressional) and S-wave (shear) velocities were measured using a stress and temperature-controlled isostatic cell (Figure 6). Each stack was enclosed in a heat-shrink Viton jacket with O-ring seals and a lateral pore-line tap, and positioned between the top and bottom broadband temperature transducers (T_1 and T_2). The jacketed assembly was secured to the sample holder, with the pore line routed to the cell head for fluid control (Figure 6). The top transducer serves as the source, and the bottom transducer as the receiver, both connected to the external control system via 4-pin Lemo© feedthrough connectors. The confining cell was filled with mineral oil and pressurized using an ISCO© pump to apply constant isostatic stress independent of temperature. The

https://doi.org/10.3390/pr14091437

ultrasonic measurements were acquired using the pulse-transmission method, in which a one MHz piezoelectric P/S-wave transducer generates an acoustic pulse that propagates through the sample. The transmitted signal is recorded using a Tektronix TBS-2000 series oscilloscope. Velocities were measured at multiple confining pressures under both dry and water-saturated conditions.

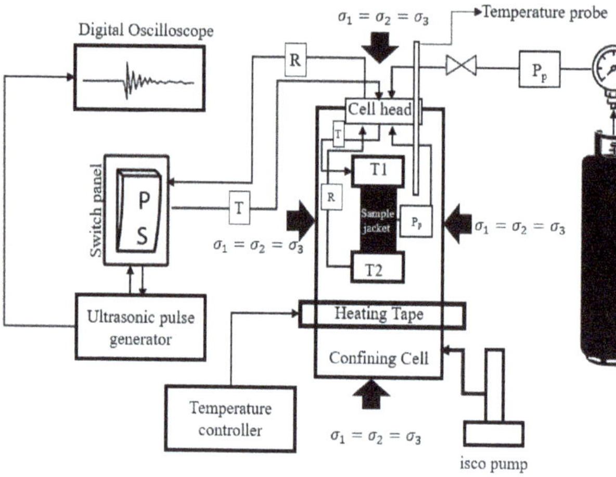

Figure 6. Diagram of the stress–temperature-dependent acoustic measurements system. This represents the first time that a stress-dependent measurement system has been constructed. The arrows indicate the direction of the stress.

3. Results

3.1. Thermal Conductivity and Acoustic Data for Fontainebleau Sandstone

The thermal conductivity of Fontainebleau sandstone increased nonlinearly with confining pressure, and the sample with the lowest porosity exhibited the highest thermal conductivity (Figure 7). Figure 8 shows the measured velocity. The initial lack of stress dependence in both measurements is attributed to the compliance associated with the low initial confining pressure.

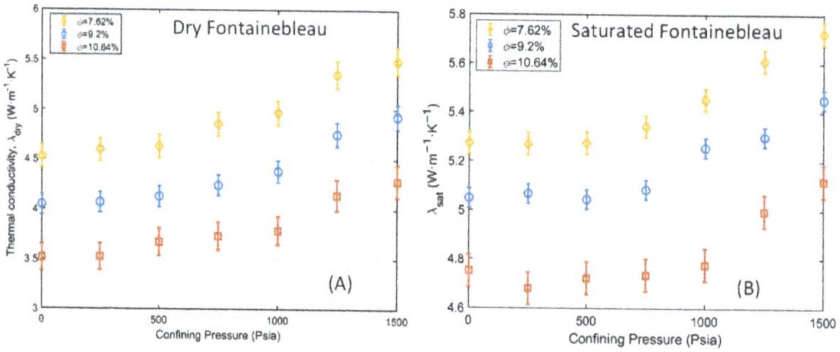

Figure 7. (**A**) Measurements of dry thermal conductivity for Fontainebleau sandstone. (**B**) Brine saturated thermal conductivity. The lack of stress dependence at low stress is interpreted as compliance effects in the samples and equipment.

https://doi.org/10.3390/pr14091437

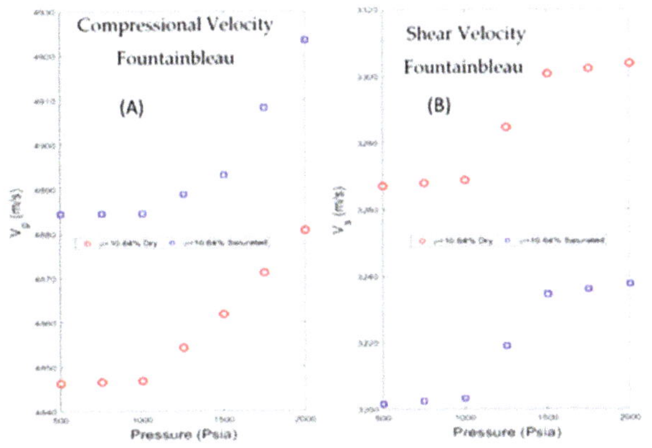

Figure 8. (**A**) is the confining pressure dependence of the compressional wave velocity (V_P). (**B**) is the shear wave velocity (V_s) for the Fontainebleau sandstone (10.6% porosity). The velocity is measured under dry (red circles) and water-saturated (blue squares) conditions.

3.2. Thermal Conductivity and Acoustic Data for the Sand Packs

The thermal conductivity cross-plots in Figure 9 demonstrate the similarity of these effects across all samples. In the dry condition (Figure 10A), the thermal conductivity of the sand packs increases with decreasing grain size and confining pressure, after the first two or three stress points where the response is likely dominated by end effects and sample conformance. Water saturation increases the thermal conductivity to nearly twice the dry value (Figure 7A,B). Figure 11A,B show that the compressional wave velocities for the sand packs increase with both confining pressure and saturation, but decrease with grain size. In Figure 12, the shear velocity shows a similar stress dependence but almost no dependence on saturation, and increasing grain size increases the velocity. However, the compressional velocity decreased with increasing grain size. Under saturated conditions, P-wave velocities increase, elevating velocities across all grain sizes compared to the dry state.

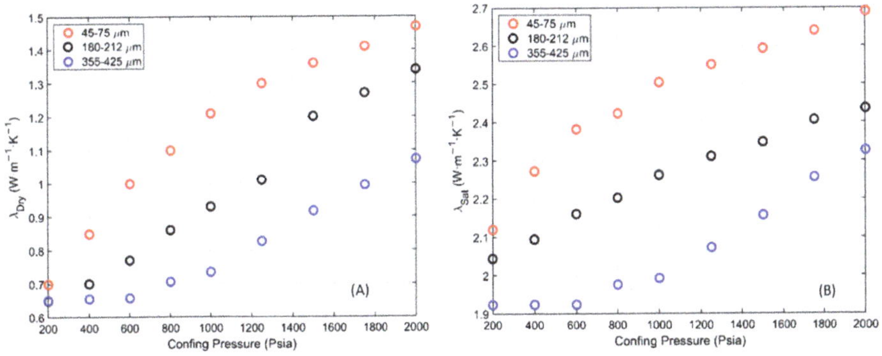

Figure 9. (**A**) is the measured thermal conductivity of saturated and dry sand packs for different grain sizes as a function of confining pressure. (**B**) shows that the larger grain sizes have lower thermal conductivities. The thermal conductivity also increased with stress.

https://doi.org/10.3390/pr14091437

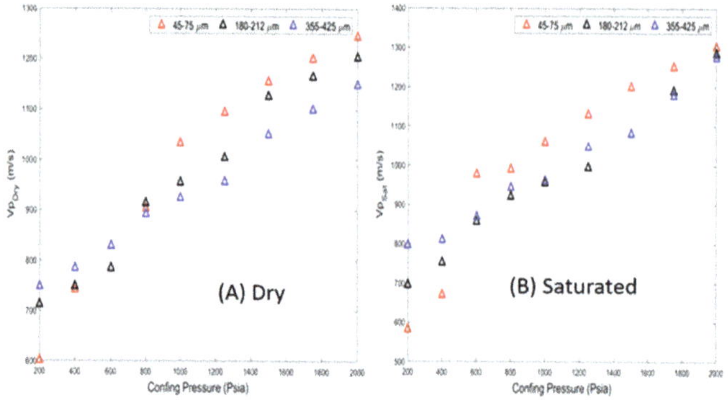

Figure 10. Measured compressional velocities of dry (**A**) and saturated sand packs (**B**) of different grain sizes as a function of confining pressure. Smaller grain sizes have higher compressional velocities.

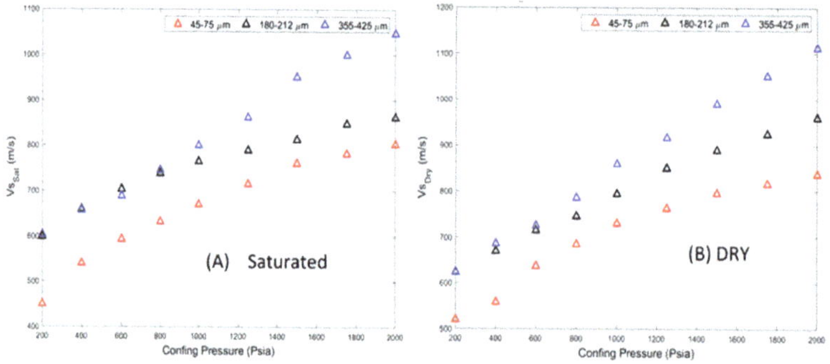

Figure 11. Measured shear velocities of saturated (**A**) and dry (**B**) sand packs of different grain sizes as a function of confining pressure. Larger grain sizes have higher shear velocities.

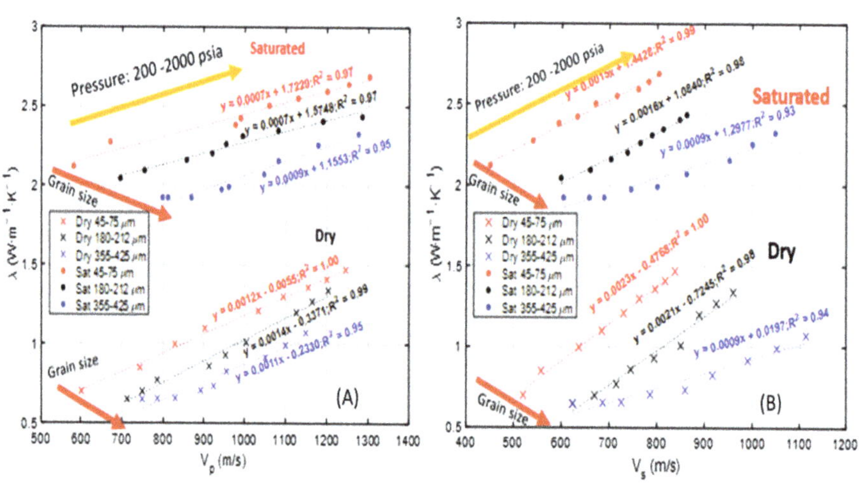

Figure 12. Stress-dependent linear regression of thermal conductivity with V_p (**A**) and V_s (**B**) for different grain size sand packs. There are excellent correlations between velocity and thermal conductivity, reflecting common trends between contact moduli (velocities) and contact resistance (thermal conductivity).

https://doi.org/10.3390/pr14091437

The thermal conductivity and acoustic velocity cross-plot (Figure 12) for the sand packs reveals a strong grain-size dependence. This dependence reflects a balance between grain contact length (favoring fine grains for thermal conductivity and V_p) and grain contact stiffness (favoring coarse grains for V_s). The extremely high correlation coefficients demonstrate the close connection between these effects. Both the modulus and the thermal conductivity are related to the contact percent among framework grains, which increases as a function of decreasing grain size. As grain size decreases in the Brazos bank sand, measured grain aspect ratios increase, resulting in a corresponding increase in grain contact area. The relationship between increasing velocity and contact percentage has been reported previously [50].

4. Discussion

4.1. Formulation of the Staged Differential Effective Medium (SDEM) Model

The Staged Differential Effective Medium (SDEM) framework was introduced for conductive inclusions and later extended [49,50] to model elastic moduli. The method introduces an interpolation parameter. L, where $L = 0$ corresponds to the Reuss bound and $L = 1$ describes the Voight bound. The staged approximation allows the model to track rock properties starting from a critical porosity to the zero-porosity endpoint [55]. Porosity is modified in sequential integration steps. The first stage uses $L = 0$ (suspension) up to the critical porosity, after which a value of $L > 0$ is applied to reflect the properties of the framework grains as they become load-bearing. This SDEM model incorporates critical porosity behavior and maintains consistency with Gassmann fluid substitution. L corresponds to measurable microstructural attributes, such as contact length and grain-fabric evolution [50]. This staged formulation allows the predicted properties to evolve with changing porosity, rather than imposing a single mixing relation across the full porosity range [51]. Starting from a percolating phase, the ordering of the integration steps is determined by the length scales present in the model. The ordering goes from small to large length scales. Thermal conductivity, like bulk modulus, depends on grain-to-grain contact geometry and is similarly bounded by the Reuss (series or iso-heat flux) and Voigt (parallel or iso-temperature gradient) limits. Substituting thermal conductivity for the bulk modulus in the SDEM model is justified because both are derived from mixture theory. Given this assumption, as shown in Equation (6), the thermal conductivity, λ, is then given by simply substituting the thermal conductivity for the modulus in earlier work [50]:

$$\lambda = \lambda_h \frac{1 + \left(\frac{\phi_0 - \phi}{\phi_0}\right) \cdot \left(\frac{\lambda_i - \lambda_h}{\lambda_h}\right) \cdot L}{1 - \left(\frac{\phi_0 - \phi}{\phi_0}\right) \cdot \left(\frac{\lambda_i - \lambda_h}{\lambda_i}\right) \cdot (1 - L)} \tag{6}$$

where ϕ_0 is the initial porosity, ϕ is the current porosity, λ_i is the thermal conductivity of the inclusion (W/m·K), λ_h is the host thermal conductivity (W/m·K), and L is the microstructural fitting parameter. The two-stage model for the effective saturated thermal conductivity (λ_{sat}) In the Reuss average form is:

$$\frac{\lambda_{sat}}{\lambda_i - \lambda_{sat}} = \frac{\lambda_{dry}}{\lambda_i - \lambda_{dry}} + \frac{\phi_c}{\phi} \cdot \frac{\lambda_L}{\lambda_i - \lambda_L} \tag{7}$$

$$\lambda_L = \left(\frac{\phi_c}{\lambda_c} + \frac{1 - \phi_c}{\lambda_i}\right)^{-1} \tag{8}$$

Equations (7) and (8) are particularly convenient forms for calculating the influence of varying the thermal conductivity of the saturating phase.

https://doi.org/10.3390/pr14091437

4.2. Modeling Fontainebleau Sandstone

This SDEM model also enables the calculation of thermal conductivity of saturated samples (λ_{sat}) from dry thermal conductivity measurements, analogous to Gassmann's fluid substitution. Figure 13 shows the thermal conductivity data measured for dry and saturated samples at 1250 psi. The SDEM model uses a quartz matrix thermal conductivity of 7.69 W/m·K for the host phase (λ_h), and a critical porosity of 40%. For both the dry and saturated thermal conductivity data, the best fit to a single L value is 0.32.

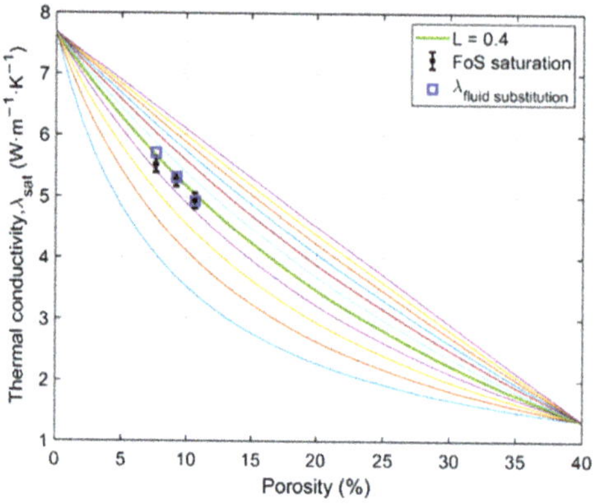

Figure 13. Measured thermal conductivity of dry and saturated Fontainebleau sandstones plotted against porosity, along with SDEM model predictions for different L values. Solid colored curves represent SDEM models for various geometric parameters L. The critical porosity, ϕ_c = 40%. The squares are the predicted saturated values calculated by fluid substitution from the dry measurements. The best fit for all the measurements gives L = 0.32.

As shown in Figure 14, plotting the thermal conductivity and the bulk modulus on a normalized plot allows a direct comparison between the two measurements. The thermal and elastic properties exhibit similar trends governed by a similar model parameter. The data for the two measurements cluster between the SDEM model curves for L = 0.3 to 0.5, demonstrating that both heat flow and elastic deformation respond to similar but not identical physical properties. The contact modulus dominates the velocity, while thermal conductivity is related to contact thermal resistance. This cross-property calibration capability demonstrates that the model can be used to jointly interpret both types of measurements. This underscores the SDEM model's value as a unified framework without assuming specific pore geometries such as ellipsoids or cracks, as assumed in prior effective medium theories [39,44].

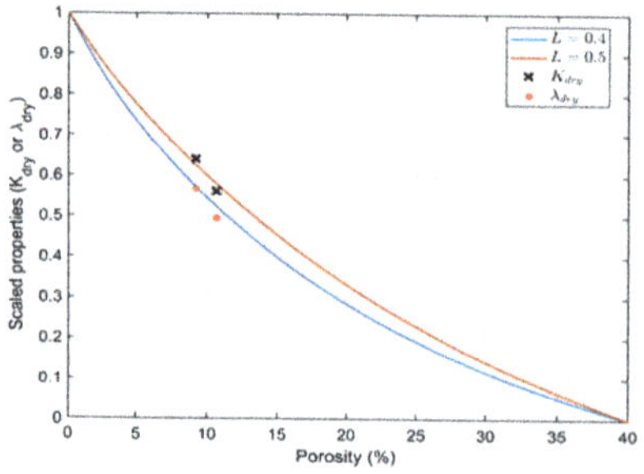

Figure 14. Normalized properties for dry thermal conductivity (λ) and bulk modulus (K) in the SDEM model for Fountainbleu sandstone.

https://doi.org/10.3390/pr14091437

4.3. SDEM Modeling of Sand Packs

To examine the impact of porosity changes, the largest grain size sand packs were compressed isostatically. As shown in Table 2, the sand packs were compressed isostatically in steps of 3500 psi up to 10,500 psi.

Table 2. Resultant change in axial strain and porosity at the end of each isostatic compaction stage.

Stage	Porosity (%)	Axial Strain (%)	Stress (psi)
1	37.0	0	0
2	28.6	12.4	3500
3	24.0	19.5	7000
4	16.7	31.2	10,500

The SDEM modeling results demonstrate that the compaction of the coarse quartz sand (355–425 μm) increases both thermal conductivity (λ) and bulk modulus (K). A single parameter, L~0.12, consistently fits λ and K for both the dry and saturated sand-pack samples (Figures 15 and 16). As porosity decreases, λ and K increase along a constant L value. At the highest compaction level, the model shows a modest change in the associated L value, a discrepancy likely arising from the significant grain crushing that occurs at this high stress. The measured λ_{sat} values for the coarse sand closely match the SDEM model predictions, confirming that saturation effects are well described by fluid substitution rather than requiring a change in the geometric parameter L. These results show that the SDEM framework tracks the evolution of both the thermal conductivity and the bulk modulus during compaction. This offers a unified tool for interpreting both thermal and elastic responses.

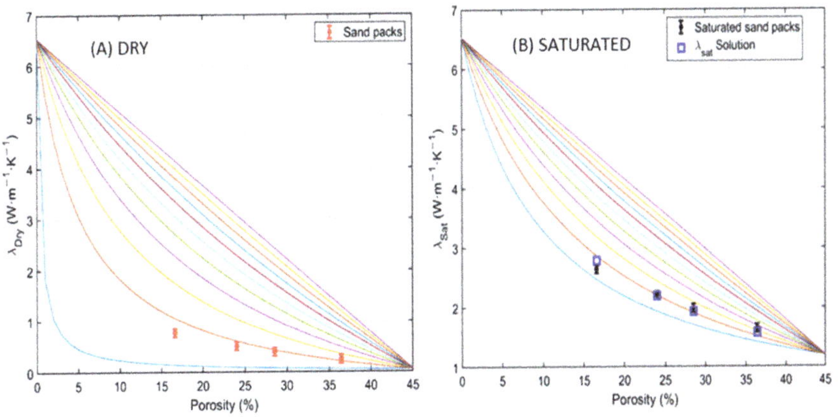

Figure 15. The SDEM model curves for compaction of coarse quartz sand for dry (**A**) and saturated conditions. (**B**) is both the fluid substitution model (square symbols) and the data. The critical porosity is assumed to be 45% for the sandpacks.

https://doi.org/10.3390/pr14091437

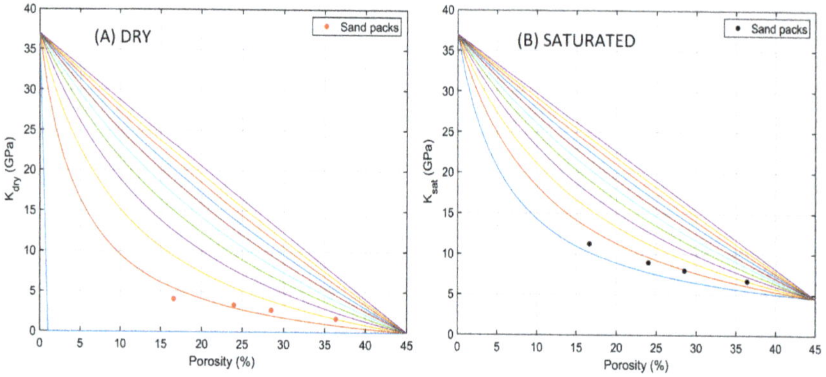

Figure 16. The SDEM model curves for the bulk modulus as a function of compaction of coarse quartz sand under dry (**A**) and saturated conditions. (**B**) is the saturated data. The critical porosity is assumed to be 45%. The best fit for sand packs is L= 0.12, significantly lower than the value found for Fontainebleau sandstone.

Comparison of Thermal Conductivity Models

In this section, the following models are compared to the SDEM model. All models are expected to fit the thermal conductivity data across the entire porosity range, up to at least the critical porosity.

Wiener bounds [23]:

$$\lambda_W^U = \phi_i\lambda_i + \phi_h\lambda_h \tag{9}$$

$$\lambda_{HS}^{mean} = \frac{1}{2}\left(\lambda_{HS}^L + \lambda_{HS}^U\right) \tag{10}$$

$$\lambda_W^L = \left(\frac{\phi_i}{\lambda_i} + \frac{\phi_h}{\lambda_h}\right)^{-1} \tag{11}$$

Hashin–Shtrikman upper and lower bounds [33]:

$$\lambda_{HS}^L = \lambda_f + \frac{3\lambda_f(\lambda_s - \lambda_f)(1-\phi)}{3\lambda_f + (\lambda_s - \lambda_f)\phi} \tag{12}$$

$$\lambda_{HS}^{mean} = \frac{1}{2}\left(\lambda_{HS}^L + \lambda_{HS}^U\right) \tag{13}$$

$$\lambda_{HS}^U = \lambda_s + \frac{3\lambda_s(\lambda_f - \lambda_s)\phi}{3\lambda_s + (\lambda_f - \lambda_s)(1-\phi)} \tag{14}$$

Maxwell–Eucken [4]:

$$\lambda = \lambda_h \frac{\lambda_i + 2\lambda_h + 2\phi_i(\lambda_i - \lambda_h)}{\lambda_i + 2\lambda_h - \phi_i(\lambda_i - \lambda_h)} \tag{15}$$

Self-consistent/Bruggeman (EMT) [37,43]:

$$0 = \phi_h \frac{\lambda_h - \lambda}{\lambda_h + \lambda} + \phi_i \frac{\lambda_i - \lambda}{\lambda_i + 2\lambda} \tag{16}$$

Geometric mean [40]:

$$\lambda = \lambda_i^{\phi_i}\lambda_h^{\phi_h} \tag{17}$$

https://doi.org/10.3390/pr14091437

λ_i is the inclusion thermal conductivity, λ_h is the host thermal conductivity, and ϕ_i and ϕ_h are the porosity of the inclusion and host respectively. Figure 17 illustrates how the different models (SDEM, Wiener Average, HS Average, Geometric Mean, and Maxwell–Eucken models) compare with the measured dry and saturated thermal conductivity.

As shown in Figure 18, the geometric-mean curve matches the Fontainebleau experimental data for both the dry and saturated cases. For the saturated case, the models make predictions similar to those in the dry case. All measurements again lie within the upper and lower bounds, but the test models consistently overestimated λ_{sat}. The SDEM with an L = 0.32 yields predictions that again cluster closely around the measured data at all porosities.

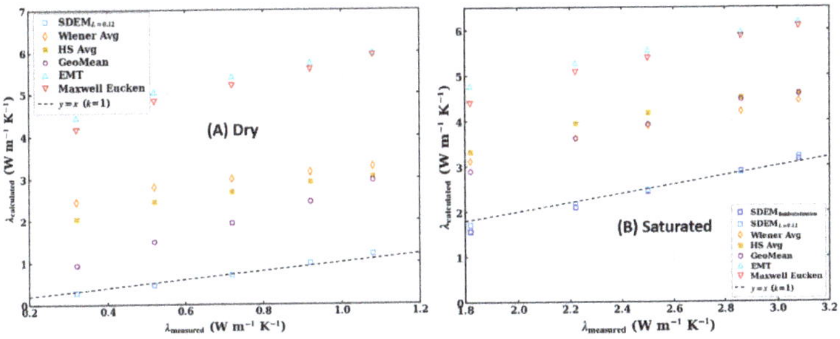

Figure 17. Comparison between measured and calculated dry (**A**) and water-saturated (**B**) thermal conductivity of the sand pack samples. The SDEM model uses a single fitted parameter of 0.12, which fits both the dry and saturated samples. All of the other models overpredict the data.

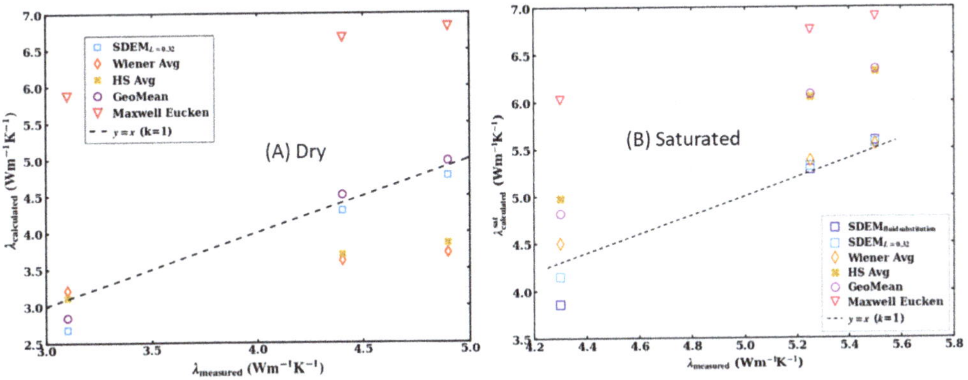

Figure 18. Comparison between measured and calculated dry (**A**) and water-saturated (**B**) thermal conductivity of the Fontainebleau sandstone samples. The SDEM model uses a single fitted parameter, 0.32, across all data.

Figure 17 shows the comparison between the models and the measured dry and saturated thermal conductivity for the sand packs. All the alternative models predict a thermal conductivity that is too high. The geometric-mean model overpredicts the saturated value. In contrast, SDEM with L = 0.12 yields a prediction that closely matches the data, including the saturation dependence.

The matrix properties determine the different L values for the sand packs and Fontainebleau and correspond to differences in the packing and distribution of thermal pathways within the material. The lower value for the sand packs is closer to a suspension

https://doi.org/10.3390/pr14091437

(a lower Wiener bound), which is expected. Fontainebleau, with its higher L value, is closer to the upper bound, consistent with its higher level of cementation.

5. Conclusions

This paper introduces a high-pressure divided-bar apparatus, designed to measure thermal conductivity under controlled confining stresses up to 2000 psi while simultaneously acquiring ultrasonic P- and S-wave velocities. This dual-measurement capability provides insight into the coupled evolution of thermal and elastic properties during compaction.

New experimental datasets on Fontainebleau sandstone and Brazos sandpacks reveal how compaction-driven increases in grain coordination and contact area enhance both thermal conductivity and bulk modulus. These results constitute one of the first systematic demonstrations of how grain size and porosity jointly affect thermal conductivity and velocity. Correlating thermal conductivity with the acoustic velocities measured at identical stress conditions both properties reflect a similar evolving rock frame. This insight enables the use of sonic-log data to infer thermal conductivity profiles, once the relationship is established.

An important contribution is the development and validation of a Staged Differential Effective Medium (SDEM) model that accurately predicts effective thermal conductivity. Across two diverse lithologies, the SDEM model reproduces observed thermal conductivity trends with a single parameter, L, that remains valid under both dry and saturated conditions. This enables a Gassmann-style fluid-substitution capability for thermal conductivity, establishing the SDEM as a powerful and generalizable tool for thermal-conductivity modeling. This modeling also allows the application of the concept of critical concentration to the analysis of thermal conductivity. The classical mixing laws all show systematic biases and fail to capture these fabric effects.

In summary, this work contributes:

- An advanced experimental platform for high-accuracy thermal measurements.
- Develops close links between thermal and elastic transport through the application of an SDEM model
- Introduces the concept of critical concentration models to the prediction of thermal conductivity.

These findings strengthen the theoretical foundation of rock thermal transport, improve prediction accuracy beyond classical approaches, and open new pathways for integrating thermal, electrical, and acoustic observations in subsurface characterization.

Future work will include the impact of changes in mineralogy. As the framework mineralogy changes, so do the grain contact % and the thermal conductivity of the starting material. In feldspathic sands, for example, at a low concentration of feldspar, increases in the contact % among grains will impact the measured conductivity more than the small decrease that should result from the mineralogic change. As feldspar increases, contact % will continue to rise, which should increase thermal conductivity, but the starting thermal conductivity of the sand's solid components will be lower. This complexity will require measuring multiple sand packs with progressively more varied mineralogy and conducting detailed image analysis to document the structural changes in the material. As such, a detailed inclusion of mineralogic changes is beyond the scope of this paper.

Author Contributions: Conceptualization, M.T.M.; methodology, Y.C., M.T.M., G.C.U. and D.M.; Software, Y.C.; formal analysis, Y.C.; investigation, Y.C., M.T.M., L.H., G.C.U. and D.M.; writing—original draft, Y.C.; writing—review and editing, M.T.M., L.H. and G.C.U.; supervision, M.T.M. and

https://doi.org/10.3390/pr14091437

Lori Hathon; project administration, M.T.M. All authors have read and agreed to the published version of the manuscript.

Funding: This research received no external funding.

Data Availability Statement: The original contributions presented in this study are included in the article. Further inquiries can be directed to the corresponding author.

Conflicts of Interest: The authors declare no conflicts of interest.

References

1. Carslaw, H.S.; Jaeger, J.C. *Conduction of Heat in Solids*, 2nd ed.; Oxford University Press: Oxford, UK, 1959.
2. García-Noval, C.; Álvarez, R.; García-Cortés, S.; García, C.; Alberquilla, F.; Ordóñez, A. Definition of a thermal conductivity map for geothermal purposes. *Geotherm. Energy* **2024**, *12*, 17. [CrossRef]
3. Preux, C.; Malinouskaya, I. Thermal conductivity model function of porosity: Review and fitting using experimental data. *Oil Gas Sci. Technol. Rev. IFP Energies Nouv.* **2021**, *76*, 66. [CrossRef]
4. Clauser, C.; Huenges, E. Thermal conductivity of rocks and minerals. In *Rock Physics & Phase Relations: A Handbook of Physical Constants*; Ahrens, T.J., Ed.; American Geophysical Union: Washington, DC, USA, 1995; Volume 3, pp. 105–126.
5. Raymond, J.; Langevin, H.; Comeau, F.-A.; Malo, M. Temperature dependence of rock salt thermal conductivity: Implications for geothermal exploration. *Renew. Energy* **2022**, *184*, 26–35. [CrossRef]
6. Mokheimer, E.M.; Hamdy, M.; Abubakar, Z.; Shakeel, M.R.; Habib, M.A.; Mahmoud, M. A comprehensive review of thermal enhanced oil recovery: Techniques evaluation. *J. Energy Resour. Technol.* **2019**, *141*, 030801. [CrossRef]
7. Horai, K.-I. Thermal conductivity of rock-forming minerals. *J. Geophys. Res.* **1971**, *76*, 1278–1308. [CrossRef]
8. Brigaud, F.; Vasseur, G. Mineralogy, porosity and fluid control on thermal conductivity of sedimentary rocks. *Geophys. J. Int.* **1989**, *98*, 525–542. [CrossRef]
9. Hermans, T.; Nguyen, F.; Robert, T.; Revil, A. Geophysical methods for monitoring temperature changes in shallow low enthalpy geothermal systems. *Energies* **2014**, *7*, 5083–5118. [CrossRef]
10. Raymond, J. Colloquium 2016: Assessment of subsurface thermal conductivity for geothermal applications. *Can. Geotech. J.* **2018**, *55*, 1209–1229. [CrossRef]
11. Beck, A.E.; Jaeger, J.; Newstead, G. The measurements of the thermal conductivities of rocks by observations in boreholes. *J. Geophys. Res.* **1971**, *76*, 3391–3401. [CrossRef]
12. Clauser, C.; Huenges, E. Thermal properties of rocks and minerals. In *Rock Physics & Phase Relations: A Handbook of Physical Constants*; Ahrens, T.J., Ed.; American Geophysical Union: Washington, DC, USA, 1995.
13. Hyndman, R.D.; Langseth, M.G.; von Herzen, R.P. A review of Deep Sea Drilling Project geothermal measurements through Leg 71. *Initial Rep. Deep Sea Drill. Proj.* **1984**, *78*, 813–823.
14. Gehlin, S. Thermal Response Test: Method Development and Evaluation. Ph.D. Thesis, Luleå University of Technology, Luleå, Sweden, 2002.
15. Cheng, L.; Zhang, Y.; Zhang, A.; Sun, Y.; Cao, Y.; Xu, Z. Investigation of measurement methods for soil thermal conductivity based on optical frequency domain reflection technology. *Int. J. Therm. Sci.* **2024**, *198*, 108889. [CrossRef]
16. Dalla Santa, G.; Peron, F.; Galgaro, A.; Cultrera, M.; Bertermann, D.; Müller, J.; Bernardi, A. An updated ground thermal properties database for ground source heat pump applications. *Energy* **2020**, *193*, 116727.
17. Parker, W.J.; Jenkins, R.J.; Butler, C.P.; Abbott, G.L. Flash method of determining thermal diffusivity, heat capacity, and thermal conductivity. *J. Appl. Phys.* **1961**, *32*, 1679–1684. [CrossRef]
18. Sass, J.H.; Lachenbruch, A.H.; Munroe, R.J.; Greene, G.W.; Moses, T.H. Heat flow in the western United States. *J. Geophys. Res.* **1971**, *76*, 6376–6413. [CrossRef]
19. Jaeger, J.C. The use of complete temperature-time curves for the determination of thermal conductivity, with particular reference to rocks. *Aust. J. Phys.* **1959**, *12*, 203–214. [CrossRef]
20. Popov, Y.A.; Pribnow, D.F.C.; Sass, J.H.; Williams, C.F.; Burkhardt, H. Characterization of rock thermal conductivity by high-resolution optical scanning. *Geothermics* **1999**, *28*, 253–276. [CrossRef]
21. Hammerschmidt, U.; Meier, V. New transient hot bridge sensor to measure thermal conductivity, thermal diffusivity, and volumetric specific heat. *Int. J. Thermophys.* **2006**, *27*, 840–865. [CrossRef]
22. Schön, J.H. *Physical Properties of Rocks: A Workbook*; Elsevier: Amsterdam, The Netherlands, 2011.
23. Wiener, O. *Theorie des Mischkörpers für das Feld der Stationären Strömung*; B.G. Teubner: Leipzig, Germany, 1912; pp. 509–598.
24. Fuchs, S.; Förster, A.; Förster, H.-J. Evaluation of common mixing models for calculating bulk thermal conductivity of sedimentary rocks: Correction charts and new conversion equations. *Geothermics* **2013**, *47*, 40–52. [CrossRef]

 https://doi.org/10.3390/pr14091437

25. Goutorbe, B.; Lucazeau, F.; Bonneville, A. Comparison of several BHT correction methods: A case study on an Australian dataset. *Geophys. J. Int.* **2006**, *164*, 104–118. [CrossRef]
26. Hartmann, A.; Rath, V.; Clauser, C. Thermal conductivity from core and well log data. *Int. J. Rock Mech. Min. Sci.* **2005**, *42*, 1042–1055. [CrossRef]
27. Waqas, U.; Rashid, H.M.A.; Ahmed, M.F.; Rasool, A.M.; Al-Atroush, M.E. Damage characteristics of thermally deteriorated carbonate rocks: A review. *Appl. Sci.* **2022**, *12*, 2752. [CrossRef]
28. Mielke, P.; Bär, K.; Sass, I. Determining the relationship of thermal conductivity and compressional wave velocity of common rock types as a basis for reservoir characterization. *J. Appl. Geophys.* **2017**, *140*, 135–144. [CrossRef]
29. Abdulagatova, Z.; Abdulagatov, I.M.; Emirov, V.N. Effect of temperature and pressure on the thermal conductivity of sandstone. *Int. J. Rock Mech. Min. Sci.* **2009**, *46*, 1055–1071. [CrossRef]
30. Labus, K.; Labus, M.; Bujok, P.; Wysocka, M. Influence of temperature on thermal conductivity of selected rocks. *J. Therm. Anal. Calorim.* **2018**, *134*, 1779–1787.
31. Vosteen, H.-D.; Schellschmidt, R. Influence of temperature on thermal conductivity, thermal capacity and thermal diffusivity for different types of rock. *Phys. Chem. Earth* **2003**, *28*, 499–509.
32. Fuchs, S.; Förster, H.-J.; Norden, B.; Balling, N.; Mielke, R.; Heckenbach, E.; Förster, A. The thermal diffusivity of sedimentary rocks: Empirical validation of a physically based α–φ relation. *J. Geophys. Res. Solid Earth* **2021**, *126*, e2020JB020595. [CrossRef]
33. Hashin, Z.; Shtrikman, S. A variational approach to the theory of the effective magnetic permeability of multiphase materials. *J. Appl. Phys.* **1962**, *33*, 3125–3131. [CrossRef]
34. Maxwell, J.C. *A Treatise on Electricity and Magnetism*; Clarendon Press: Oxford, UK, 1873.
35. Eucken, A. Allgemeine Gesetzmäßigkeiten für das Wärmeleitvermögen verschiedener Stoffarten und Aggregatzustände. *Forsch. Geb. Ingenieurwes. A* **1940**, *11*, 6–20. [CrossRef]
36. Levy, F. A modified Maxwell–Eucken equation for calculating the thermal conductivity of two-component solutions or mixtures. *Int. J. Refrig.* **1981**, *4*, 223–225. [CrossRef]
37. Berryman, J.G. Mixture theories for rock properties. In *Rock Physics & Phase Relations: A Handbook of Physical Constants*; Ahrens, T.J., Ed.; American Geophysical Union: Washington, DC, USA, 1995; pp. 205–228.
38. Norris, A.N. A differential scheme for the effective moduli of composites. *Mech. Mater.* **1985**, *4*, 1–16. [CrossRef]
39. Mavko, G.; Mukerji, T.; Dvorkin, J. *The Rock Physics Handbook*, 3rd ed.; Cambridge University Press: Cambridge, UK, 2020.
40. Woodside, W.; Messmer, J.H. Thermal conductivity of porous media. I. Unconsolidated sands. *J. Appl. Phys.* **1961**, *32*, 1688–1699. [CrossRef]
41. Fuchs, S.; Förster, H.-J.; Braune, K.; Förster, A. Calculation of thermal conductivity of low-porous, isotropic plutonic rocks of the crust at ambient conditions from modal mineralogy and porosity: A viable alternative for direct measurement? *J. Geophys. Res. Solid Earth* **2018**, *123*, 8602–8614. [CrossRef]
42. Ghanbarian, B.; Daigle, H. Thermal conductivity in porous media: Percolation-based effective-medium approximation. *Water Resour. Res.* **2016**, *52*, 295–314. [CrossRef]
43. Zimmerman, R.W. Thermal conductivity of fluid-saturated rocks. *J. Pet. Sci. Eng.* **1989**, *3*, 219–227. [CrossRef]
44. Pimienta, L.; Sarout, J.; Esteban, L.; Delle Piane, C. Prediction of rocks thermal conductivity from elastic wave velocities, mineralogy and microstructure. *Geophys. J. Int.* **2014**, *197*, 860–874. [CrossRef]
45. El Sayed, A.M.A.; El Sayed, N.A. Thermal conductivity calculation from P-wave velocity and porosity assessment for sandstone reservoir rocks. *Geothermics* **2019**, *82*, 91–96. [CrossRef]
46. Gegenhuber, N.; Schön, J. New approaches for the relationship between compressional wave velocity and thermal conductivity. *J. Appl. Geophys.* **2012**, *76*, 50–55. [CrossRef]
47. Orlander, T.; Adamopoulou, E.; Jerver Asmussen, J.; Marczyński, A.A.; Milsch, H.; Pasquinelli, L.; Lykke Fabricius, I. Thermal conductivity of sandstones from Biot's coefficient. *Geophysics* **2018**, *83*, D173–D185. [CrossRef]
48. Esteban, L.; Pimienta, L.; Sarout, J.; Delle Piane, C.; Haffen, S.; Géraud, Y.; Timms, N.E. Study cases of thermal conductivity prediction from P-wave velocity and porosity. *Geothermics* **2015**, *53*, 255–269. [CrossRef]
49. Myers, M. Pore modeling: Extending the Hanai–Bruggeman equation. In Proceedings of the SPWLA 30th Annual Logging Symposium, Houston, TX, USA, 11–14 June 1989.
50. Myers, M.; Hathon, L. Staged differential effective medium (SDEM) models for acoustic velocity. In *SEG Technical Program Expanded Abstracts 2009*; Society of Exploration Geophysicists: Tulsa, OK, USA, 2009; pp. 2130–2134.
51. Villarroel, A.; Myers, M.; Hathon, L. Integrating Thomas–Stieber with a staged differential effective medium model for saturation interpretation of thin-bedded shaly sands. In Proceedings of the SPWLA 63rd Annual Logging Symposium, Stavanger, Norway, 25–29 June 2022.
52. Myers, M.; Hathon, L. Staged differential effective medium (SDEM) models for the acoustic velocity in carbonates. In *Proceedings of the ARMA US Rock Mechanics/Geomechanics Symposium, Chicago, IL, USA, 24–27 June 2012*; American Rock Mechanics Association: Alexandria, VA, USA, 2012.

53. Thiry, M.; Bertrand-Ayrault, M.; Grisoni, J.-C. Groundwater silicification and leaching in sands: Example of the Fontainebleau Sand (Oligocene) in the Paris Basin. *Geol. Soc. Am. Bull.* **1988**, *100*, 1283–1290. [CrossRef]
54. *ASTM E1225-13*; Standard Test Method for Thermal Conductivity of Solids Using the Guarded Comparative-Longitudinal Heat Flow Technique. ASTM International: West Conshohocken, PA, USA, 2013.
55. Nur, A.; Mavko, G.; Dvorkin, J.; Galmudi, D. Critical porosity: A key to relating physical properties to porosity in rocks. *Lead. Edge* **1998**, *17*, 357–362. [CrossRef]

https://doi.org/10.3390/pr14091437

MDPI AG
Grosspeteranlage 5
4052 Basel
Switzerland
Tel.: +41 61 683 77 34

Processes Editorial Office
E-mail: processes@mdpi.com
www.mdpi.com/journal/processes

www.ingramcontent.com/pod-product-compliance
Lightning Source LLC
LaVergne TN
LVHW070453061126
851858LV00011B/226
* 9 7 8 3 7 2 5 8 8 1 5 1 2 *